Current Topics in
219 Microbiology
and Immunology

Editors

R.W. Compans, Atlanta/Georgia
M. Cooper, Birmingham/Alabama · H. Koprowski,
Philadelphia/Pennsylvania · F. Melchers, Basel
M. Oldstone, La Jolla/California · S. Olsnes, Oslo
M. Potter, Bethesda/Maryland · H. Saedler, Cologne
P. K. Vogt, La Jolla/California · H. Wagner, Munich

Springer-Verlag Berlin Heidelberg GmbH

Toxoplasma gondii

Edited by U. Gross

With 31 Figures and 14 Tables

Springer

PD Dr. Uwe Gross
Institute for Hygiene and Microbiology
University of Würzburg
Josef-Schneider-Str. 2
D-97080 Würzburg
Germany

Cover illustration: Immunolocalization of the bradyzoite-specific antigen BAG1 within tissue cysts of Toxoplasma gondii. Top picture shows a light microscopic image of a peroxidase-stained cyst within the brain of a chronically infected mouse. Left insert: a confocal laserscan of a fluorescence-labeled cell culture-derived cyst presenting the positively stained bradyzoites (orange) within the cyst while the surrounding tachyzoites are negative (green). Right insert: electron micrograph of the periphery of a tissue cyst showing the gold particles (pseudocolored) over the cytoplasm of the bradyzoites. Cover illustration provided by D.J.P. Ferguson and U. Gross

Cover design: Springer-Verlag Heidelberg, Design & Production

ISBN 978-3-642-51016-8   ISBN 978-3-642-51014-4 (eBook)
DOI 10.1007/978-3-642-51014-4

© Springer-Verlag Berlin Heidelberg 1996
Originally published by Springer-Verlag Berlin Heidelberg New York in 1996
Softcover reprint of the hardcover 1st edition 1996

Library of Congress Catalog Card Number 15-12910

SPIN: 10517627   27/3140-5 4 3 2 1 0 – Printed on acid-free paper

Preface

Toxoplasma gondii has been recognized as one of the most successful protozoan parasites infecting 10%–25% of the world's human population. For a long time, toxoplasmosis has been known as a disease severely affecting newborns who have acquired the parasite in utero. Post-natally acquired toxoplasmosis is usually an asymptomatic infection that persists lifelong. However, in immunocompromised patients, such as those suffering from the acquired immune deficiency syndrome (AIDS), or those who receive organ transplants or aggressive cancer chemotherapy, persistent *Toxoplasma gondii* infections can reactivate, and this can eventually lead to fatal outcomes.

The emergence of toxoplasmosis in immunocompromised patients has increased worldwide interest in this parasite. For example, the European Union has launched a program for collaborations on research of *Toxoplasma gondii* and/or HIV. Several workshops focusing on these diseases have been held, such as "*Toxoplasma gondii* Research in Europe". By bringing experts together from around the world, this volume of *Current Topics in Microbiology and Immunology* offers comprehensive reviews and unpublished data about current knowledge of the interaction of *Toxoplasma gondii* and its human and animal hosts.

The extreme success of *Toxoplasma gondii* infection is due to its ability to invade any nucleated cell of human individuals and most warm-blooded animals. Once infected, *Toxoplasma gondii* multiplies and resides inside the cell. The differentiation from the rapidly dividing tachyzoite stage to the dormant bradyzoite stage is one important prerequisite for lifelong persistence. This stage conversion seems to depend on the hosts' immune response; for example, in immunocompromised patients, reactivation occurs and is associated with reconversion from bradyzoites to tachyzoites. The conditions that are associated with this close interaction between *Toxoplasma gondii* and the host are still not well understood, but recent advances in molecular biology, cell biology, and immunology will help to solve this mystery and to understand the pathogenesis of the disease that is caused by this fascinating parasite.

It has been shown that *Toxoplasma gondii* has a clonal population structure that seems to be associated with clinically overt toxoplasmosis, a finding that eventually may lead to the identification of virulence-associated genes. Gene cloning and the generation of gene knockout mutants will help to understand the function of certain antigens and specific parasite organelles (such as rhoptries, micronemes, dense granules, and the multimembraneous plastid-like component). The identification of host genes, which show correlation with disease, opens new doors to understanding infectious diseases. Immunological investigations, including the use of knockout mice, have demonstrated the importance of IFN-γ and the cytokine network in controlling infection, and these results are equally informative for other intracellular parasites. Although eradication of *Toxoplasma gondii* is still not possible with antimicrobial agents, recent progress in molecular biology will help to develop new therapeutic strategies. Finally, *Toxoplasma gondii* research has evolved through the development of in vivo and in vitro models and genetic manipulation of the parasite and its host, and can serve as a model for other protozoan parasites.

Acknowledgements. I wish to express my thanks to André Capron for his support in initializing this volume. I am extremely indebted to Christiana Cooper for helping with the final editing. Special thanks are also due to Christoph A. Jacobi, Sören Schubert, and Michael Weig for helpful discussions and managing or converting computer programs . Finally, I wish to express my gratitude to Doris M. Walker from Springer-Verlag for excellent cooperation in the realization of this volume.

Würzburg   U. GROSS

Contents

C   The Host: Immunogenetics and Immune Response

D   Diagnosis and Treatment of Toxoplasmosis

List of Contributors

(Their addresses can be found at the beginning of their respective chapters.)

ALEXANDER, J. 183
AMBROISE-THOMAS, P. 155
AMICHAY, D. 127
AMMASSARI, A. 209
ANTINORI, A. 209
APPLEFORD, P. 67
BISWAS-HUGHES, G. 67
BLUETHMANN, H. 183
BOHNE, W. 81
BRUN-PASCAUD, M. 223
CANDOLFI, E. 141
CESBRON-DELAUW, M.-F. 59, 165
CHAU, F. 223
CINGOLANI, A. 209
DARDÉ, M.L. 27
DECOSTER, A. 199
DE LUCA, A. 209
DEROUIN, F. 223
DUBREMETZ, J.F. 55, 76
DUTTON, S. 67
ESTES, R. 95
FACCHETTI, P. 165
FARBER, J.M. 127
FISCHER, H.G. 175
FOURMAUX, M.N. 55
GARCIA-RÉGUET, N. 55
GAZZINELLI, R.T. 127
GHIOTTO, F. 165
GROSS, U. 81, 235
GRUNWALD, E. 127
GUO, Z.-G. 17
HADDING, U. 175
HOWE, D.K. 3
HUNTER, C.A. 113
JEBBARI, H. 183
JOHNSON, A.M. 17

JOHNSON, J. 95
KIEN, T.T. 141
KUTICIC, V. 261
LACROIX, C. 223
LECORDIER, L. 59
MACK, D. 95
MASLO, C. 223
McLEOD, R. 95
McNEIL, G. 67
MERCEREAU-PUIJALON, O. 55
MERCIER, C. 59
MURRI, R. 209
PARMLEY, S.F. 81
PELLOUX, H. 155
PISTOIA, V. 165
POHL, F. 235
PRIGIONE, I. 165
REICHMANN, G. 175
REMINGTON, J.S. 113
ROBERTS, C.W. 183
ROMAND, S. 223
ROOS, D.S. 247
SATOSKAR, A. 183
SHARTON-KERSTEN, T. 127
SHER, A. 127
SIBLEY, L.D. 3
SMITH, J.E. 67
SOÊTE, M. 76
SUBAUSTE, C.S. 113
SUZUKI, Y. 113
THOUVENIN, M. 141
TOMAVO, S. 45
VILLARD, O. 141
WIKERHAUSER, T. 261
YANG, S. 81
ZHANG, Y.W. 67

A

Population Structure of *Toxoplasma gondii*: Implications of Clonality

Genetic Basis of Pathogenicity in Toxoplasmosis

L.D. Sibley and D.K. Howe

1 Introduction

Toxoplasma gondii is among the most prevalent chronic parasitic infections in humans, infecting from 10% to 25% of the world's population (DUBEY and BEATTIE 1988). While infections are often benign, toxoplasmosis has emerged as an important opportunistic pathogen in immunocompromised patients. The identification of virulence factors is complicated by the fact that disease is rarely overt in the healthy host. This problem is amplified by the unusual population structure of *T. gondii* that results in coinheritance of many unlinked loci. Despite these complications, it is important to identify specific parasite components that contribute to pathology as they provide predictive markers of disease progression and may identify potential targets for intervention. The recent advent of genetic tools for use in protozoan parasites enables direct molecular identification of virulence determinants. This new-found technology also obligates investigators to a higher standard in establishing the molecular basis of virulence. This review provides a framework for the application of molecular genetics to investigate pathogenicity of toxoplasmosis.

Department of Molecular Microbiology, Washington University School of Medicine, St. Louis, MO 63110, USA

2 Toxoplasmosis

Toxoplasma gondii is a protozoan parasite that infects most warm-blooded vertebrates and causes disease in agricultural animals and humans (Dubey 1977; Dubey and Beattie 1988). Due to its extremely high prevalence and long-term chronicity of infection, *T. gondii* is well suited to take advantage of any compromise in the host immune status. Newly acquired infections are an important cause of spontaneous abortion in domestic animals (Dubey and Beattie 1988) and of congenital disease in humans (Desmonts and Couvreur 1974; Wong and Remington 1994). *T. gondii* is also an important opportunistic pathogen due to reactivation of chronic infections in immunocompromised hosts including organ transplant, cancer chemotherapy (Israelski and Remington 1993), and AIDS patients (Luft and Remington 1992).

The majority of infections with *T. gondii* do not lead to clinically overt disease. In the case of congenital infection, about 25% of maternal infections acquired in the first trimester and about 65% of maternal infections acquired in the third trimester lead to congenital infection (Desmonts and Couvreur 1974). Even among cases in which congenital infection occurs, the clinical outcome is quite variable, with infections acquired in early gestation generally being more severe (Desmonts and Couvreur 1974). In immunocompromised patients, AIDS cases being the most represented group, only about 30% of patients with chronic infection go on to develop the severe central nervous system (CNS) pathology that accompanies reactivation (Luft and Remington 1992). The reasons for this partial penetrance are not understood but presumably reflect a combination of host genetics, immune status, tissue burden of parasites, and the genetic make-up of the parasite.

One of the intriguing mysteries of toxoplasmosis is how readily the balance of subclinical infection can be disturbed leading to overt disease. In the majority of primary infections, toxoplasmosis is benign, causing mild flu-like symptoms before subsiding into a long-term chronic state that typically remains subclinical (Frenkel 1988). The transition between acute and chronic infection is accompanied by stage conversion whereby the parasite changes from a rapidly replicating tachyzoite form that is lytic to a slow-growing bradyzoite contained within long-lasting tissue cysts (Frenkel 1988). Tissue cysts mature slowly, eventually lysing to reestablish the chronic infection (Frenkel and Escajadillo 1987). In mice, tissue cysts are present for the life of the infected host and serological studies support a similar persistence in others animals including humans (Krahenbuhl and Remington 1982). Toxoplasmosis is controlled by a vigorous cell-mediated immune response capable of killing infected cells and parasites (Krahenbuhl and Remington 1982). The continued presence of this aggressive response is thought to prevent relapse and hence curtail pathology in the chronically infected host.

Circumstances which upset this delicate balance are of interest for two primary reasons. In strictly practical terms, the accurate prediction of which infections will remain subclinical versus those which may cause overt disease would greatly increase the efficiency of treatment thus reducing both economic and health losses. Of broader significance, these breakdowns in surveillance provide important clues about the normal regulation of parasitic infections that may provide targets for direct intervention.

3 Factors Influencing the Severity of Disease

The outcome of primary infection with *T. gondii* depends on a number of factors that combine to influence the severity of disease. In mice, inoculation with tachyzoites or bradyzoites leads to an acute infection that can culminate in death at high inocula (DEROUIN and GARIN 1991). The cause of death during acute infection is related to high parasitemia and subsequent inflammation and necrosis in the lungs, liver, and CNS (MCLEOD et al. 1989a; DEROUIN and GARIN 1991). Parasitemia and mortality are proportional to the inoculum size (ARAUJO et al. 1976), but both the parasite life-cycle stage and the route of injection also greatly influence the outcome of infection in mice (BROWN and MCLEOD 1994). Lower challenge doses, particularly of strains which are inherently less virulent for mice, lead to acute infections that are readily controlled by a vigorous immune system and which develop into long-term chronic infections (SUMYUEN et al. 1995). Several inbred strains of mice also succumb to chronic infection, typically due to encephalitis that is associated with diminished cell-mediated immunity and a decrease in production of Th1-type cytokines (interferon-γ, interleukin-2) (SUZUKI et al. 1991; HUNTER et al. 1992; GAZZINELLI et al. 1993).

Susceptibility to toxoplasmosis varies widely with different species of hosts. Mice, rabbits, and hamsters are all relatively susceptible, as are animals which evolved in the absence of significant feline predation, such as Australian marsupials, Madagascar lemurs, and New World monkeys (FRENKEL 1988). Among domesticated animals, abortion caused by toxoplasmosis occurs in pigs, sheep, and goats (DUBEY and BEATTIE 1988), while cattle and horses are relatively resistant to infection and have very low prevalence rates of *T. gondii* infection (DUBEY 1992). Susceptibility to toxoplasmosis varies with immune status with young animals being more susceptible than adults of the same species (DUBEY and BEATTIE 1988). This trend is particularly acute in neonatal animals and in part explains the severe pathology that can occur with congenital infections.

Given the importance of cell-mediated immunity in controlling toxoplasmosis, it is perhaps not surprising that the genotype of the host plays a major role in mediating resistance. In mice, in which the genetic dissection of host resistance is best described, there are at least five separate loci that control susceptibility versus resistance to acute infection (MCLEOD et al. 1989b; WILLIAMS et al. 1978). Additional genes must control resistance to chronic infection as those strains which are resistant to acute infection are often susceptible to chronic infection and develop severe CNS disease (SUZUKI et al. 1993). This complexity reflects both the underlying biological diversity of the parasite and the complexity of immune mechanisms involved in control of infection. Genetic factors are likely to play a role in infection of nonmurine hosts, yet these components have not been identified. In humans, there is no obvious genetic predisposition to toxoplasmosis and although infection prevalence rates vary considerably among different peoples, these differences are thought to reflect exposure rather than susceptibility. One of the complications in establishing such patterns is the extremely varied genetic make-up of human populations and the corresponding large numbers of samples necessary to make statistically valid correlations between host haplotypes and disease.

It is also likely that the genetic composition of the parasite plays a role in toxoplasmosis in animals and humans. This prediction is based on paradigms established in bacterial and viral pathogens in which it is clear that specific genes contribute directly to disease progression. In parasites, awareness of such genetic relationships is just emerging, coincident with the development of experimental genetic tools, which are essential for the identification of genes involved in pathogenesis. While it could be argued that the present day lack of clearly defined genetic determinants in parasite infections is evidence for their minor role, this argument is flawed by the prior lack of genetic tools needed to make such determinations.

4 Classical and Molecular Genetics

Toxoplasma gondii is an obligate intracellular parasite that propagates mitotically as a haploid cell throughout most of its life cycle (CORNELISSEN et al. 1984). Mating occurs exclusively in intestinal epithelial cells of the cat leading to the formation of oocysts (DUBEY and FRENKEL 1972). Oocysts are shed in the feces and go on to sporulate into haploid progeny that are the result of a single round of meiosis (DUBEY and FRENKEL 1972). Experimental crosses conducted in cats have established that *T. gondii* does not have a predetermined mating type (CORNELISSEN and OVERDULVE 1985). Instead, coinfection of cats with different parental clones results in recovery of both recombinant and parental genotypes due to mating and self-fertilization, respectively (PFEFFERKORN and PFEFFERKORN 1980). The basic parameters of meiosis and a rudimentary genetic linkage map have been established based on restriction fragment length polymorphism (RFLP) markers (SIBLEY et al. 1992). While it is possible to map a given phenotype to a specific chromosome by linkage, this strategy is limited by the labor- and time-intensive nature of RFLP mapping. A more substantial limitation is the relatively low recombination rate (1 centimorgan=300 kb), which predicts that genes will not be readily obtainable by linkage mapping alone.

The recent advent of molecular genetics has rapidly expanded our repertoire of available tools for experimental investigations in *T. gondii*. There are presently at least four dominant selectable markers (cat, dhfr, trp, ble) that have been used for DNA transformation and two independent systems for transient expression (CAT, β-Gal) (reviewed in Roos et al. 1994; Boothroyd et al. 1995; Sibley et al. 1995). The availability of molecular genetics enables the direct cloning of genes by complementation, expression of heterologous genes or altered genes and cloning of genes by insertional mutagenesis and marker-rescue strategies (Donald and Roos 1995). Integration of DNA into the genome is predominantly nonhomologous; however, several strategies have successfully been used for allelic replacement or gene knockouts (Kim et al. 1993; Donald and Roos 1994). The feasibility of performing such reverse genetics provides a powerful system for testing the role of specific genes in virulence and pathogenicity.

5 Population Genetic Structure

One of the tools needed for genetic analysis are polymorphic markers that can be used for analyzing population structures, for linkage mapping, and for establishing correlations between parasite genotype and geographic distribution, host range, and disease severity. In *T. gondii*, such markers have been difficult to define due to the low diversity between strains: all strains from animals and humans isolated from around the world are grouped into a single species, *T. gondii*. While this initial distinction was based on morphological grounds, modern molecular analyses fully support such a unified taxonomy. The major antigens of tachyzoites are highly conserved with similar alleles being found in all strains (Handman et al. 1980; Couvreur et al. 1988). Sequence analysis of the major tachyzoite surface antigens SAG1 (Burg et al. 1988; Bülow and Boothroyd 1991) and SAG2 (Parmley et al. 1994) indicate the presence of only two separate alleles among a wide collection of independent strains. This result is a reflection of the low allelic diversity of *T. gondii* strains which is also revealed by a wide variety of both RFLP (Sibley and Boothroyd 1992; Howe and Sibley 1995) and isoenzyme markers (Dardé et al. 1992).

With the availability of polymorphic DNA and isoenzyme markers, it has recently been possible to define the population genetic structure of *T. gondii*. Despite the presence of a sexual phase in the life cycle, the population structure of *T. gondii* is overwhelmingly clonal. This clonality is manifested by the overabundance of a limited number of genotypes that are widespread and by the absence of many possible recombinant genotypes (Tibayrenc et al. 1990). Combining the RFLP and isoenzyme analysis of approximately 125 separate strains,

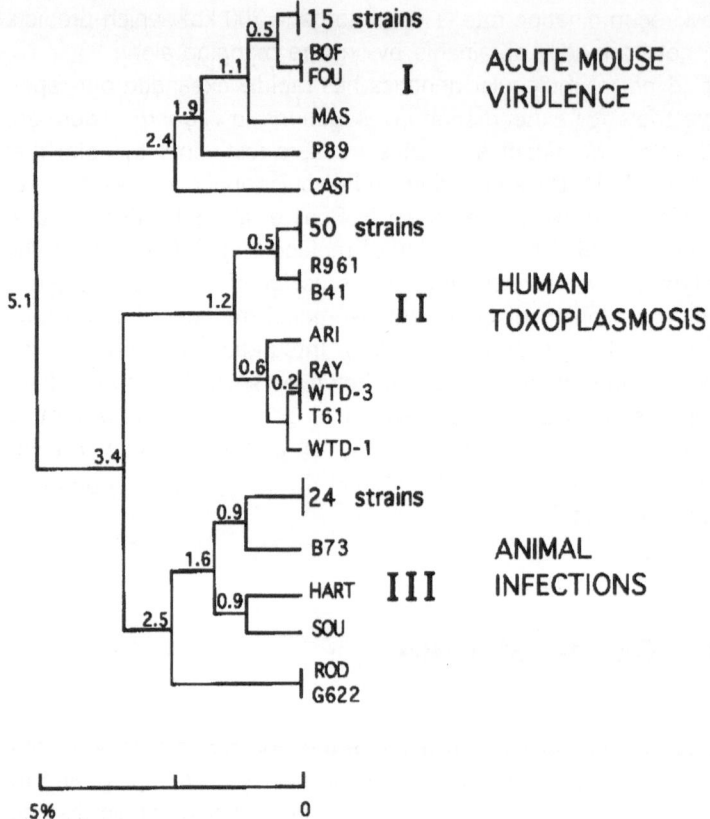

Fig. 1. Phylogenetic analysis of 106 *Toxoplasma gondii* strains based on restriction fragment length polymorphisms. *T. gondii* has a population structure that is highly clonal, comprised of three main lineages. Each lineage has specific biological phenotypes that are associated with it. Scale represents percent nucleotide divergence based on six independent single-copy markers. (From Howe and Sibley 1995)

it is clear that *T. gondii* is comprised of three predominant, clonal lineages (Dardé et al. 1992; Howe and Sibley 1995). This conclusion is supported by both phylogenetic analyses which group the strains into three major branches and by statistical analyses on the frequency of predominant genotypes which indicate that each branch represents a separate clonal lineage (Fig. 1; Howe and Sibley 1995).

The vast majority of *T. gondii* strains that have been analyzed were collected in North America and Europe. The rare isolates obtained from South America, Asia, Africa, and Australia also fit neatly into one of the three clonal lineages, indicating that strains from these regions are not significantly different (Howe and Sibley 1995). Nonetheless, we cannot rule out the possibility that additional lineages exist, particularly given the large populations of Africa, India, and China from which few isolates are available. Despite this limitation, there is no indication that the lineages are geographically separated and some other mechanism must underlie their genetic separation.

Among bacterial and fungal pathogens that show clonal population structures, this pattern is often accounted for by local outbreaks or periodic epidemics that are dominated by one or several genotypes (SMITH et al. 1993). In order for epidemically clonal populations to occur, rapid transmission between hosts is necessary. Epidemic spread is unlikely to explain the pattern of clonality in *T. gondii*: infections are not spread by close contact, but require ingestion of infected tissues or oocysts shed from cats. Instead, clonality likely persists due to the limited opportunity for meiosis in the wild combined with transmission through the food chain by carnivorous feeding or scavenging that propagates mitotically dividing forms. In order for meiosis to take place, coinfection of a cat with separate strains must happen during the limited time period when fertilization occurs (5–15 days after primary infection). The extent to which infected animals harbor more than a single strain in the wild is unknown, although experimental studies have demonstrated that in mice infection with a second strain is possible despite the substantial immunity conferred by primary infection (REIKVAM and LORENTZEN-STYR 1976). Double infection is also possible in cats, and following challenge with the second strain, the number of resulting oocysts shed is substantially reduced (FRENKEL and SMITH 1982). It has not been determined if any of these oocysts contain recombinants between the two separate strains and, although this possibility seems remote, it is biologically feasible.

5.1 What Is Meant by Clonality?

Population clonality is an often misunderstood phenomenon, therefore it is worth reflecting on exactly what is meant by the term clonality. Simply put, clonality indicates that sexual recombination, meiosis in the case of *T. gondii*, occurs much less frequently in the population than would be expected. Instead, most isolates arise by mitotic expansion of a limited number of genotypes. A consequence of this phenomenon is that many possible recombinant genotypes are not found. In the case of RFLP markers that have been used to distinguish *T. gondii* strains, there are over 1700 different possible combinations of alleles at six independent loci, yet only 15 combinations were found in a survey of over 100 independent strains (HOWE and SIBLEY 1995). The absence of the remaining genotypes is unlikely to reflect any inherent problem with their survival as randomly assorted genotypes are readily produced during experimental crosses between strains (SIBLEY et al. 1992). Instead, the observed absence of the many possible genotypes from natural populations is a direct indication that sexual recombination is rare.

Population clonality does not imply that no difference exists between individual members of a clonal lineage. In fact, small differences are expected to arise by mutation or genetic changes that can occur in mitotically propagating lines. Identification of these differences is most easily made using multicopy probes that have a higher sensitivity for detecting small differences and which are often rapidly evolving (CRISTINA et al. 1991; HOWE and SIBLEY 1994; CRISTINA et al. 1995). Such differences are not inconsistent with the concept of clonality nor do they refute its existence. The population structure, whether clonal or not, is determined by evaluating genetic recombination at independent loci among different lineages (TIBAYRENC and AYALA 1991; TIBAYRENC et al. 1990, 1991). Such recombinations are the direct evidence of meiosis and their absence is an indication of clonal descent. A logical extension is that finding one or several exceptions to the clonal pattern also does not refute its predominance at the population level. As might be expected for an organism with a fully developed sexual cycle, genetic exchange does occasionally occur in *T. gondii*. In a survey of 106 strains, only four strains were found to have extensively mixed genotypes due to meiotic recombination (HOWE and SIBLEY 1995). The existence of these recombinant strains demonstrates that the three lineages of *T. gondii* are not separate species, as they fulfill the canon of undergoing mating in naturally occurring populations. Despite these examples, the population structure of *T. gondii* remains predominantly clonal.

5.2 Implications of Clonality

There are a number of predications about biological diversity that stem directly from a clonal population structure, such as that seen in *T. gondii*. First, members of the same clonal lineage will share common phenotypes. Second, members of separate clonal lineages may differ significantly in biological traits such as growth, infectivity, and pathogenesis. Finally, the genetic differences between separate clonal lineages are likely to underlie any differences in biology including the development of disease (TIBAYRENC and AYALA 1991). Although not all genetic differences can be expected to directly impart important biological phenotypes, even those which are not causal provide important surrogate markers that may be useful for predicting disease.

The correlation between parasite genotype and disease association is borne out in the case of toxoplasmosis. Analysis of approximately 40 chronic animal infections and 60 human cases of toxoplasmosis reveals striking correlations between biological phenotypes and specific parasite lineages (summarized in Fig. 1). Type I strains, which are acutely virulent in mice, show a greater prevalence in human congenital toxoplasmosis than in chronic animal infections. Type II strains, which have a propensity for producing high cyst burdens and chronic pathology in mice, are most often associated with human toxoplasmosis both in congenitally infected and immunocompromised patients. Finally, type III strains are abundant in animals, yet rarely seen in human toxoplasmosis.

One complication is the current impossibility of determining the genotypes of *T. gondii* strains that cause chronic, subclinical infections in humans. Consequently, it is not clear if the higher prevalences of type I and type II strains in human toxoplasmosis reflect differences in primary infection or in the capacity of strains to cause disease. There is no discernible pattern in animal prevalences based on whether particular species are consumed by humans as food or not, suggesting the difference is not simply a matter of transmission frequency. The high ratio of type II/III strains in human toxoplasmosis is not simply a reflection of the higher cyst-forming capacity of type II strains, as animals infected by carnivorous feeding or scavenging do not display a similar bias. In summary, these findings support a role for the parasite genotype in disease progression and severity of toxoplasmosis.

6 The Genetic Basis of Acute Virulence in Mice

One of the most discrete phenotypes of *T. gondii* is the acute virulence of different parasite strains to mice. Strains are readily grouped into two broad phenotypes based on their ability to kill mice following i.p. challenge with tachyzoites. Acutely virulent strains have an LD_{100} of a single organism: infection always leads to death that is typically rapid (6–10 days). Nonvirulent strains have LD_{50} values that range from 10^2 to $>10^5$ and animals usually succumb between days 10 and 20 postinfection. Acute virulence is apparent independent of the mouse strain used; however, some nonvirulent strains (i.e., ME49) show enhanced virulence to inbred versus outbred mice (HOWE and SIBLEY, unpublished). Reports of strains increasing in virulence with passage are widespread; however, these reports are largely anecdotal and have not been documented based on carefully conducted titrations of LD_{50}. In our experience, in which such LD_{50} assays have routinely been conducted, nonvirulent strains may increase in virulence following rapid passage but do not cross this barrier of $LD_{50} >10^2$. For example, with repeated passage after isolation from chronic infection, the virulence of ME49 increases from $LD_{50} >10^5$ to an LD_{50} of 10^2: during this transition, its multilocus genotype as determined by RFLP analysis remains identical (SIBLEY, unpublished). A further example of this stability is the behavior of two commonly used, nonvirulent laboratory strains, PLK (KASPER and WARE 1985) and CEP (PFEFFERKORN et al. 1977). These strains were isolated more than 10 years ago and they have been maintained by laboratory passage ever since, yet both their genotypes and phenotypes remain unaltered (SIBLEY and BOOTHROYD 1992; SIBLEY, unpublished). The acute virulence of *T. gondii* strains for mice is strongly associated with a particular widespread clonal lineage designated type I (SIBLEY and BOOTHROYD 1992). Type I strains have been isolated from domestic animals (pig, cow, goat) and from human cases of toxoplasmosis from the US, France, Zaire, Holland, Austria, and Brazil, indicating they are widespread globally. Virulent strains do not require prolonged passage to manifest their ability to kill mice at low inocula (DUBEY 1980). These isolates share a highly similar genotype that differs significantly from the remaining two lineages, types II and III, which comprise exclusively nonvirulent strains (based on the above definition) (SIBLEY and BOOTHROYD 1992; HOWE and SIBLEY 1995). Virulence can be predicted based on the allele present at the SAG1 which shows a near perfect correlation with acute virulence in mice (SIBLEY and BOOTHROYD 1992; HOWE and SIBLEY 1995). SAG1 is located on the distal end of chromosome VIII, while adjacent markers, for example SAG2, do not show such a correlation (HOWE and SIBLEY, unpublished). The correlation indicates that a genetic factor unique to type I strains regulates the acute virulence of toxoplasmosis in mice. The adaptive value of such a phenotype is uncertain, but type I strains reach higher parasitemia in mice and are more persistent in rats (REMINGTON et al. 1961), which could lead to greater transmission via carnivorous feeding or scavenging. Elevated parasitemia may also explain the higher prevalence of type I strains in human congenital toxoplasmosis due to either increased likelihood of congenital transmission or greater severity of the resulting infection.

The availability of animal models for both acute and chronic toxoplasmosis provides an experimental system for dissecting the genetic contributions of parasite and host on the development of toxoplasmosis. Given the tight correlation between SAG1 and acute virulence, it seems likely that a locus involved in this trait is located on the distal end of chromosome VIII. Analysis of loci on other chromosomes does not identify similar correlations with acute virulence, therefore the association with chromosome VIII is not merely due to linkage disequilibrium (HOWE and SIBLEY 1995). Despite this strong correlation, SAG1 alone does not appear to be sufficient for virulence as at least two natural recombinant isolates, which have the SAG1 allele normally associated with virulence, were not acutely virulent when tested in mice (HOWE and SIBLEY, unpublished). This finding suggests that there is either a second gene or a closely linked locus that is required for the expression of acute virulence. The relatively low recombination rate in T. gondii is consistent with that gene lying anywhere within a 500 kb region of SAG1 on chromosome VIII.

Fortunately, there are several experimental approaches which are feasible for identifying the molecular basis of Toxoplasma virulence in mice. The availability of an RFLP linkage map for T. gondii makes it possible to analyze specific loci for their association with biological traits. A genetic cross between type I and type III strains would allow mapping of loci controlling acute virulence based on segregation with specific genetic markers. Such classical genetic studies will directly address whether acute virulence is mediated by a single locus or if it is multigenic and will provide preliminary data on the location of virulence genes. A second independent approach is based on DNA transfection of a cosmid library from the virulent lineage into nonvirulent recipients followed by selection for acute virulence in mice (HOWE and SIBLEY, unpublished). One limitation of these experiments is that virulence must be mediated by a single dominant locus in order to be rescued by this strategy. A further use of transfection would be to delete genes thought to be involved in virulence from type I strains and then test their resulting phenotype. Although it is not possible to select for such avirulent clones on a population level, it is reasonable to test individual loci once they have been implicated.

In addition to acute virulence, it may be possible to map and identify additional genes involved in chronic pathogenesis. For example, type II strains are most commonly associated with reactivation of chronic infections in AIDS patients. Type II strains produce high cyst burdens in mice (SUZUKI et al. 1989) and preliminary evidence suggests that while type II strains cause chronic pathology, at least one type III strain is much less prone to do so (SUZUKI and JOH 1994). Obviously, it needs to be determined whether this phenotype is true of type II versus III strains in general, but the clonal nature of these two lineages predicts that this phenotype is also genetically determined.

7 Perspectives

The availability of classical genetics and the recent explosion in molecular genetics in T. gondii provide the tremendous potential of being able to identify and characterize virulence determinants. There are three fundamental steps for this process. First it is necessary to establish correlations between genetic polymorphisms and disease. One of the challenges will be to distinguish between genetic polymorphisms that are merely predictive of disease potential versus those that contribute directly to development of disease. Second, a molecular understanding of the genes involved in pathogenesis requires a system for testing the role of specific genes in virulence. Where the phenotype of interest is well defined, it may be possible to test the role of specific genes in vitro (for example, cell invasion). Finally, the true test of whether a particular gene is involved in virulence requires testing in appropriate animal models.

While these criteria place a large burden on the investigator, they offer the potential for identifying specific virulence factors that may provide targets for intervention as well as attenuated strains for vaccine development.

Acknowledgements. Supported by the National Institutes of Health, The American Foundation for AIDS Research, and The Burroughs Wellcome Fund.

References

Araujo FG, Williams DM, Grumet FC, Remington JS (1976) Strain-dependent differences in murine susceptibility to Toxoplasma. Infect Immun 13:1528– 1530
Boothroyd JC, Black M, Kim K, Pfefferkorn ER, Seeber F, Sibley LD, Soldati D (1995). Forward and reverse genetics in the study of the obligate, intracellular parasite Toxoplasma gondii. In: Adolph K (ed) Methods in molecular genetics. Academic, New York, pp 3–29

Brown C, McLeod R (1994) Mechanisms of survival of mice during acute and chronic Toxoplasma gondii infection. Parasitol Today 10:290–292

Bülow R, Boothroyd JC (1991) Protection of mice from fatal Toxoplasma infection by immunization with P30 antigen in liposomes. J Immunol 147:3496–3500

Burg JL, Perlman D, Kasper LH, Ware PL, Boothroyd JC (1988) Molecular analysis of the gene encoding the major surface antigen of Toxoplasma gondii. J Immunol 141:3584–3591

Cornelissen AWCA, Overdulve JP (1985) Sex determination and sex differentiation in coccidia: gametogony and oocyst production after monoclonal infection of cats with free-living and intermediate host stages of Isospora (Toxoplasma) gondii. Parasitol 90:35–44

Cornelissen AWCA, Overdulve, JP, Van der Ploeg, M (1984) Determination of nuclear DNA of five Eucoccidian parasites, Isospora (Toxoplasma) gondii, Sarcocystis cruzi, Eimeria tenella, E. acervulina, and Plasmodium berghei, with special reference to gametogenesis and meiosis in I. (T.) gondii. Parasitol 88:531–553

Couvreur G, Sadak A, Fortier B, Dubremetz JF (1988) Surface antigens of Toxoplasma gondii. Parasitol 97:1–10

Cristina N, Oury P, Ambroise-Thomas P, Santoro F (1991) Restriction fragment length polymorphisms among Toxoplasma gondii strains. Parasitol Res 77:266–268

Cristina N, Dardé ML, Boudin C, Tavernier G, Pestre-Alexandre M, Ambroise-Thomas P (1995) A DNA fingerprinting method for individual characterization of Toxoplasma gondii strains: combination with isoenzymatic characters for determination of linkage groups. Parasitol Res 81:32–37

Dardé ML, Bouteille B, Pestre-Alexandre M (1992) Isoenzyme analysis of 35 Toxoplasma gondii isolates: biological and epidemiological implications. J Parasitol 78:786–794

Derouin F, Garin YJF (1991) Toxoplasma gondii: blood and tissue kinetics during acute and chronic infections in mice. Exp Parasitol 73:460–468

Desmonts G, Couvreur J (1974) Congenital toxoplasmosis: a prospective study of 378 pregnancies. N Engl J Med 290:1110–1116

Donald RGK, Roos DS (1994) Homologous recombination and gene replacement at the dihydrofolate reductase-thymidylate synthase locus in Toxoplasma gondii. Mol Biochem Parasitol 63:243–253

Donald RGK, Roos DS (1995) Insertional mutagenesis and marker rescue in a protozoan parasite: cloning of the uracil phosphoribosyl transferase locus from Toxoplasma gondii. Proc Natl Acad Sci 92:5749–5753

Dubey JP (1977). Toxoplasma, Hammondia, Besnoitia, Sarcocystis, and other tissue cyst-forming coccidia of man and animals. In: Kreier JP (ed) Parasitic Protozoa. Academic, New York, pp 101–237

Dubey J (1980) Mouse pathogenicity of Toxoplasma gondii isolated from a goat. Am J Vet Res 41:427–429

Dubey JP (1992) Isolation of Toxoplasma gondii from a naturally infected beef cow. J Parasitol 78:151–153

Dubey JP, Beattie CP (1988) Toxoplasmosis of animals and man. CRC, Boca Raton

Dubey JP, Frenkel, JF (1972) Cyst-induced toxoplasmosis in cats. J Protozool 19:155–177

Frenkel JK (1988) Pathophysiology of toxoplasmosis. Parasitol Today 4:273–278

Frenkel JK, Escajadillo A (1987) Cyst rupture as a pathogenic mechanism of toxoplasmic encephalitis. Am J Trop Med Hyg 36:517–522

Frenkel JK, Smith DD (1982) Immunization of cats against shedding of Toxoplasma oocysts. J Parasitol 68:744–748

Gazzinelli RT, Eltoum I, Wynn TA, Sher A (1993) Acute cerebral toxoplasmosis is induced by in vivo neutralization of TNF-α and correlates with the down-regulated expression of inducible nitric oxide synthase and other markers of macrophage activation. J Immunol 151:3672–3681

Handman E, Goding JW, Remington JS (1980) Detection and characterization of membrane antigens of Toxoplasma gondii. J Immunol 124:2578–2583

Howe DK, Sibley LD (1994) Toxoplasma gondii: analysis of different laboratory stocks of RH strain reveals genetic heterogeneity. Exp Parasitol 78:242–245

Howe DK, Sibley LD (1995) Toxoplasma gondii is comprised of three clonal lineages: correlation of parasite genotype with human disease. J Infect Dis 172:1561–1566

Hunter CA, Roberts CW, Alexander J (1992) Kinetics of cytokine mRNA production in the brains of mice with progressive toxoplasmic encephalitis. Eur J Immunol 22:2317–2322

Israelski DM, Remington JS (1993) Toxoplasmosis in the non-AIDS immunocompromised host. Curr Clin Top Infect Dis 13:322–356

Kasper LH, Ware PL (1985) Recognition and characterization of stage-specific oocyst/sporozoite antigens of Toxoplasma gondii by human antisera. J Clin Invest 75:1570–1577

Kim K, Soldati D, Boothroyd JC (1993) Gene replacement in Toxoplasma gondii with chloramphenicol acetyltransferase as selectable marker. Science 262:911–914

Krahenbuhl JL, Remington JS (1982). The immunology of Toxoplasma and toxoplasmosis. In: Cohen S, Warren KS (eds) Immunology of parasitic infections. Blackwell London, pp 356–421

Luft BJ, Remington JS (1992) Toxoplasmic encephalitis in AIDS. Clin Infect Dis 15:211–222

McLeod R, Eisenhauer P, Mack D, Brown C, Filice G, Spitalny G (1989a) Immune responses associated with early survival after peroral infection with Toxoplasma gondii. J Immunol 142:3247–3255

McLeod R, Skamene E, Brown CR, Eisenhauer PB, Mack DG (1989b) Genetic regulation of early survival and cyst number after peroral Toxoplasma gondii infection of AXB/BXA recombinant inbred and congenic mice. J Immunol 143:3031–3034

Parmley SF, Gross U, Sucharczuk A, Windeck T, Sgarlato GD, Remington JS (1994) Two alleles of the gene encoding surface antigen P22 in 25 strains of Toxoplasma gondii. J Parasitol 80:293–301

Pfefferkorn ER, Pfefferkorn LC, Colby ED (1977) Development of gametes and oocysts in cats fed cysts derived from cloned trophozoites of Toxoplasma gondii. J Parasitol 63:158–159

Pfefferkorn LC, Pfefferkorn ER (1980) Toxoplasma gondii: Genetic recombination between drug resistant mutants. Exp Parasitol 50:305–316

Reikvam S, Lorentzen-Styr AM (1976) Virulence of different strains of Toxoplasma gondii and host response in mice. Nature 261:508–509

Remington JS, Jacobs L, Melton M (1961) Congenital transmission of toxoplasmosis from mother animals with acute and chronic infections. J Infect Dis 108:163–173

Roos DS, Donald RGK, Morrissette NS, Moulton AL (1994) Molecular tools for genetic dissection of the protozoan parasite Toxoplasma gondii. Meth Cell Biol 45:28–61

Sibley LD, Boothroyd JC (1992) Virulent strains of Toxoplasma gondii comprise a single clonal lineage. Nature 359:82–85

Sibley LD, LeBlanc AJ, Pfefferkorn ER, Boothroyd JC (1992) Generation of a restriction fragment length polymorphism linkage map for Toxoplasma gondii. Genetics 132:1003–1015

Sibley LD, Howe DK, Wan KL, Khan S, Aslett M, Ajioka J (1995). Toxoplasma gondii as a model genetic system. In: Smith D, Parsons M (eds) Molecular biology of parasitic protozoa. Oxford University, Cambridge

Smith JM, Smith NH, O'Rouke M, Spratt BG (1993) How clonal are bacteria? Proc Natl Acad Sci 90:4384–4388

Sumyuen MH, Garin YJF, Derouin F (1995) Early kinetics of Toxoplasma gondii infection in mice orally infected with cysts of an avirulent strain. J Parasitol 81:327–329

Suzuki Y, Joh K (1994) Effect of the strain of Toxoplasma gondii on the development of toxoplasmic encephalitis in mice treated with antibody to interferon-γ. Parasitol Res 80:125–130

Suzuki Y, Conley FK, Remington JS (1989) Differences in virulence and development of encephalitis during chronic infection vary with the strain of Toxoplasma gondii. J Infect Dis 159:790–794

Suzuki Y, Joh K, Orellana MA, Conley FK, Remington JS (1991) A gene(s) within the H-2D region determines the development of toxoplasmic encephalitis in mice. Immunol 74:732–739

Suzuki Y, Orellana MA, Wong SY, Conley FK, Remington JS (1993) Susceptibility to chronic infection with Toxoplasma gondii does not correlate with susceptibility to acute infection in mice. Infect Immun 61:2284–2288

Tibayrenc M, Ayala FJ (1991) Towards a population genetics of microorganisms: the clonal theory of parasitic protozoa. Parasitol Today 7:228–235

Tibayrenc M, Kjellberg F, Araud J, Oury B, Breniere SF, Dardé ML, Ayala FJ (1991) Are eukaryotic microorganisms clonal or sexual? A population genetics vantage. Proc Natl Acad Sci 88:5129–5133

Tibayrenc M, Kjellberg F, Ayala FJ (1990) A clonal theory of parasitic protozoa: the population structures of Entamoeba, Giardia, Leishmania, Naegleria, Plasmodium, Trichomonas, and Trypanosoma and their medical and taxonomic consequences. Proc Natl Acad Sci 87:2414–2418

Williams DM, Grumet FC, Remington JS (1978) Genetic control of murine resistance to Toxoplasma gondii. Infect Immun 19:416–420

Wong S, Remington JS (1994) Toxoplasmosis in pregnancy. Clin Infect Dis 18:853–862

DNA Polymorphisms Associated with Murine Virulence of *Toxoplasma gondii* Identified by RAPD-PCR

Z.-G. Guo and A.M. Johnson

1 Virulence in *Toxoplasma gondii* Strains

There are large differences in virulence among the various strains and isolates of *T. gondii*, although only one species of the genus *Toxoplasma* has been observed to exist so far. This virulence diversity among isolates may have considerable impact on epidemiology, immunology, pathology and the para-site-host relationship. Evidence that there is more than one strain of *T. gondii* is based on the observation of differences in virulence among different isolates for laboratory animals (KRAHENBUHL and REMINGTON 1982). Virulence can be es-timated as either the time taken before infected animals die and the percentage of mortality, or the number of parasites needed to kill infected animals (KAUFMAN et al. 1959; DUBEY and FRENKEL 1973). Some strains such as the extremely virulent RH strain have lost the ability to form oocysts in the cat, and as few as ten tachyzoites can be lethal for a mouse within a week when injected i.p. Other strains such as the avirulent S-1 strain form cysts in the brains of mice injected with 1000 oocysts (equivalent to 8000 tachyzoites) and these mice survive without ill effects (DUBEY and FRENKEL 1973). Early studies showed that differences in virulence correlated with tachyzoite generation time in tissue culture (KAUFMAN 1958; KAUFMAN et al. 1959). Virulent strains rapidly destroy the cells while avirulent strains grow slowly causing minimal cell damage. The pathogenicity of some avirulent strains has been observed to increase following frequent i.p. passage of the tachyzoites in mice (FRENKEL 1956). It seems that

Molecular Parasitology Unit, Department of Cell and Molecular Biology, University of Technology, Sydney, P.O. Box 123, Broadway, NSW 2007, Australia

this change in virulence may be host dependent as avirulent strains passaged in mice show increased pathogenicity whereas the same strains passaged in chick embryos fail to display the change (KASPER and BOOTHROYD 1993). It is also well recognised that different host species and host strains vary in their susceptibility to *T. gondii* infection. A suspected virulent parasite strain which is lethal for mice may exhibit little clinical effect on adult rats even when challenged with large numbers of the parasite strain. Differences in susceptibility among different inbred and outbred strains of mice are well recognised (McLEOD et al. 1984), and the route of challenge of the test mice also has a significant outcome on the definition of virulence given to the parasite strain (JOHNSON 1984). However, within the parasite itself, it would seem that different *T. gondii* strains exhibit a large range of variation in virulence for a given host species.

2 Characterisation of Strain Variation in *Toxoplasma gondii*

Initial efforts to explain the differences in murine virulence concentrated on the protein or antigenic structure of the different strains (reviewed in JOHNSON 1989, 1990). Restriction fragment length polymorphisms (RFLPs) have been used to analyse several strains of *T. gondii*. SIBLEY and BOOTHROYD (1992) demonstrated a correlation between RFLP patterns and strain virulence for mice, finding that ten virulent strains had an identical genotype at the three *T. gondii* genomic loci investigated (*SAG1*, 850 and BS), therefore comprising a single clonal lineage, while 18 avirulent strains had moderately polymorphic RFLP patterns different from those of the virulent strains. Conversely, the RFLP patterns generated by using different DNA probes in other studies (CRISTINA et al. 1991a, 1991b; PARMLEY et al. 1994) did not correlate with virulence. In addition, using two repetitive probes (TGR1E and TGR6), CRISTINA et al. (1995) revealed, in another RFLP study, that three virulent strains are closely related, giving similar RFLP patterns. However, the virulent strain MAS had completely different RFLP patterns from those of the other three virulent strains. Also, a recent study (ASAI et al. 1995) on the isozyme forms of the nucleoside triphosphate hydrolase (NTPase) of *T. gondii*, found RFLPs among virulent *T. gondii* strains. It seems that virulent strains, like avirulent strains, may be polymorphic when different strains are included in analyses comparing larger numbers of different loci. Recently, comparison of the 315 bp sequence of the 3' region of the major surface antigen *SAG1* gene from nine *T. gondii* strains revealed polymorphisms that distinguished six virulent strains from three avirulent strains (RINDER et al. 1995). However, relationships established by polymorphisms in the sequences of a defined genomic locus may be limited when different loci are analysed. For example, the level of heterogeneity in the 18S ribosomal DNA gene sequence of *T. gondii* strains was too low to establish genetic relatedness among eight *T. gondii* strains (LUTON et al. 1995). These results suggested that strains of *T. gondii* have evolved only relatively recently. Therefore, in order to establish relationships likely to accurately represent those for a much wider range of *T. gondii* isolates, we should perhaps use tests that compare as large a number of different parasite DNA loci as possible. One recently developed test that does this is the random amplified polymorphic DNA polymerase chain reaction (RAPD PCR) .

3 Random Amplified Polymorphic DNA Polymerase Chain Reaction

In 1990, two groups independently and almost simultaneously described a novel DNA polymorphism assay. WILLIAMS et al. (1990) described genetic mapping applications and termed the new technique the RAPD, for random amplified polymorphic DNA, while WELSH and colleagues concentrated on genome fingerprinting and termed their assay arbitrary primer-PCR (AP-PCR; WELSH and McCLELLAND 1990; WELSH et al. 1991). The technique is based on random amplification of DNA fragments by the use of a single short primer (commonly 10-mers) with an arbitrary sequence. The single primer will support DNA amplification from the genomic template if binding sites on opposite strands of the template exist within a distance that can be traversed by DNA polymerase (up to several thousand nucleotides) (INNIS et al. 1990). Separation of the PCR products on agarose gels and visualisation of the DNA bands with ethidium bromide staining can generate fingerprint patterns, and specific fragments may be present as unique bands. The method detects abundant polymorphisms which can be used for genetic mapping applications, genetic diagnosis and for genetic comparison of a large range of organisms (WILLIAMS et al. 1993). The nature of the amplified fragments is highly dependent on the primer sequence and on the DNA sequence of the genome being assayed. Genomic polymorphisms at one or both primer binding sites may result in disappearance of the amplified bands. RAPD PCR also has the additional advantages of technical simplicity, speed, high resolution and the requirement for only small amounts of DNA (about 20 ng DNA per reaction).

4 Major Findings by Using RAPD PCR on *Toxoplasma gondii* Strains

The RAPD PCR has shown its potential in the detection of polymorphisms randomly distributed in the genome among very closely related organisms including parasitic protozoa without the need for predetermined sequence information (GUO and JOHNSON 1995a,b; MACKENSTEDT and JOHNSON 1995; STEVENS and TIBAYRENC 1995). Our previous results of the analysis of 11 *T. gondii* strains by RAPD PCR using seven primers suggested that polymorphic DNA can be easily detected among *T. gondii* isolates (GUO and JOHNSON 1995a). The six murine virulent *T. gondii* strains formed one group and the five avirulent strains formed another. These results suggested that *T. gondii* may actually comprise two major clonal lineages, correlated with their virulence. In order to confirm these initial findings on the genetic relatedness of *T. gondii* strains, we have just completed another study involving 18 primers and 35 parasite strains in order to identify an even larger number of significant polymorphisms between virulent and avirulent strains. The two *Toxoplasma* clonal lineages found are consistent with the fact that strains with similar virulence types form two defined subsets which have probably evolved independently following their initial separation. The results also show that the populations in both lineages appear to consist of a range of strains with a similar level of genetic diversity, which is in contrast to a previous suggestion that virulent strains comprise a single clonal lineage while avirulent strains are moderately polymorphic (SIBLEY and BOOTHROYD 1992), but consistent with the findings of CRISTINA et al. (1995) of genetic diversity among virulent strains. During our RAPD PCR studies on the genetic relationships among *T. gondii* strains, murine virulence- and avirulence-specific polymorphic DNA bands have been identified. For example, genomic DNA from *T. gondii* strains was amplified by RAPD PCR using two 10-mer arbitrary primers from Operon Technologies Inc (USA), B12 (5'-CCTTGACGCA-3') and B5 (5'-TGCGCCCTTC-3'). The conditions of the PCR amplification were as described previously (GUO and JOHNSON 1995a). PCR products were separated by electrophoresis in 1.5% agarose gels and visualised with ethidium bromide staining. Primer B12 was found to be able to generate a virulence-specific DNA fragment we called B12-v. The DNA band B12-v was amplified in all eight *T. gondii* virulent strains tested (RHa, RHu, ENT, PT, GT1, CT1, S48 and ts4) but B12-v could not be visually detected in the seven avirulent strains analysed (ME49, PLK, CEP, Tg51, TPR, Beverley and Fukaya) (Fig. 1a). By contrast, primer B5 was found to be able to amplify an avirulence-specific DNA band we called B5-av. The DNA band B5-av was visually detected only in the seven avirulent strains (ME49, PLK, CEP, Beverley, TPR, Tg51 and Fukaya) but could not be detected in the five virulent strains tested (RHa, ENT, PT, GT1 and CT1) (Fig. 1b). Because of their virulence or avirulence uniqueness and easy identification, B12-v and B5-av were purified from RAPD PCR products by low melting agarose gel electrophoresis and directly cloned

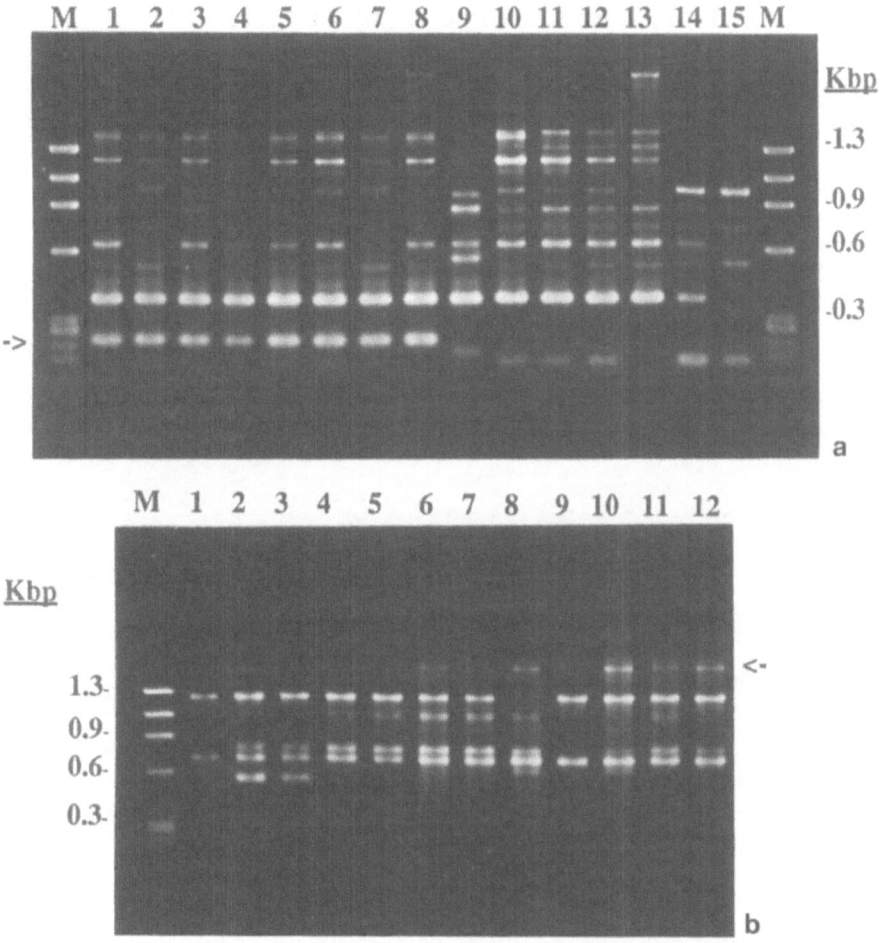

Fig. 1. a A RAPD PCR products amplified by using primer B12; lane M: DNA size markers (*Hae*III cut ØX 174 DNA); *lanes 1–15*: RAPD PCR products amplified from *T. gondii* strains RHu, RHa, ENT, PT, GT1, CT1, S48, ts4, ME49, PLK, CEP, Tg51, TPR, Beverley and Fukaya, respectively. *Marker* points to the DNA band B12-v which was detected in taxa in *lanes 1–8*. **b** RAPD PCR products amplified by using primer B5; lane M: DNA size markers (*Hae*III cut ØX 174 DNA, Promega); *lanes 1–12*: RAPD PCR products amplified for *T. gondii* strains RHa, ENT, PT, GT1, CT1, ME49, PLK, CEP, Beverley, TPR, Tg51 and Fukaya, respectively. *Marker* points to the DNA band B5-av which was detected in taxa in *lanes 6–12*

into the plasmid pUC18. To further characterise their molecular nature, B12-v and B5-av were collected from the respective purified recombinant pUC18 following double digestion with endonuclease *Bam*HI and *Kpn*I. B12-v was then radiolabelled with [32P]dCTP by using a Megapriming DNA labelling system (Amersham, UK) and B5-av was labelled with horseradish peroxidase in the ECL nucleic acid labelling system (Amersham) and hybridised to the RAPD PCR products of *T. gondii* strains generated by either primer B12 or B5, respectively, which had been separated by agarose gel electrophoresis and

1 2 3 4 5 6 7 8 9 10 11 12 13 14 15

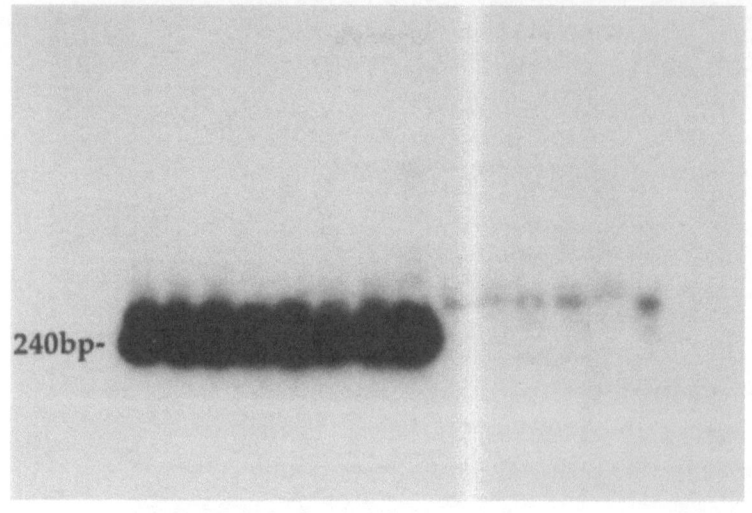

240bp-

a

1 2 3 4 5 6 7 8 9 10 11 12

·1.5kbp

b

Fig. 2. a A Southern blot of RAPD PCR products amplified by using primer B12, which was probed with a ^{32}P-labelled B12-v DNA fragment. Lanes are the same as Fig. 1a. **b** Southern blot of RAPD PCR products amplified by using primer B5, which was probed with horseradish peroxidase labelled B5-av DNA fragment. Lanes are same as Fig. 1b

Fig. 3a,b. Southern blots of genomic DNA digested with endonucleases *Sac*I and *Msp*I, which were hybridised with **a** radiolabelled B12-v fragment; **b** radiolabelled B5-av fragment. *Lanes 1–4* are digested DNA from *T. gondii* strains RHa, ENT, ME49, and TPR respectively; *lane 5* is digested DNA from the host cell line (MLA 144 gibbon lymphoma)

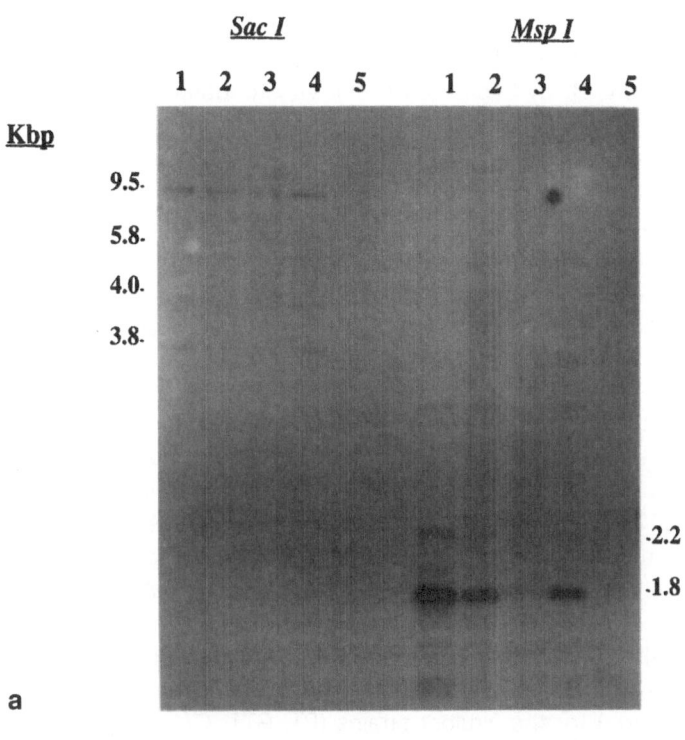

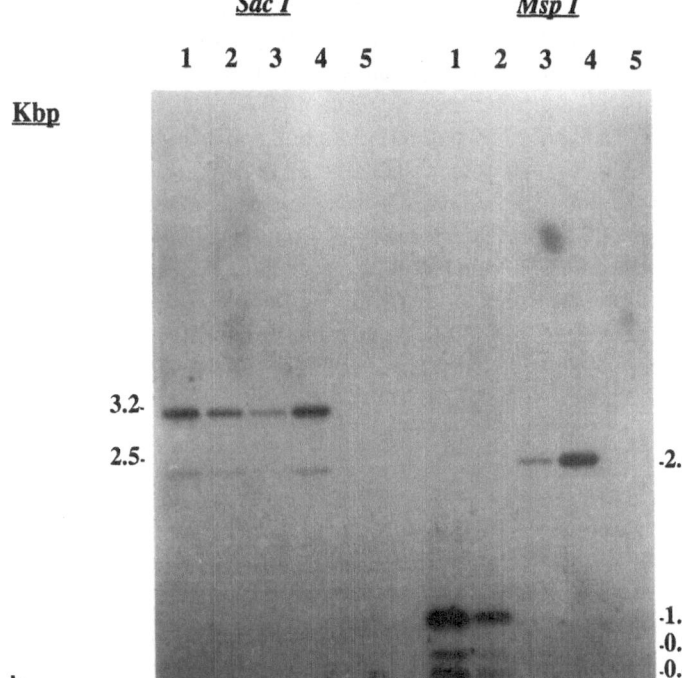

Southern blotted to nylon membrane. The results show that distinct hybridisation bands were detected exclusively in all virulent strains by using probe
B12-v (Fig. 2a). Similarly, B5-av hybridised exclusively to identical bands in all
RAPD PCR products of avirulent strains (Fig. 2b). In order to determine whether
the DNA fragments amplified by using primers B5 and B12 were present in
the genomes of virulent and/or avirulent parasite strains, *Msp*I and *Sac*I restriction digested genomic DNA from two virulent *T. gondii* strains (RHa and
ENT) and two avirulent strains (ME49 and TPR) were probed with radiolabelled
B12-v or B5-av. The genomic DNA hybridisation results show that B12-v hybridised to genomic DNA from both virulent and avirulent *T. gondii* strains
digested with *Msp*I and *Sac*I and produced identical RFLP patterns for the
virulent and avirulent strains (Fig. 3a). This result suggested that the B12-v
RAPD PCR polymorphism is not caused by a deletion in the genome between
virulent and avirulent strains but that it is likely that there are polymorphisms
in the primer B12 binding sites of the genomes of virulent and avirulent strains
which results in B12-v being amplified exclusively in virulent strains. B5-av
hybridised to genomic DNA from both virulent and avirulent *T. gondii* strains
digested with *Msp*I and *Sac*I, but produced different RFLP patterns in the
*Msp*I restriction digest of the virulent and avirulent strains (Fig. 3b). Genomic
DNA from seven other *T. gondii* strains (PT, GT1, CT1, S48, PLK, Tg51 and
Fukaya) were also tested by Southern hybridisation as described above. The
results (data not shown) for the virulent strains (PT, GT1, CT1 and S48) and
the avirulent strains (PLK, Tg51 and Fukaya) were identical to those obtained
for the virulent strains (RHa and ENT) and avirulent strains (ME49 and TPR)
shown in Fig. 3b. This further confirmed that B5-av can produce distinguishing
RFLP between virulent and avirulent *T. gondii* strains. These results are consistent with the hypothesis that there are significant DNA polymorphisms not
only in the primer B5 binding sites between virulent and avirulent *T. gondii*
strains, but also that there are virulence-specific RFLPs along the genome
between the B5 primer binding sites. In order to further characterise the RAPD
PCR virulence markers, B5-av and B12-v were sequenced (deposited as Genbank accession numbers L48960 and L48935) on an LI-COR automatic DNA
sequencer (Model 4000) with SequiTherm Cycle Sequencing (Epicenter Technologies, USA). Genbank database (FASTA) searches revealed no significant
homology with each other or with any other sequences in the Genbank database.

5 Discussion

Recently, RAPD PCR has been used to identify DNA loci of interest from
closely related organisms including species in apicomplexan genera such as
Eimeria and *Sarcocystis* (GRANSTROM et al. 1994; MacPHERSON and GAJADHAR 1994;

JOACHIM et al. 1996). DNA polymorphisms related to loci of interest have also been identified in near isogenic lines of plants by RAPD PCR (MARTIN et al. 1991). Three of the RAPD PCR products were confirmed to be tightly linked to the *Pseudomonas* resistance gene.

Based on these above studies, we believed that the RAPD PCR technique may also be suitable for identifying DNA polymorphisms in the loci of interest (especially those associated with virulence) within *Toxoplasma*. As described above, probe B5-av produced distinct RFLP that can be used to differentiate between virulent and avirulent *T. gondii* strains. We believe that this DNA fragment, and others like it that can be identified from RAPD PCR, will be of great value in discriminating murine virulent and avirulent *T. gondii* strains.

References

Asai T, Miura S, Sibley LD, Okabayashi H, Takeuchi T (1995) Biochemical and molecular charac-
    terization of nucleoside triphosphate hydrolase isozymes from the parasitic protozoan Toxo-
    plasma gondii. J Biol Chem 270:11391–11397
Cristina N, Liaud M, Santoro F, Oury B, Ambroise-Thomas P (1991a) A family of repeated DNA
    sequences in Toxoplasma gondii: cloning, sequence analysis, and use in strain characterisation.
    Exp Parasitol 73:73–81
Cristina N, Oury B, Ambroise-Thomas P, Santoro F (1991b) Restriction fragment length polymorph-
    isms among Toxoplasma gondii strains. Parasitol Res 77:266–268
Cristina N, Dardé ML, Boudin C, Tavernier G, Pestre-Alexandre M, Ambroise-Thomas P (1995) A
    DNA fingerprinting method for individual characterisation of Toxoplasma gondii strains: com-
    bination with isoenzymatic characters for determination of linkage groups. Parasitol Res 81:32–37
Dubey JP, Frenkel JK (1973) Experimental Toxoplasma infection in mice with strains producing
    oocysts. J Parasitol 59:505–512
Frenkel JK (1956) Pathogenesis of toxoplasmosis and of infections with organisms resembling
    Toxoplasma. Ann N Y Acad Sci 64:215–251
Granstrom DE, MacPherson JM, Gajadhar AA, Dubey JP, Tramontin R, and Stamper S (1994)
    Differentiation of Sarcocystis neurona from eight related coccidia by random amplified poly-
    morphic DNA assay. Mol Cell Probes 8:353–356
Guo Z-G, Johnson AM (1995a) Genetic characterization of Toxoplasma gondii strains by random
    amplified polymorphic DNA polymerase chain reaction. Parasitology 111:127–132
Guo Z-G, Johnson AM (1995b) Genetic comparison of Neospora caninum with Toxoplasma and
    Sarcocystis by random amplified polymorphic DNA-polymerase chain reaction. Parasitol Res
    81:365–370
Innis MA, Gelfand DH, Sninsky JJ, White TJ (1990) In: White TJ (ed) PCR protocols: a guide to
    methods and applications. Academic, San Diego, pp 482
Jacobs L (1972) New knowledge of Toxoplasma and toxoplasmosis. Adv Parasitol 11:631–669
Joachim A, Tenter AM, Jeffries AC, Johnson AM (1996) A RAPD-PCR derived marker can differentiate
    between pathogenic and non-pathogenic Sarcocystis species of sheep. Mol Cell Probes (in
    press)
Johnson AM (1984) Strain-dependent, route of challenge-dependent, murine susceptibility to tox-
    oplasmosis. Z Parasitenkd 70:303–309
Johnson AM (1990) Toxoplasma: biology, pathology, immunology and treatment. In: Long PL (ed)
    Coccidiosis of man and domestic animals. CRC, Boca Raton, pp 121–153
Johnson AM (1989) Toxoplasma vaccines. In: Wright IG (ed) Veterinary protozoan and hemoparasite
    vaccines. CRC, Boca Raton, pp 177–202
Kasper LH, Boothroyd JC (1993) Toxoplasma gondii: immunology and molecular biology. In: Warren
    K (ed) Immunology and molecular biology of parasitic infections. Blackwell, London, pp 269–299
Kaufman HE (1958) Toxoplasmosis: the nature of virulence. Am J Ophthalmol 46:255–260
Kaufman HE, Melton ML, Remington JS, Jacobs L (1959) Strain differences of Toxoplasma gondii.
    J Parasitol 45:189–190

Krahenbuhl JL, Remington JS (1982) The immunology of Toxoplasma and toxoplasmosis. In: Cohen S, Warren KS (eds) Immunology of parasitic infections. Blackwell, London, pp 356–421

Luton K, Gleeson M, Johnson AM (1995) rRNA gene sequence heterogeneity among Toxoplasma gondii strains. Parasitol Res 81:310–315

Mackenstedt U, Johnson AM (1995) Genetic differentiation of pathogenic and nonpathogenic strains of Entamoeba histolytica by random amplified polymorphic DNA polymerase chain reaction. Parasitol Res 81:217–221

MacPherson JM, Gajadhar AA (1994) Specific amplification of Sarcocystis cruzi DNA using a randomly primed polymerase chain reaction assay. Vet Parasitol 55:267–277

Martin GB, Williams JGK, Tanksley SD (1991) Rapid identification of markers linked to a Pseudomonas resistance gene in tomato by using random primers and near-isogenic lines. Proc Natl Acad Sci 88:2336–2340

McLeod R, Estes GR, Mack DG, Cohen H (1984) Immune response of mice to ingested Toxoplasma gondii: A model of Toxoplasma infection acquired by ingestion. J Infect Dis 149:234–244

Parmley SF, Gross U, Sucharczuk A, Windeck T, Sgarlato GD, Remington JS (1994) Two alleles of the gene encoding surface antigen P22 in 25 strains of Toxoplasma gondii. J Parasitol 80:293–301

Rinder H, Thomschke A, Dardé ML, Löscher T (1995) Specific DNA polymorphisms discriminate between virulence and non-virulence to mice in nine Toxoplasma gondii strains. Mol Biochem Parasitol. 69:123–126

Sibley LD, Boothroyd JC (1992) Virulent strains of Toxoplasma gondii comprise a single clonal lineage. Nature 359:82–85

Stevens JR, Tibayrenc M (1995) Detection of linkage of disequilibrium in Trypanosoma brucei isolated from tsetse flies and characterised by RAPD analysis and isoenzymes. Parasitology 110:181–186

Welsh J, McClelland M. (1990) Fingerprinting genomes using PCR with arbitrary primers. Nucleic Acids Res 18:7213–7218

Welsh J, Petersen C, McClelland M (1991) Polymorphisms generated by arbitrary primed PCR in the mouse: application to strain identification and genetic mapping. Nucleic Acids Res 19:303–306

Williams JGK, Hanafey MK, Rafalski JA, Tingey SV (1993) Genetic analysis using random amplified polymorphic DNA markers. Meth Enzymol 218:704–740

Williams JGK, Kubelik AR, Livak KJ, Rafalski JA, Tingey SV (1990) DNA polymorphisms amplified by arbitrary primers are useful as genetic markers. Nucleic Acids Res 18:6531–6535

Biodiversity in *Toxoplasma gondii*

M.L. Dardé

1 Introduction

Studies on the extent and nature of genetic polymorphism in medically important protozoan parasites are important in order to understand epidemiological and biological aspects of parasitic infections. Genetic variations among *Toxoplasma gondii* isolates first were suggested by differences in pathogenicity in Swiss mice, but also by the large range of clinical manifestations in humans, not entirely explained by the immune status of the host. Besides, due to the worldwide distribution, the broad host range of this species, and its capacity of sexual reproduction, a large genetic polymorphism was expected. Isoenzymatic studies (DARDÉ et al. 1988) first confirmed the existence of this genetic polymorphism, with three main zymodemes originally described. Correlations

---

Laboratoire de Parasitologie, CHU Dupuytren, 87042, Limoges Cédex, France

between genetic polymorphism and virulence in mice were observed. Further isoenzymatic analysis and the more recent application of molecular biology techniques allowed a better understanding of the significance of genetic polymorphism in *Toxoplasma* regarding aspects of epidemiology, such as population structure, geographical and zoological repartition, and regarding biological characteristics, such as pathogenicity to mice or oocyst production. The role of *T. gondii* genotype in the clinical presentation of human disease will be the next subject of extensive research.

2 Antigenic Diversity

Antigenic differences first were reported by WARE and KASPER (1987) among three cloned strains: three main surface antigens (P40, P30, P22) appeared to be different in the mouse-virulent RH strain and in two mouse-avirulent strains (P and C strains). Monoclonal antibodies against P22 and P30 did not kill P and C strains as effectively as the RH strain. Two forms of the P22 antigen corresponding to two alleles of the P22 gene could be identified (GROSS et al. 1991; PARMLEY et al. 1994). Similarly, the P30 antigenic differences suggested by WARE and KASPER (1987) could be related to *SAG1* (*P30*) gene polymorphism, which correlates with virulence in mice (SIBLEY and BOOTHROYD 1992a; RINDER et al. 1995). Morever, a monoclonal antibody against a 27 kDa antigen, present in the cytosol fraction, was also used to distinguish between mouse-virulent and -avirulent strains (BOHNE et al. 1993).

3 Diversity Detected by Genomic Analyses

The application of genetic characterization procedures confirmed the existence of differences between *Toxoplasma* isolates. Chromosome size variation was demonstrated by molecular karyotyping (SIBLEY and BOOTHROYD 1992b). A DNA polymorphism was detected by restriction fragment length polymorphism (RFLP) patterns (CRISTINA et al. 1991, 1995; SIBLEY and BOOTHROYD 1992a) with or without PCR amplification, by random amplified polymorphic DNA-polymerase chain reaction (RAPD-PCR) (GUO and JOHNSON 1995), by analysis of single-stranded (ss) rRNA gene sequences (LUTON et al. 1995) or DNA sequencing of a portion of the *P30* gene (RINDER et al. 1995). The results of these different techniques will be evaluated together with those of the isoenzyme analysis.

4 Diversity Detected by Isoenzyme Analysis

Isoenzyme analysis is a classical approach to assess the level of genetic variations in populations and to provide information on the reproductive biology or the population genetic data of a given organism. Its major advantage is the ability to examine a very large number of structural genes. The analysis of 15 enzyme systems is generally considered as a sample representative of the structural genome of a protozoan.

4.1 Isoenzyme Analysis in a Population of *Toxoplasma gondii* Isolates

The isoenzyme analysis was performed on the tachyzoite stage of 61 *Toxoplasma* isolates (Table 1), using the 15 enzyme systems already described, after isofocussing on polyacrylamide or agarose gels (DARDÉ et al. 1992). These isolates originated mainly from France (40 isolates), but also from other European countries (Denmark, England, Germany, and Holland), the Americas (USA, Argentina, Uruguay, and French Guiana), and Australia and Japan. They were isolated from human infections (41 isolates) or from animal infections (20 isolates).

As already described on a more limited sample (DARDÉ et al. 1992), only six of the 15 enzyme systems were polymorphic: acid phosphatase (ACP, EC 3.1.3.2), amylase (AMY, EC 3.2.1.1), aspartate aminotransferase (ASAT, EC 2.6.1.1), glucose phosphate isomerase (GPI, EC 5.3.1.9), glutathione reductase (GSR, EC 1.6.4.2), and propionyl esterase (PE). Two isoenzyme patterns (type I and type II) can be described for ASAT, AMY, GSR, and PE and three isoenzyme patterns (types I, II, and III) for GPI and ACP. The different combinations of the isoenzyme types allowed the identification of 11 zymodemes amongst this population of 61 *Toxoplasma* isolates (Fig. 1 and Table 2). In fact, seven zymodemes comprise only one isolate each, so that four main zymodemes (Z1, Z2, Z3, and Z4) cluster 88% of the isolates. Moreover, 29 isolates of this sample (that is nearly 50% of the isolates) belong to Z2. So, despite the expansion of the studied sample, zymodeme 2 is still markedly overrepresented in this *Toxoplasma* population, as previously described in a more limited sample (DARDÉ et al. 1992). It is possible that this overrepresentation of Z2 could be biased by the predominant human origin of the isolates (41 of 61 isolates), as discussed below. The 12 isolates of Z1 are all mouse-virulent isolates behaving like the well-known RH strain. Although 12 out of the 61 isolates belong to Z1, this does not accurately reflect the genuine proportion of this zymodeme in nature, because mouse-virulent isolates were selected for this analysis. For instance, the nine French isolates belonging to Z1 are nearly the only ones that have been isolated in France in 10 years, in contrast to hundreds of mouse-avirulent isolates.

Table 1. Isolates of *Toxoplasma gondii* characterized by isoenzyme analysis

	Geographical origin	Year of isolation	Host	Clinical history	Mouse-virulence[+]
Zymodeme 1					
CT-1	U.S.A.	1989	Cattle	Bovine intestine	V
RH	U.S.A.	1939	Human	Encephalitis	V
FOU	France	1992	Human	Lethal AT* (kidney transplant)	V
DPHT	France	1993	Human	Lethal AT (kidney transplant)	V
GPHT	France	1987	Human	Moderate CT**	V
GIL	France	1988	Human	Severe CT	V
MOR	France	1988	Human	Severe CT	V
ENT	France	1985	Human	Latent CT	V
FAJI	France	1991	Human	Latent CT	V
P***	France	1984	Human	Latent CT	V
PIL	France	1994	Human	Latent CT	V
BK	Holland	1948	Human	Lethal CT	V
ts-4	U.S.A	1976		mutant of RH strain	0
Zymodeme 2					
NTE	Germany	1990	Human	Encephalitis (AIDS)	A
BOU	France	1985	Human	Encephalitis (AIDS)	A
SUR	France	1993	Human	Disseminated AT (AIDS)	A
DAM	France	1993	Human	Disseminated AT (Aplasia)	A
JONES	England	1986	Human	Lymphadenopathies	A
CRO	France	1993	Human	CT	A
FAR	France	1991	Human	CT	A
CHAT	France	1988	Human	CT	A
FOUA	France	1988	Human	CT	A
AUDB	France	1988	Human	CT	A
CAL	France	1989	Human	Latent CT	A
REN	France	1989	Human	Latent CT	A
SZY	France	1988	Human	Latent CT	A
PRE	France	1993	Human	Latent CT	A
ROD	France	1988	Human	Latent CT	A
PON	France	1986	Human	Latent CT	A
CHAM	France	1983	Human	Lethal CT	A
PRUGNIAUD	France	1963	Human	Lethal CT	A
Tg132	Japan		Human	?	A
BEVERLEY	England	1959	Rabbit		A
S1	France	1977	Sheep	Lethal CT	A
S2	France	1979	Sheep	Lethal CT	A
S3	France	1980	Sheep	Lethal CT	A
ME49	U.S.A	1965	Sheep		A
PIG 3	Argentina	1993	Pig		A
P101	U.S.A.	1991	Pig		A
CH1	France	1986	Cat		A
CH2	France	1987	Cat		A
76K	France	1963	Guinea pig	Lethal AT	A
Zymodeme 3					
COR	France	1992	Human	Lethal AT (lymphoma)	I
NED	France	1989	Human	Latent CT	A
M7741	USA	1958	Sheep		I
C(EP)	USA	1977	Cat		A
OPA-OPA	Uruguay	1993	Pig		A
C56	USA	1961	Chicken		I

Table 1. Continued

	Geographical origin	Year of isolation	Host	Clinical history	Mouse-virulence[+]
Zymodeme 4					
ELG	France	1990	Human	Encephalitis (AIDS)	I
CHAMON	France	1988	Human	CT	A
DAS	France	1992	Human	Latent CT	A
DEG	France	1987	Human	Latent CT	A
C	France	1981	Human	Lethal CT	A
Tg96	Australia		Human		A
SQM	England	1990	Monkey	Lethal AT	A
Zymodeme 5					
MAS	France	1991	Human	Severe CT	V
Zymodeme 6					
RUB	French Guinea	1992	Human	Pneumonitis (immunocompetent patient)	V
Zymodeme 7					
CASTELLS	Uruguay	1993	Sheep	Lethal CT	I
Zymodeme 8					
TONT	France	1992	Human	Severe CT	V
Zymodeme 9					
SSI 119	Denmark	1968	Pig		A
Zymodeme 10					
P89	USA	1991	Pig		I
Zymodeme 11					
P80	USA	1991	Pig		A

[+] V: virulent to mice; A: avirulent; I: intermediate virulence; 0: no persistence in mice
* AT: acquired toxoplasmosis
** CT: congenital toxoplasmosis
*** P refered as PT by Sibley and Boothroyd (1992a)

4.2 Isoenzymes in Different Lines or Stocks of the Same Strains

Two strains (S1, S2) maintained in our laboratory and in another French laboratory in different conditions (stabilates or more or less frequent serial passages) were found to keep their isoenzyme characteristics. Different culture conditions of the same strains (human fibroblast cell culture or passages in vivo in mouse sarcoma cells) did not induce a shift in the isoenzyme pattern of differentiating enzymes; but a new isoenzyme form of lactate dehydrogenase (LDH) appeared for all the isolates cultivated in vitro in human fibroblasts (Dardé et al. 1990). This phenotypic variation could reflect the metabolic adaptability of *T. gondii* to new host cell environments or it could represent the appearance of bradyzoites in cell culture (Soête et al. 1993, Bohne et al., this volume).

Five laboratory stocks of the RH strain, maintained respectively in Würzburg and Göttingen (Germany), in Helsinki (Finland), in the USA (RH-88, A. Sher),

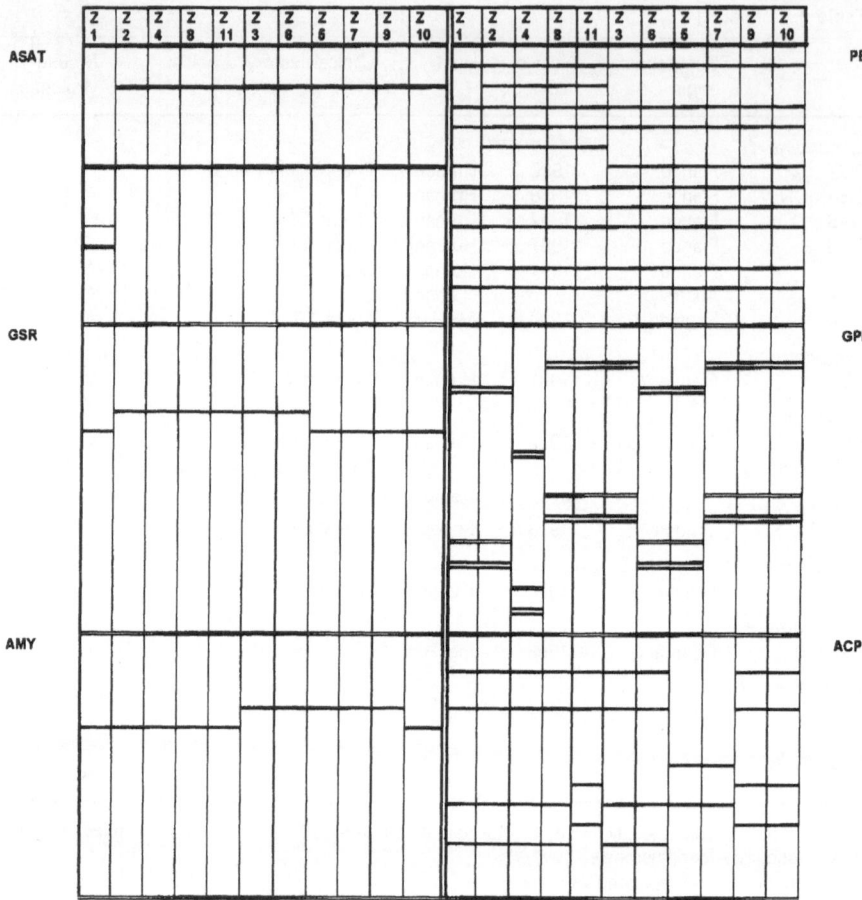

Fig. 1. Diagrammatic representation of isoenzyme patterns obtained for the eleven zymodemes, with six polymorphic enzyme systems

Table 2. Zymodemes in a population of 61 *T. gondii* isolates as defined by isoenzyme types of six variable enzyme systems

Zymodeme	Isoenzyme type						Number of isolates
	ASAT	GSR	AMY	GPI	PE	ACP	
1	I	I	I	I	I	I	12
2	II	II	I	I	II	I	29
3	II	II	II	II	I	I	6
4	II	II	I	III	II	I	7
5	II	I	II	I	I	II	1
6	II	II	II	I	I	I	1
7	II	I	II	II	I	II	1
8	II	II	I	II	II	I	1
9	II	I	II	II	I	III	1
10	II	I	I	II	I	III	1
11	II	II	I	II	II	III	1

and in Limoges (France), and the ts-4 strain, a temperature sensitive mutant derived from the RH strain (PFEFFERKORN and PFEFFERKORN 1976), exhibited identical isoenzyme patterns. These results contrast with the genetic heterogeneity demonstrated by RFLP analysis of five laboratory stocks of the RH strain and of the ts-4 strain using repetitive DNA probes (HOWE and SIBLEY 1994) and by ssrRNA gene sequence analysis of seven different lines of the RH strain (LUTON et al. 1995). From this point of view, isoenzyme phenotypes behave more like single-copy loci, like, for example, *SAG1*, and *SAG2* (HOWE and SIBLEY 1994).

4.3 Clustering Analysis of Electrophoretic Data

The expansion in the number of zymodemes makes a numerical taxonomy indispensable to establish relatedness between the different genotypes. A similarity coefficient for a pairwise comparison was calculated according to Jaccard's formula ($S=a/a+b+c$; in which a is the number of common bands between two isolates and b and c are the number of bands peculiar to each isolate). The group average clustering strategy was applied to the similarity matrix. The dendrogram constructed from the data obtained for the 11 zymodemes demonstrated three principal clusters (Fig. 2). The first group included the isolates of Z3, Z7, Z9 and Z10, the second group comprised Z2, Z4, Z8 and Z11. Finally, Z1 was closely related to the isolates of Z5 and Z6.

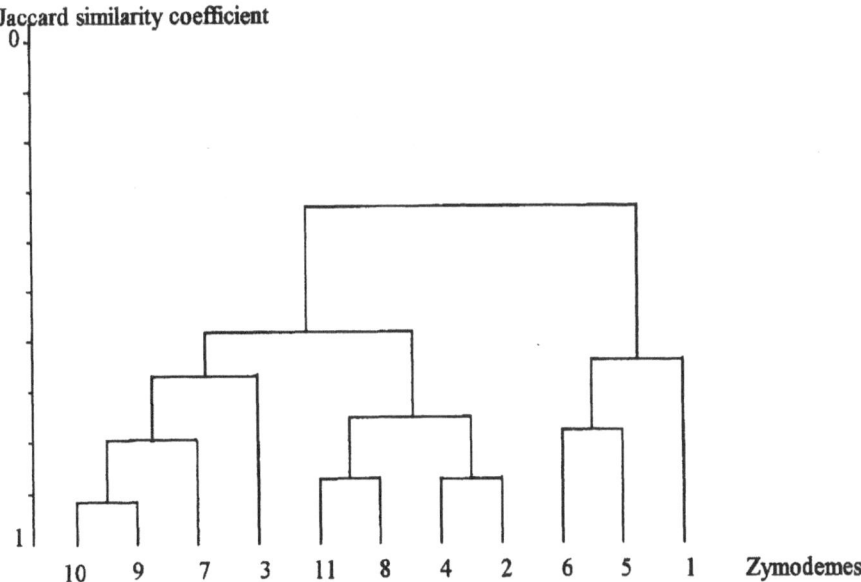

Fig. 2. Dendrogram of the 11 zymodemes constructed with the data of the isoenzyme types of six polymorphic enzymes, according to the group average clustering strategy

5 Intraspecies Level of Polymorphism in *Toxoplasma gondii*

Given the ubiquitous nature of the parasite, its worldwide distribution, and its possible sexual reproduction, an extensive heterogeneity would be expected. However, the genetic polymorphism detected among the population of 61 *Toxoplasma* isolates by isoenzyme analysis is still relatively limited, even if new isoenzyme patterns were detected. Only six out of 15 (40%) enzyme loci were polymorphic and only two or three different isoenzyme types were detected for these enzymes. Some zymodemes, e.g., Z2, Z4 and Z8, differed only by the isoenzyme type of GPI.

This low degree of polymorphism was also observed by SIBLEY and BOOTHROYD (1992a) using single-copy loci to distinguish between 28 strains. Similarly, the analysis of ssrRNA gene sequences of seven strains showed that *T. gondii* has accumulated very few differences compared to other Apicomplexans (LUTON et al. 1995). However, the level of discrimination provided by isoenzyme electrophoresis or by single-copy gene analysis is relatively low. The use of techniques such as DNA fingerprinting with multicopy probes, which provides a high level of genotypic discrimination, allows individual identification of different strains. A comparative study, performed on 14 *Toxoplasma* isolates with isoenzyme and RFLP analysis using two endonucleases (*Sal*I and *Pst*I) and two moderately repetitive DNA probes (TGR1E and TGR6), showed that an individual zymodeme comprised isolates that could be differentiated by RFLP analysis with that kind of repetitive probes (CRISTINA et al. 1995).

6 Correlation Between Genetic Markers

Several isolates or strains were separately studied by various genetic markers. A good correlation between the results of these different methods was observed (Table 3). For instance, 12 isolates studied by isoenzyme analysis were also studied in an RFLP analysis with five different genetic markers (SIBLEY and BOOTHROYD 1992a); these isolates were almost identical for the six polymorphic enzyme loci and for these five molecular markers (*SAG1* locus and loci hybridizing to probes 850 and BS, derived from a repetitive element dispersed in the *Toxoplasma* genome). The only slight differences were observed with the multicopy BS probe which results in hyperpolymorphic patterns. An RFLP analysis using repetitive DNA probes detected an important polymorphism among 14 isolates belonging to five different zymodemes (CRISTINA et al. 1995). However, when a numerical taxonomy method (similarity index) was applied to these RFLP data, a classification of *T. gondii* isolates into three main groups was observed that also correlated with zymodemes: the first

Table 3. Correlations between genetic markers

	Isoenzyme type						SAG1	850	BS			P22
	ASAT	GPI	AMY	PE	GSR	ACP			Mspl	Ddel	Hhal	
CT-1	I	I	I	I	I	I	1	1	1	1	1	
RH	I	I	I	I	I	I	1	1	1	1	1	1
P	I	I	I	I	I	I	1	1	1	1	1	
ENT	I	I	I	I	I	I	1	1	1	1	1	
ME49	II	I	I	II	II	I	2	2	2	2	2	2
S1	II	I	I	II	II	I	2	2	2	2	2	
BEV	II	I	I	II	II	I	2	2	2	2	2	2
Tg132	II	I	I	II	II	I	2	2	5	2	5	
Tg96	II	III	I	II	II	I	2	2	2	2	4	
CEP	II	II	II	I	II	I	2	1	3	3	3	1
M7741	II	II	II	I	II	I	2	1	4	3	3	
C56	II	II	II	I	II	I	2	1	4	3	3	1

Results of the analysis of *SAG1*, 850 and BS loci were collected from Sibley and Boothroyd (1992b) and of *P22* loci from Parmley et al. (1994).

group clustered the virulent strains (Z1–Z5), the second group clustered isolates belonging to Z2 and Z4 in a kind of superfamily, and the third corresponded to Z3 isolates.

This clustering into three main groups (A, B and C) is in agreement with a classification based on the combination of different alleles of the three genetic loci *P22* (*SAG2*), 850, and *SAG1* (PARMLEY et al. 1994). An analysis of six polymorphic single-copy genetic loci identified three types of *Toxoplasma* strains (I, II and III; HOWE and SIBLEY 1995). Taken together, the results of the studies of PARMLEY et al. (1994), SIBLEY and BOOTHROYD (1992a), HOWE and SIBLEY (1995) and isoenzyme analysis suggest that group A could be equated to genetic type I and to zymodeme 1, group B to type II and to Z2–Z4, and group C to type III and zymodeme 3.

7 Relationships Between Genetic Markers and Biological Characteristics

7.1 Pathogenicity to Mice

Most of the studies concerning genetic polymorphism among *Toxoplasma* strains have tried to establish relationships between pathogenicity in mice and genotypes. The isoenzyme analysis of 61 *Toxoplasma* isolates confirms the correlation between genetic markers and pathogenicity to mice: most of the mouse-virulent strains tested in this sample (12/15) belong to the same zymodeme (Z1). However, three highly mouse-virulent isolates exhibited clearly different isoenzyme patterns (Z5, Z6, and Z8).

The strains giving rise to chronic infections in mice are found in the seven other zymodemes. In fact, among the so-called nonpathogenic or mouse-avirulent isolates, different kinds of pathogenicity to mice can be observed: mice infected with certain nonpathogenic strains may live a normal life span, whereas those infected with other strains slowly die off during the first few months after infection with manifestations of subacute encephalitis (DARDÉ et al. 1988, SUZUKI et al. 1989). Zymodemes 2 and 4 contain only mouse-avirulent strains, even if two of the Z4 isolates initially killed the inoculated mice and became nonvirulent only after a certain number of passages in mice with a very small inoculum. Z3 isolates were originally described as nonvirulent isolates. However, under our laboratory conditions, at least two of them (M7741 and C56) are responsible for the death of most of the inoculated mice 10–13 days after inoculation or for the development of subacute encephalitis. Similarly, the Z10 isolate, which is genetically related to Z3, was described by DUBEY et al. (1995) as more virulent for mice than other "avirulent" isolates that also originated from pigs (such as the P101 isolate belonging to Z2 or the P80 isolate belonging to Z11). Therefore, this heterogenous behavior in mice could reflect the genetic subdivisions observed in the mouse-avirulent group. It should be noted that three of these Z3 strains (CEP, M7741, and C56) with an intermediate pathogenicity also exhibited an intermediate genetic pattern of P22 and 850 alleles (PARMLEY et al., 1994; SIBLEY and BOOTHROYD 1992a). Due to the influence of host or environmental factors on the expression of virulence, it could not be ascertained that this subtle difference in virulence reflects genetic differences between isolates. The course of infection depends on the potency of the parasite inoculum, the route and kinetics of infection, and the genetic background of the congenic mouse strains (BROWN et al. 1995). Pathogenicity in a given host will also be increased by frequent and rapid passages (FRENKEL 1973), as is the case for most of the laboratory strains.

7.2 Oocyst Production

FRENKEL et al. (1976) demonstrated that frequent and rapid passages in mice could eventually induce the loss of oocyst production capacity in some *T. gondii* strains. They concluded that host-induced selection leads to a loss of alleles in isolates passaged in mice. However, the capacity of oocyst production was absent even for some strains studied soon after their isolation in mice, i.e., after less than ten passages (DARDÉ et al. 1992). Therefore, besides pathogenicity in mice, capacity of sexual reproduction can be additional evidence of the biological diversity among *T. gondii* strains.

Oocyst production has never been observed for strains belonging to Z1, except for the CT-1 isolate (DARDÉ et al. 1992), which originated from the intestinal tissue of a beef cow (DUBEY 1992). Although present, the oocyst production of the CT-1 isolate is markedly reduced in comparison to strains belonging to other zymodemes.

8 Epidemiological Implications of the Genetic Analysis

The isoenzyme analysis shows that the main zymodemes do not seem to be specific for a given geographic area. For example, Z1 originates from both the USA and France. For sampling reasons, Z2 isolates come mainly from France, but they are also found in other European countries, in Asia (Japan) or in North and South America (USA, Argentina). Z3 isolates originate from the USA, Uruguay, and France, and Z4 isolates from various regions of France, England, and Australia.

However, this set of isolates still may be too limited to draw definitive conclusions. The exchange of meat and animals (pigs, sheep) between these countries could explain the isolation of the same zymodeme in different regions of the world. In this respect, particular genotypes, such as the Z6 isolate which originated from the deep forest of French Guinea where domestic cats are missing, and where wild felines are the only possible definitive hosts, suggest that separate epidemiologic reservoirs which harbor distinct sets of *T. gondii* could exist in some isolated geographic area.

The large host range of *Toxoplasma* and its transmission through carnivorism or ingestion of oocysts makes host specificity of zymodemes unlikely (Table 1), unless immune mechanisms of the host could preferentially eliminate some *Toxoplasma* genotypes. Isoenzyme analysis showed that 21 of 32 (66%) Z2–Z4 isolates are from human origin, compared to only two of six (30%) Z3 isolates. This finding is consistent with a genetic analysis of a large collection of *Toxoplasma* isolates (Howe and Sibley 1995). However, it should be noted that only a very small sample of the spectrum of possible hosts has been analysed.

9 Population Genetic Analysis

Population genetic data obtained using techniques such as DNA sequencing, RFLP patterns and enzyme electrophoresis have recently been applied to answer fundamental questions about the predominant mode of reproduction of parasitic protozoa. Although the studied *Toxoplasma* sample, which results from accumulation of a disparate collection of isolates originating from different geographic areas and host species, cannot be considered as a random sample of *Toxoplasma* isolates circulating in nature, isoenzyme and genomic analyses support data suggesting a mainly clonal population structure for *Toxoplasma* (TIBAYRENC et al. 1991; TIBAYRENC 1993; SIBLEY and BOOTHROYD 1992a; PARMLEY et al. 1994). The arguments for clonality proposed by TIBAYRENC et al. (1991) are easily recovered in the studies concerning a large sample of *Toxoplasma* isolates: (1) the isolation of identical multilocus genotypes over large geographic areas and at intervals of several years, (2) the small number of different genotypes and the overrepresentation of the observed genotypes by comparison with panmixic expectations, providing evidence of a high linkage disequilibrium and (3) the correlation between independant sets of genetic markers, such as the different enzyme loci and the molecular markers presented in the RFLP studies.

Genetic studies proposed three main clonal lineages of *T. gondii* (HOWE and SIBLEY 1995; PARMLEY et al. 1994). Isoenzyme analysis also supports data suggesting that each of the main zymodemes could be considered as a clonal entity and that the *Toxoplasma* population is basically clonal, as proposed by TIBAYRENC et al. (1991). This clonal structure, suggesting a predominant uniparental cycle (self-fertilization in cat, transmission through intermediate hosts by carnivorism, absent or reduced oocystogenesis for Z1 isolates) does not totally exclude occasional sexual reproduction with cross-fertilization: the isoenzyme patterns obtained for Z9, Z10 and Z11 suggest a possible recombination in the definitive host (Table 2). Experimental mixed infection has been performed (PFEFFERKORN and PFEFFERKORN 1976), but this seems to be rare in nature. Strong immunity conferred by toxoplasmic infection could explain that both cats and intermediate hosts usually harbor only one genotype of *Toxoplasma* isolates, providing a considerable potential for self-fertilization in cats. This clonal population structure is important in establishing correlations between particular strains and clinical presentation of toxoplasmosis.

10   Reliability with Pathogenicity in Humans?

In human toxoplasmosis, the role of the infecting *Toxoplasma* strain is not easily assessed although it is clear that infections can have dramatically different outcomes. There is overwhelming evidence that the immunological status of the host plays a major part in the clinical presentation of this opportunistic infection.

10.1   Acquired Toxoplasmosis

A population of parasite isolates could be defined from asymptomatic, immunocompetent, pregnant women who have given birth to infected children. The corresponding *Toxoplasma* strain isolated from the infected child is likely to be identical to the strain that originally had infected the mother. Each of these isolates belong to one of the main zymodemes (Z1, Z2, Z3, and Z4). Therefore, an absolute relationship between asymptomatic infection and a given zymodeme does not seem to exist.

Two isolates with unusual genotypes belonging to Z5 and Z6 gave rise to uncommon clinical presentations in immunocompetent patients. The Z5 isolate originated from a French case of congenital toxoplasmosis. The mother acquired infection at 18 weeks of pregnancy without any clinical symptoms and was treated with spiramycin. Congenital infection of the fetus was antenatally diagnosed, followed by therapeutic abortion: *Toxoplasma* was isolated from blood, brain and liver of the infected fetus. The mother presented symptomatic toxoplasmosis with fever and persisting lymphadenopathy 1 month after abortion. It remains to be answered if the unusual outcome of this acquired toxoplasmosis in the pregnant women is entirely due to a different immune response or if the unusual genetic characteristics of the *Toxoplasma* isolate played a role in this outcome. The only isolate of Z6 originates from a case of severe toxoplasmosis (pneumonitis, renal impairment, and pericarditis) acquired by an immunocompetent soldier in the deep forest of French Guinea. Four other soldiers of the same group also became infected during the same journey: one of them died from toxoplasmosis and the three others developed a severe pulmonary toxoplasmosis (PINON M., personal communication). It could be hypothesized that humans are highly susceptible to this type of isolate, which is found in an environment unusual for humans. By contrast, the usual *Toxoplasma* genotypes, which have coevolved with humans in a domestic cycle, are well adapted to their host and do not give rise to severe clinical manifestations. Similarly, New World monkeys confronted with a strain belonging to a zymodeme circulating in Europe or North America (SQM isolate, Z4) developed severe toxoplasmosis during acute epidemics in zoos (CUNNINGHAM et al. 1992). These two isolates could provide evidence that the genetic characteristics of the isolate plays a part in the clinical outcome. It must be remembered that these two isolates are also unusual mouse-virulent isolates, as they do not belong to Z1.

The number of isolates originating from cases of acquired toxoplasmosis in immunodeficient patients is not large enough to draw any significant conclusion (only four AIDS isolates). However, once again, reactivation of toxoplasmic infection appeared to be possible with each of the main zymodemes.

10.2 Congenital Toxoplasmosis

Isoenzyme analysis shows that each zymodeme could be responsible for congenital toxoplasmosis. The main factor for the severity of congenital infection seems to be the term of pregnancy at the time of contamination rather than the zymodeme of the infecting strain: independent of latter, latent toxoplasmosis in the child is mainly observed after infection in the last trimester, and symptomatic or lethal toxoplasmosis after infection in the first or second trimester.

11 Conclusion

These studies of *Toxoplasma* polymorphism and its implications are still in their infancy compared, for instance, to what has been discovered about *Plasmodium* or *Entamoeba histolytica*. However, major evidence is accumulating concerning the role of *Toxoplasma* polymorphism regarding some of its biological characteristics and new insights on the reproductive biology of this species have been proposed. Nonetheless, extensive research is still needed before definitive conclusions can be drawn, especially about the relationship between strain genotype and the human clinical manifestations of toxoplasmosis.

Acknowledgements. This work was supported in part by the European Union (BMH1-CT92-1535; EC concerted action on *Toxoplasma* strain differences and AIDS). I also want to acknowledge all the scientists who kindly provided *Toxoplasma* isolates.

References

Bohne W, Gross U, Heesemann J (1993) Differentiation between mouse-virulent and -avirulent strains of Toxoplasma gondii by a monoclonal antibody recognizing a 27-kilodalton antigen. J Clin Microbiol 31:1641–1643

Brown CR, Hunter A, Estes RG, Beckmann E, Forman J, David C, Remington JS (1995) Definitive identification of a gene that confers resistance against Toxoplasma cyst burden and encephalitis. Immunol 85:419–428

Cristina N, Oury B, Ambroise-Thomas P, Santoro F (1991) Restriction fragment length polymorphism among Toxoplasma gondii strains. Parasitol Res 77:266–268

Cristina N, Dardé ML, Boudin C, Tavernier G, Pestre-Alexandre M, Ambroise-Thomas P (1995) A DNA fingerprinting method for individual characterization of Toxoplasma gondii strains: combination with isoenzymatic characters for determination of linkage groups. Parasitol Res 81:32–37

Cunningham AA, Buxton D, Thomson KM (1992) An epidemic of toxoplasmosis in a captive colony of squirrel monkeys (Saimiri sciureus). J Comp Path 107:207–209

Dardé ML, Bouteille B, Pestre-Alexandre M (1988) Isoenzymic characterization of seven strains of Toxoplasma gondii by isoelectrofocalisation in polyacrylamide gels. Am J Trop Med Hyg 39:551–558

Dardé ML, Bouteille B, Pestre-Alexandre M (1990) Comparison of isoenzyme profiles of Toxoplasma gondii tachyzoites produced in different culture conditions. Parasitol Res 76:367–371

Dardé ML, Bouteille B, Pestre-Alexandre M (1992) Isoenzyme analysis of 35 Toxoplasma gondii isolates and the biological and epidemiological implications. J Parasitol 78:786–794

Dubey JP (1992) Isolation of Toxoplasma gondii from a naturally infected beef cow. J Parasitol 78:151–153

Dubey JP, Thulliez P, Powell EC (1995) Toxoplasma gondii in Iowa sows: comparison of antibody titers to isolation of T. gondii by bioassays in mice and cats. J Parasitol 81:48–53

Frenkel JK (1973) Toxoplasmosis: parasite life cycle, pathology and immunology. In: Hammond DM, Long PL(eds) The Coccidia Eimeria, Isospora, Toxoplasma and related genera. University Park, Baltimore, pp 343–410

Frenkel JK, Dubey JP, Hoff RL (1976) Loss of stages after continuous passage of Toxoplasma gondii and Besnoitia jellisoni. J Protozool 23:421–424

Gross U, Müller WA, Knapp S, Heesemann J (1991) Identification of a virulence-associated antigen of Toxoplasma gondii by use of a mouse monoclonal antibody. Infect. Immun. 59:4511–4516

Guo ZG, Johnson AM (1995) Genetic comparison of Neospora caninum with Toxoplasma and Sarcocystis by random amplified polymorphic DNA-polymerase chain reaction. Parasitol Res 81:365–370

Howe DK, Sibley LD (1994) Toxoplasma gondii: analysis of different laboratory stocks of the RH strain reveals genetic heterogeneity. Exp Parasitol 78:242–245

Howe DK, Sibley LD (1995) Toxoplasma gondii is comprised of three clonal lineages: correlation of parasite genotype with human disease. J Infect Dis 172:1561–1566

Luton K, Gleeson M, Johnson AM (1995) rRNA gene sequence heterogeneity among Toxoplasma gondii strains. Parasitol Res 81:310–315

Parmley SF, Gross U, Sucharczuk A, Windeck T, Sgarlato GD, Remington JS (1994) Two alleles of the gene encoding surface antigen P22 in 25 strains of Toxoplasma gondii. J Parasitol 80:293–301

Pfefferkorn ER, Pfefferkorn LC (1976) Toxoplasma gondii: isolation and primary characterization of temperature-sensitive mutants. Exp Parasitol 39:365–376

Rinder H, Thomschke A, Dardé ML, Löscher T (1995) Specific DNA polymorphisms discriminate between virulence and non-virulence to mice in nine Toxoplasma gondii strains. Mol Biochem Parasitol 69: 123–126

Sibley LD, Boothroyd JC (1992a) Virulent strains of Toxoplasma gondii comprise a single clonal lineage. Nature 359:82–85

Sibley LD, Boothroyd JC (1992b) Construction of a molecular karyotype for Toxoplasma gondii. Mol Biochem Parasitol 51:291–300

Soête M, Fortier B, Camus D, Dubremetz JF (1993) Toxoplasma gondii: kinetics of bradyzoite-to-chyzoite interconversion in vitro. Exp Parasitol 76:259–264

Suzuki Y, Conley FK, Remington JS (1989) Differences in virulence and development of encephalitis during chronic infection vary with the strain of Toxoplasma gondii. J Infect Dis 159:790–794

Tibayrenc M (1993) Entamoeba, Giardia, and Toxoplasma: clones or cryptic species? Parasitol Today 9:102–105

Tibayrenc M, Kjellberg F, Arnaud J, Oury B, Brénière SF, Dardé ML, Ayala F (1991) Are eukaryotic organisms clonal or sexual ? A population genetics vantage. Proc Natl Acad Sci 88:5129–5133

Ware P, Kasper LH (1987) Strain-specific antigens of Toxoplasma gondii. Infect Immun 55:778–783

B

The Parasite:
From Antigens to Genes

The Major Surface Proteins
of *Toxoplasma gondii*: Structures and Functions

S. Tomavo

1 Introduction

The obligate intracellular protozoan *Toxoplasma gondii* can penetrate into a variety of eukaryotic cells. This broad host and tissue specificity, which is an unique feature of this parasite among the Apicomplexa parasites such as *Plasmodium*, *Eimeria*, and *Sarcocystis*, indicates that ligands and receptors mediating initial attachment of *T. gondii* to host cells are likely to be highly conserved and ubiquitously distributed. *Toxoplasma* enters host cells by an active process called invasion, in which special organelles, designated rhoptries and micronemes, are believed to be involved in cell penetration. Although the precise mechanism of invasion is not understood yet, it is generally agreed that the attachment or adhesion of *T. gondii* to host cells is a critical step in invasion. It would be most helpful if receptors and ligands on *Toxoplasma* and the host cell membrane were identified and their implications in cell invasion established. This review will focus essentially on structures of the major surface proteins of *Toxoplasma* and will discuss recent progress toward the identification of their functions in host cell recognition and adhesion.

INSERM U.415, Institut Pasteur, 1, rue du Prof. A. Calmette, 59019 Lille, France

Table 1. The surface of *Toxoplasma* tachyzoites is dominated by five major proteins

Proteins (kDa)	Encoding Sequences Known[1]	GPI-anchor[2]	N-glycosylation[3]	Functions[4]
30	*SAG1*	Yes	No	Attachment factor
22	*SAG2*	Yes	No	NK
43	*SAG3*	Yes	No	Attachment factor
35	NK	Yes	No	NK
23	NK	Yes	Yes	NK

NK: not known

[1]SAG means surface antigen: SIBLEY et al. (1991), BURG et al. (1988), PRINCE et al. (1990), CESBRON-DELAUW et al. (1994)

[2]NAGEL and BOOTHROYD 1989; TOMAVO et al. (1989, 1992a)

[3]ODENTHAL-SCHNITTLER et al. (1993)

[4]GRIMWOOD and SMITH (1992), MINEO et al. (1993), TOMAVO (this review)

2 Identification of the Major Surface Proteins

Due to the ease of both in vitro and in vivo propagation, most of the work thus far has been focused on the rapidly replicating tachyzoite stage. HANDMAN et al. (1980), KASPER et al. (1982, 1983) and COUVREUR et al. (1988) have pioneered the characterization of tachyzoite surface proteins by generating monoclonal antibodies that identify five major surface proteins of 22, 23, 30, 35 and 43 kDa (designated P22, P23, P30, P35 and P43) according to their mobilities in SDS-polyacrylamide gel electrophoresis (Table 1). A monoclonal antibody specific for P30 has been used to purify this protein by immunoprecipitation and to subsequently show that P30 represents 5% of the total amount of tachyzoite proteins (KASPER et al. 1983). It appears that most of the surface proteins electrophorese slower under reduced gel conditions; in addition, P35 and P30 proteins comigrate under unreduced conditions. The posttranscriptional modifications (such as N-, O-glycosylation, and membrane anchoring) that occur with these proteins have been studied. Indeed, in addition to the peptide backbone, these structures could play a role as attachment or adhesion factors.

3 Glycosyl-Phosphatidylinositol Anchors

Although some of the plasma membrane proteins of the parasitic protozoa use transmembrane polypeptide anchors, all five known surface proteins of *Toxoplasma* have glycosyl-phosphatidylinositol (GPI) structures that serve to anchor these molecules to the plasma membrane (Fig. 1; NAGEL and BOOTHROYD 1989; TOMAVO et al. 1989). This type of anchor is more frequently used in

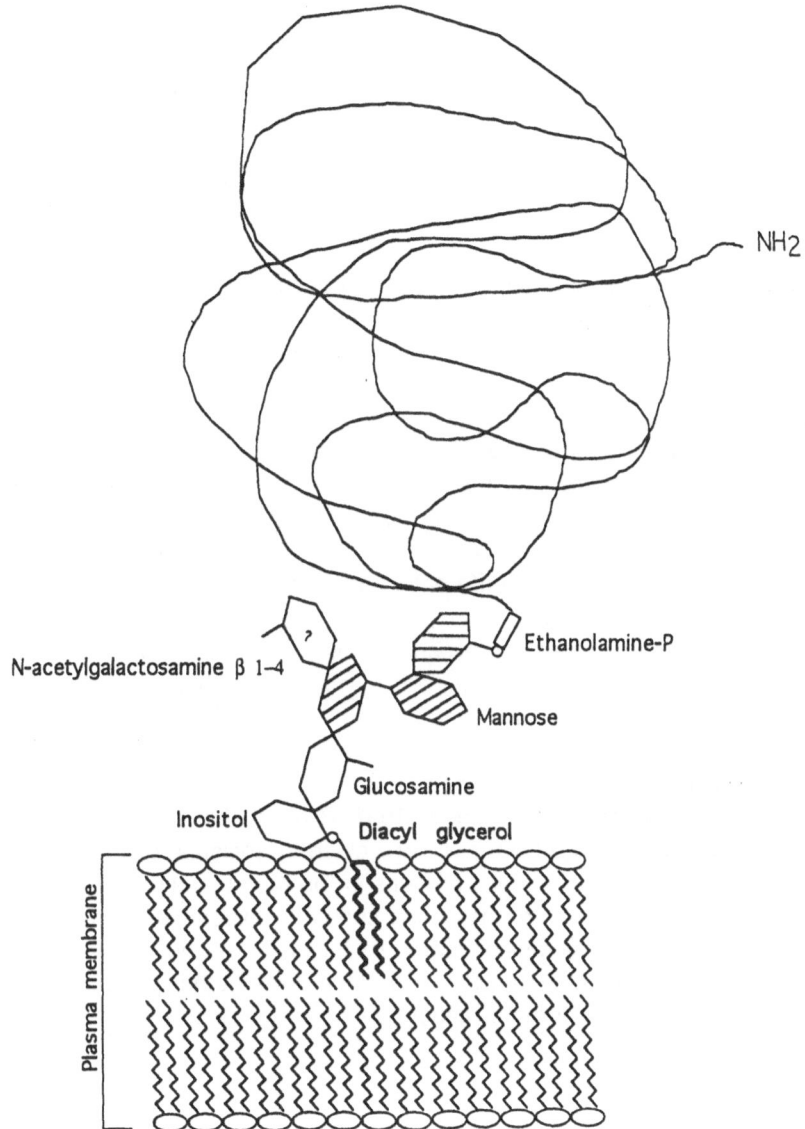

Fig. 1. Structure of glycolipid anchors of *Toxoplasma gondii* P30 and P23 proteins. All glycosidic linkages of the GPI backbone are in the α-configuration (ToMAVO et al. 1989, 1992a,b, 1993), and the β-configuration of GalNAc is not yet determined

Toxoplasma than in higher eukaryotes. Indeed, no other type of plasma membrane anchoring is known in *T. gondii*. However, genetically engineered P30, containing a transmembrane anchor instead of the GPI anchor, is efficiently targeted to the parasite surface, indicating that a GPI anchor is not absolutely necessary to localize a protein on the surface of *T. gondii* (SEEBER and BOOTHROYD, personal communication). The partial structure of the GPI anchors of these proteins has been determined using metabolic labeling (tritiated glucosamine, mannose, galactose, palmitic and myristic acids, and inositol). The sensitivity of these proteins to a phosphatidylinositol phospholipase C clearly demonstrates that these proteins possess GPI anchors. The release of these proteins from the surface of live parasites, which causes a cross-reacting determinant (CRD) of the soluble forms to be accessible to anti-CRD serum of trypanosomes, also demonstrates that they possess GPI anchors. Further investigations led to the identification of the evolutionarily conserved linear GPI core structure (Fig. 1), consisting of ethanolamine-PO_4-6Manα1-2Manα1-6Manα1-4GlcNAcα1-6-inositol on P30 and P23 proteins (Tomavo et al. 1992a, 1993). Four major glycolipids have been isolated either in living parasites or in a parasite extract. These glycolipids have the same GPI core structure and can serve as preassembled precursors of GPI anchors linked to these proteins (Tomavo et al. 1992b). As reported for the Thy-1 molecule (Homans et al. 1988), N-acetylgalactosamine has been identified on GPI anchors of both proteins and glycolipids, suggesting the presence of side chain modifications on the core structure (Fig. 1). The pathway of GPI biosynthesis in *T. gondii* consists of sequential addition of different monosaccharides to phosphatidylinositol until the transfer of ethanolamine phosphate takes place (Tomavo et al. 1992b), as previously described for trypanosomes (Masterson et al. 1989; Menon et al. 1990). A molecule specific for the *Toxoplasma* anchors, which can be used for drug design or specific targeting, has not been found so far. The role of GPI anchors in *T. gondii* is not known; however, in other systems it can be more than a simple anchor. For example, the GPI anchor can be involved in signal transduction (Stefanova et al. 1991; Brown 1993). One possible function of the GPI anchor might be to allow a closer association of the proteins with themselves and other surface proteins in the membrane (Ferguson 1994). Evidence for this in *T. gondii* has been shown in experiments with the above-mentioned transmembrane-anchored P30, which does not show the observed association of GPI-anchored P30 with itself and/or other proteins (Seeber and Boothroyd, personal communication).

4   N-Glycosylation

The presence of N- and O-glycosylated structures on the surface molecules of tachyzoites has been controversial since the work of Handman et al. (1980). Based on the inability of living parasites and radioiodinated proteins to bind to lectins, which are generally used to identify surface glycoproteins of most eukaryotic cells, these authors concluded that there were no such post-transcriptional modifications on the surface of *T. gondii* tachyzoites. In contrast, Mauras et al. (1981) and Johnson et al. (1981) have detected concanavalin A binding to both living tachyzoites and isolated proteins. Using a combination

of metabolic labeling with tritiated monosaccharides, gel filtration chromatography and enzymatic digestions (peptidase-N-glycanase F, endoglycosidases
H, F and exoglycosidases), N-linked glycans in the hybrid- or complex-type
form (composed of N-acetylgalactosamine, N-acetylglucosamine and mannose)
have been identified on one surface protein, P23 (ODENTHAL-SCHNITTLER et al.
1993; SCHWARZ and TOMAVO 1993; SCHWARZ et al. 1993). Even if *T. gondii* micro-
somes are able to synthesize a glycosylated lipid-bound high-mannose structure, known as N-glycosylation precursor in eukaryotic cells (ODENTHAL-SCHNITTLER
et al. 1993), N-linked structures are not a predominant modification occurring
on tachyzoite surface proteins. This is also true for *Plasmodium falciparum,*
in which a complete inability to synthesize N-linked glycans has been documented (DIECKMAN-SCHUPPERT et al. 1992). It has been shown that *Plasmodium
falciparum* modifies its proteins with O-glycosylated structures instead of N-
glycosylation. The presence of O-glycans on *T. gondii* surface proteins still
needs to be further explored.

5 Determination of Primary Sequences

The primary sequences of three major *T. gondii* surface proteins, P30, P22
and P43, derived from cDNA or gene sequences have been published (BURG
et al. 1988; PRINCE et al. 1990; CESBRON-DELAUW et al. 1994). Similar to the yeast
nomenclature, the proteins are designated as SAG1, 2 and 3 (SIBLEY et al.
1991). As described in other GPI-anchored proteins, all three sequences contain
processing signals in their translated product. Firstly, they contain an NH_2-ter-
minal signal sequence for entry into the lumen of the endoplasmic reticulum
and, secondly, they contain a GPI signal sequence at the COOH-terminal.
These COOH-terminal hydrophobic tails are 26, 14 and 25 amino acids long,
confirming that these molecules contain a GPI anchor. During the trafficking
of the newly synthesized polypeptide through the endoplasmic reticulum, the
hydrophobic tails are likely to be cleaved and replaced with a GPI anchor by
a transaminidase (CROSS 1990). Such a transaminidase has not been isolated
yet in any eukaryotic cell. Further analysis of these sequences revealed one
putative N-linked site (N-A-S) in SAG1. However, this putative N-linked motif
is not used by the N-glycosylation machinery of *T. gondii*. N-linked motifs are
not present in the sequences of SAG2 and SAG3 proteins. Data bank searches
for homology to known molecules, which would help to identify biological
functions, were unsuccessful; however, surprisingly, we found significant re-
semblance between SAG1 and SAG3 proteins in the number (12) and distribution of cysteine and tryptophan residues (CESBRON-DELAUW et al. 1994).
Additionally, these proteins share two identical peptides, Q-Y-C-S-G and G-A-
T-L-T-I, and their overall hydrophobic/philic profiles are remarkably similar.
These observations suggest that SAG1 and SAG3 may have the same folding

pattern and, therefore, probably similar functions. The genes encoding P35 and P23 surface proteins remain to be cloned.

6 Using Mutants as an Approach to Study Biological Functions of *Toxoplasma* Surface Proteins

6.1 Chemical Mutagenesis

Mutants unable to express SAG1 or mutants expressing altered SAG1 have been generated by ethylnitrosourea mutagenesis followed by selection of resistance to antibody-complement lysis (KASPER 1987). One mutant, designated B-mutant, contains a point mutation in the *SAG1* gene (stop codon instead of a leucine at position 203). As a consequence, the SAG1 protein was not detectable by western blot and immunofluorescence assay (IFA). A second mutant, designated A-mutant, expresses an altered SAG1 protein (a serine instead of a cysteine at amino acid 295), resulting in decreased mobility compared to wild-type SAG1 on unreduced SDS-PAGE gel. This conformationally altered SAG1 is not recognized by antisera to native SAG1 (KIM et al. 1994). A third mutant, designated C-mutant, which has a point mutation in the *SAG1* gene, has also been described. Despite the ability of SAG1 antisera to inhibit wild-type parasite invasion of host cells (GRIMWOOD and SMITH 1992; MINEO et al. 1993), these SAG1 mutants appear to invade host cells with no significant difference compared to wild-type *T. gondii* (MINEO et al. 1993). Since these mutants are able to infect host cells, it can be concluded that alternative molecules must exist to allow host cell invasion. A SAG2-mutant (P22 surface protein) has also been obtained using chemical mutagenesis (KASPER et al. 1982), and this mutant is also able to invade host cells. However, a major concern with using the SAG1 and SAG2 mutants is that chemical mutagenesis may also have induced other mutations outside the *SAG1* and *SAG2* genes. This could explain the absence of a clear-cut phenotype for these mutants. The recent development of transfection systems in *T. gondii* (KIM et al. 1993; SOLDATI and BOOTHROYD 1993) allows the use of reverse genetics to create recombinant parasites with a perfect gene replacement (gene knockout).

6.2 Reverse Genetics

We have engineered a SAG3 knockout mutant by transfecting tachyzoites of *T. gondii* (RH strain) with a linearized plasmid containing a *Toxoplasma* selection cassette (5' *TUB*-promoter-chloramphenicol acetyltransferase-*SAG1* untranslated region 3') contained within the 5' (2.8 kb) and 3' (2.2 kb) untranslated flanking sequences of the *SAG3* gene. Four mutants, lacking the SAG3 protein due to a perfect replacement of the *SAG3* gene, which has been proven by southern and northern blot analyses, have been obtained (WYLS et al., unpublished). Two of these mutants, through further analysis, showed less virulence in vivo (LD$_{50}$ in mice was 500- to 1000-fold greater than with the wild-type RH strain). Their ability to invade different eukaryotic cells in vitro was reduced to 40%–60% compared to the wild type. These findings show that the SAG3 protein is an attachment/virulence factor expressed on the surface of *T. gondi* and that a lack of the SAG3 protein is not lethal for the parasite. The fact that a significant proportion (about 50%) of the SAG3 knockout mutants still invaded host cells suggests the existence of other alternative attachment/virulence factors in these mutants. Such conclusions have been drawn from studies on other protozoan parasites like *Plasmodium, Leishmania,* and *Trypanosoma* (WILSON and HARDIN 1990; SCHENKMAN et al. 1991; GALINSKI et al. 1992). In addition, the striking homology between SAG3 and SAG1 proteins clearly points out that redundancy exists in parasite molecules which are involved in attachment and invasion of host cells and possibly in virulence as well. Studies are now in progress for the generation of double mutants devoid of both SAG3 and SAG1 proteins.

7 Other Attachment Factors

Recent investigations indicate that, in addition to the implication of the SAG1 protein (GRIMWOOD and SMITH 1992; MINEO et al. 1993) and the SAG3 protein involved in attachment (described in this review), *T. gondii* attachment may also be mediated either by a parasite laminin and host cell laminin receptor or by parasite surface lectin-like molecules (FURTADO et al. 1992; KASPER and MINEO 1994). FURTADO et al. (1992) have shown that laminin, but not fibronectin, increase parasite attachment to the murine macrophage cell line J774 in a dose-dependent manner. The cyclic Y-I-G-S-R, a laminin-derived peptide (which inhibits laminin binding to the 32/67 kDa laminin protein of host cells), is able to block parasite attachment mediated by laminin. Hence, an antiserum to the 32/67 kDa protein also inhibits the attachment of *T. gondii* to host cells. Other candidates for attachment may be lectins, since several studies have shown that bovine serum albumin (BSA)-glucosamide binds to extracellular tachyzoites with high affinity (ROBERT et al. 1991). In addition, the neoglycoprotein BSA-glucosamide has been shown to competitively block in vitro infection of human fibroblast cells with tachyzoites (MINEO et al. 1993). As discussed above, lectins are not able to bind to surface molecules of live tachyzoites. It appears that neoglycoproteins involved in attachment are detected in the parasite's internal organelles, such as micronemes and rhoptries. Interestingly, DE CARVALHO et al. (1991) have shown that gold-labeled lectin and neoglycoprotein localize sugar residues and sugar-binding sites on the parasite rhoptries. It should be mentioned that the precise mechanism with which these lectins, laminin, and surface proteins of *T. gondii* interact with ligands of host cells remains to be explained.

8 Conclusions and Future Directions

The last decade has seen considerable advances in the understanding of the interaction of attachment and adhesion between host cells and Apicomplexa parasites (especially *Plasmodium* and *Toxoplasma*). Attachment and adhesion are prerequisites for the penetration of these parasites into host cells. However, the detailed processes and the molecular mechanisms involved in invasion are still not completely understood. Informative and innovative experiments will certainly be performed in the future, because of our ability to generate gene knockout mutants and to complement them with wild-type, altered, or foreign molecules. These mutants will be extremely useful for further examinations of the structures within the attachment/adhesion factors, which promote the interactions between *Toxoplasma* and host cells. Knockout mutants will also be extremely important when considering new ways to control the parasite, for example, by inhibiting invasion.

Acknowledgements. The author would like to thank Wyls V and Toursel C for their contributions in the generation of *SAG3* mutants and Dubremetz JF and Boothroyd JC for helpful comments on this review. This work has been supported by INSERM, CNRS and the Pasteur Institute of Lille.

References

Burg JL, Perelman D, Kasper LH, Ware PL, Boothroyd JC (1988) Molecular analysis of the gene encoding the major surface antigen of Toxoplasma gondii. J Immunol 141:3584–3591
Brown D (1993) The tyrosine kinase connection: how GPI-anchored proteins activate T cells. Curr Opin Immunol 5:349–354
Cesbron-Delauw MF, Tomavo S, Beauchamps P, Fourmaux MP, Camus D, Capron A, Dubremetz JF (1994) Similarities between the primary structure of two distinct major surface proteins of Toxoplasma gondii. J Biol Chem 269:16217–16222
Couvreur G, Sadak A, Fortier B, Dubremetz JF (1988) Surface antigens of Toxoplasma gondii. Parasitology 97:1–10
Cross GAM (1990) Glycolipid anchoring of plasma membrane proteins. Ann Rev Cell Biol 6:1–39
de Carvalho L, Souto-Padron T, de-Souza W (1991) Localization of lectin-binding sites and sugar-binding proteins in tachyzoites of Toxoplasma gondii. J Parasitol 77:156–161
Dieckmann-Schuppert A, Bender S, Odenthal-Schnittler M, Bause E, Schwarz RT (1992) Apparent lack of N-glycosylation in the asexual intraerythrocytic stage of Plasmodium falciparum. Eur J Biochem 205:815–825
Ferguson MAJ (1994) What can GPI do for you? Parasitol Today 10:48–52
Furtado GC, Slowik M, Kleinman HK, Joiner KA (1992) Laminin enhances binding of Toxoplasma gondii tachyzoites to J774 murine macrophage cells. Infect Immun 60:2337–2342

Galinski MR, Medina CC, Ingravallo P, Barnwell JW (1992) A reticulum-binding protein complex of Plasmodium vivax merozoites. Cell 69:1213–1226

Grimwood J, Smith JE (1992) Toxoplasma gondii: the role of a 30-kDa surface protein in host cell invasion. Exp Parasitol 74:106–111

Handman E, Goding JW, Remington JS (1980) Detection and characterization of membrane antigens of Toxoplasma gondii. J Immunol 124:2578–2583

Homans SW, Ferguson MAJ, Dwek RA, Rademacher TW, Anand R, Williams AF (1988) Complete structure of the glycosyl phosphatidylinositol membrane anchor of rat brain Thy-1 glycoprotein. Nature 333:269–272

Johnson AM, McDonald PJ, Neoh SH (1981) Molecular weight analysis of the major polypeptides and glycopeptides of Toxoplasma gondii. Biochem Biophys Res Commun 100:934–943

Kasper LH (1987) Isolation and characterization of a monoclonal anti-P30 antibody resistant mutant of Toxoplasma gondii. Parasite Immunol 9:433–445

Kasper LH, Mineo JR (1994) Attachment and invasion of host cells by Toxoplasma gondii. Parasitol Today 10:184–188

Kasper LH, Crabb JH, Pfefferkorn ER (1982) Isolation and characterization of a monoclonal antibody-resistant antigenic mutant of Toxoplasma gondii. J Immunol 129:1694–1699

Kasper LH, Crabb JH, Pfefferkorn ER (1983) Purification of a major membrane protein of Toxoplasma gondii by immunoabsorption with a monoclonal antibody. J Immunol 130:2407–2412

Kim K, Soldati D, Boothroyd JC (1993) Gene replacement in Toxoplasma gondii with chloramphenicol acetyltransferase as selectable marker. Science 262:911–914

Kim K, Bülow R, Kampmeier J, Boothroyd JC (1994) Conformational appropriate expression of the Toxoplasma antigen SAG1 (p30) in CHO cells. Infect Immun 62:203–209

Masterson W, Doering TL, Hart GW, Englund PT (1989) A novel pathway for glycan assembly: biosynthesis of the glycosyl phosphatidylinositol anchor of trypanosome variant surface glycoprotein. Cell 56:793–800

Mauras G, Dodeur M, Laget P, Senet JM, Bourrillon R (1981) Partial resolution of the sugar content of Toxoplasma gondii membrane. Biochem Biophys Res Commun 100:934–943

Menon AK, Schwarz RT, Major S, Cross GAM (1990) Cell-free synthesis of glycosyl phosphatidylinositol precursors for the glycolipid membrane anchor of Trypanosoma brucei variant surface glycoproteins: structural characterization of putative biosynthetic intermediates. J Biol Chem 265:9033–9042

Mineo JR, McLeod R, Mack D, Smith J, Khan IA, Ely KH, Kasper LH (1993) Antibodies to Toxoplasma gondii major surface protein (SAG-1, P30) inhibit infection of host cells and are produced in murine intestine after peroral infection. J Immunol 150:3951–3964

Nagel SD, Boothroyd JC (1989) The major surface antigen, P30, of Toxoplasma gondii is anchored by a glycolipid. J Biol Chem 264:5569–5574

Odenthal-Schnittler M, Tomavo S, Becker D, Dubremetz JF, Schwarz RT (1993) Evidence for N-linked glycosylation in Toxoplasma gondii. Biochem J 291:713–721

Prince JB, Auer KL, Huskinson J, Parmley SF, Araujo FG, Remington JS (1990) Cloning, expression, and cDNA sequence of surface antigen P22 from Toxoplasma gondii. Mol Biochem Parasitol 43:97–106

Robert R, de la Jarrige PL, Mahaza C, Cottin J, Marot-Leblond A, Senet JM (1991) Specific binding of neoglycoproteins to Toxoplasma gondii tachyzoites. Infect Immun 59:4670–4673

Schenkman S, Diaz C, Nussenzweig V (1991) Attachment of Trypanosoma cruzi trypomastigotes to receptors at restricted cell surface domains. Exp Parasitol 72:76–86

Schwarz RT, Tomavo S (1993) The current status of the glycobiology of Toxoplasma gondii: glycosyl phosphatidylinositols, N- and O-linked glycans. Res Immunol 144:24–31

Schwarz RT, Tomavo S, Odenthal-Schnittler M, Striepen B, Becker D, Dubremetz JF (1993) Recent advances in the glycobiology of Toxoplasma gondii. In: Smith J (ed) Toxoplasmosis. NATO ASI Series H. Cell biology; vol. 78. Springer Berlin,Heidelberg, New York, pp. 109–121

Sibley LD, Pfefferkorn ER, Boothroyd JC (1991) Proposed nomenclature for Toxoplasma gondii. Parasitol Today 7:327–328

Stefanova I, Horejsi V, Ansotegui IJ, Knapp W, Stockinger H (1991) GPI-anchored cell-surface molecules complexed to protein tyrosine kinases. Science 254:1016–1019

Soldati D, Boothroyd JC (1993) Transient transfection and expression in the obligate intracellular parasite Toxoplasma gondii. Science 260:349–352

Tomavo S, Schwarz RT, Dubremetz JF (1989) Evidence for glycosyl-phosphatidylinositol anchoring of Toxoplasma gondii major surface antigens. Mol Cell Biol 9:4576–4580

Tomavo S, Dubremetz JF, Schwarz RT (1992a) A family of glycolipids from Toxoplasma gondii. Identification of candidate glycolipid precursor(s) for Toxoplasma gondii glycosylphosphatidyl-inositol membrane anchors. J Biol Chem 267:11721–11728

Tomavo S, Dubremetz JF, Schwarz RT (1992b) Biosynthesis of glycolipid precursors for glycosyl-phosphatidylinositol membrane anchors in a Toxoplasma gondii cell-free system. J Biol Chem 267:21446–21458

Tomavo S, Dubremetz JF, Schwarz RT (1993) Structural analysis of glycosyl-phosphatidylinositol membrane anchor of the Toxoplasma gondii tachyzoite surface glycoprotein gp23. Biol Cell 78:155–162

Wilson ME, Hardin KK (1990) The major Leishmania donovani chagasi surface glycoprotein in tunicamycin-resistant promastigotes. J Immunol 144:4825–4834

Toxoplasma gondii Microneme Proteins: Gene Cloning and Possible Function

M.N. Fourmaux[1], N. Garcia-Réguet[1],
O. Mercereau-Puijalon[2], and J.F. Dubremetz[1]

1 Introduction

Micronemes are small apical organelles found in variable amounts in the invasive stages of all Apicomplexa. They are believed to play a role in recognition of the host cell by exocytosing their contents at the time of invasion, but very little direct evidence of this exocytosis has been obtained (ENTZEROTH et al. 1992). Rather, in several models, molecules known to be stored in micronemes have been shown to possess adhesive domains or even binding properties, explaining the host cell specificity of the organism (ADAMS et al. 1990; TOMLEY et al. 1991; ESCHENBACHER et al. 1993). This is especially true in *Plasmodium* sp., in which erythrocyte and hepatocyte binding molecules have been extensively characterized in recent years (ADAMS et al. 1992; CERAMI et al. 1992; CHITNIS et al. 1994; SIM et al. 1994; ROBSON et al. 1995).

2 *Toxoplasma gondii* Microneme Proteins

Three different *Toxoplasma gondii* microneme proteins have been described so far (ACHBAROU et al. 1991). Their characteristics are as follows:

---

[1] Unité 42 INSERM, 369 rue J Guesde, 59650 Villeneuve d'Ascq, France
[2] Unité d' Immunologie Moléculaire des Parasites, Institut Pasteur, 25, rue du Docteur Roux, 75015 Paris, France

- MIC1 is a 60 kDa protein that migrates at 50 kDa under nonreducing conditions, suggesting the presence of internal disulfide bridges. It has a pI of 6.5 and can be labeled with tritiated glucosamine.
- MIC2 is a 120 kDa protein. It has a a pI of 5 and undergoes partial processing to a 116 and a 110 kDa molecule upon host cell invasion (ACHBAROU, unpublished observations).
- MIC3 is a 90 kDa protein of pI 6.75, made of two 38 kDa monomers of pI 6.7 and 6.75. These monomers are linked by intermolecular disulfide bridge(s). MIC3 monomers are synthesized as 40 kDa precursors that are processed to 38 kDa final products within about 30 min after synthesis.

MORRISSETTE et al. (1994) have obtained monoclonal antibodies directed against MIC2 and MIC3. They have found by immunofluorescence that MIC3 is differentially located depending on the replicating stage of the parasite, being also detected in the perinuclear region in recently divided parasites but found only in the apex at other stages. These findings suggest that microneme protein synthesis is regulated during the cell cycle of *T. gondii*.

MIC1 has been shown to bind to host cells in vitro: when a T. gondii tachyzoite lysate is incubated with Vero cells or HFF, MIC1 is one of the major parasite proteins that binds to the cells and that can be detected by western blotting of the incubated cells (ACHBAROU, unpublished). MIC2 does not possess this property whereas MIC3, which is fully insoluble in physiological medium, has not been studied in this respect.

3 Cloning of Genes Encoding Microneme Proteins

The gene encoding the MIC1 protein has been cloned and sequenced (FOURMAUX et al., unpublished). It is a single copy gene with three introns, all located in the first half of the coding sequence. The intron sizes are 320, 391 and 263 bp. The cDNA has an open reading frame (ORF) of 1368 bp that encodes a polypeptide of 456 amino acids, with a putative signal sequence of 16 amino acids and no other hydrophobic domain. Two polyadenylation sites have been found at 400 bp and 570 bp after the stop codon. The 5' untranslated mRNA is 144 nucleotides long, as has been determined by rapid amplification of cDNA ends (RACE). An interesting feature of the deduced polypeptide sequence is that it contains a tandemly repeated domain of 88 amino acids, with six conserved cysteine residues in each domain. The two domains share 22% identity and present a perfect conservation of the positions at the cysteine residues. Moreover, part of the molecule has homologies with the thrombospondin-related adhesive protein (TRAP) of *Plasmodium falciparum* (ROBSON et al. 1988). The TRAP protein has recently been shown to bind to human hepatocytes (ROBSON et al. 1995) and is therefore probably involved in host cell binding by *Plasmodium* sporozoites, as is the circumsporozoite protein (CSP) that also possesses a thrombospondin-related domain (Cerami et al. 1992). Both TRAP and CSP are located on the surface and in micronemes of *Plasmodium* sporozoites (Fine et al. 1984; Rogers et al. 1992). As *Eimeria* also possesses a microneme protein with a homology to TRAP (Tomley et al. 1991), this suggests that Apicomplexa express a family of microneme proteins that may derive from a common ancestor and that are involved in binding to the host cell surface. The differences in sequence would reflect the receptor-ligand specificity of host cell invasion of the different genera. This TRAP-like family seems to be expressed in Apicomplexa parasites which invade nucleated cells. In contrast, the erythrocyte-binding antigen (EBA) family (Adams et al. 1992), which has been found in the micronemes of the genus *Plasmodium*, seems to be restricted to erythrocyte-invading Apicomplexa parasites.

The gene coding for the MIC3 protein has also been cloned from a cDNA library (Garcia-Reguet et al., unpublished), and the sequence data obtained so far encompass a 1089 bp ORF, encoding a polypeptide of 363 amino acids, and a 3′ untranslated region of 409 bp preceding the polyA tail. Southern blot experiments suggest the presence of a single copy gene. As the protein is a heterodimer of two isoforms, the difference between these two is likely to be due to posttranslational modifications, which so far are unknown. The deduced amino acid sequence indicates that MIC3 is also a cysteine-rich protein with a putative transmembrane sequence. It also contains two epidermal growth factor (EGF)-like domains as a tandem repeat in the COOH-terminal. EGF-like domains have not previously been found in microneme proteins of Apicomplexa, but they have been found in surface proteins of *Plasmodium* sp. merozoites (Blackman et al. 1991) and ookinetes (Kaslow et al. 1988). We do not know whether any relationship exists between these molecules and MIC3.

Finding these EGF-like domains in *T gondii* micronemes suggests the existence of a second receptor-ligand structure in the same organelle, which makes them especially interesting. If micronemes contain at least two different types of putative host cell ligands, either alternate or complementary routes of host cell recognition/binding could then be envisioned for this organism that can invade almost any cell type. These preliminary data make the study of *T. gondii* micronemes a highly promising field. Further investigations can hopefully show: (1) whether additional proteins are expressed in these organelles, (2) how and when exocytosis occurs, and (3) whether the binding proteins really play a part in invasion, either at initial recognition or at the moving junction formation.

References

Achbarou A, Mercereau-Puijalon O, Autheman JM, Fortier B, Camus D, Dubremetz JF (1991) Characterization of microneme proteins of Toxoplasma gondii. Mol Biochem Parasitol 47:223–233

Adams JH, Hudson DE, Torii M, Ward GE, Wellems TE, Aikawa M, Miller LH (1990) The Duffy receptor family of Plasmodium knowlesi is located within the micronemes of invasive malaria merozoites. Cell 63:141–153

Adams JH, Sim BKL, Dolan SA, Fang XD, Kaslow DC, Miller LH (1992) A family of erythrocyte binding proteins of malaria parasites. Proc Natl Acad Sci 89:7085–7089

Blackman MJ, Ling, IT, Nicholls SC, and Holder AA (1991) Proteolytic processing of the Plasmodium falciparum merozoite surface protein-1 produces a membrane-bound fragment containing two epidermal growth factor-like domains. Mol Biochem Parasitol 49:29–34

Cerami C, U Frevert, Sinnis P, Takacs B, Clavijo P, Santos MJ, Nussenzweig V (1992) The basolateral domain of the hepatocyte plasma membrane bears receptors for the circumsporozoite protein of Plasmodium falciparum sporozoites. Cell 70:1021–1033

Chitnis C, Miller L (1994) Identification of the erythrocyte binding domain of Plasmodium vivax and Plasmodium knowlesi proteins involved in erythrocyte invasion. J Exp Med 180:497–506

Entzeroth R, Kerchkoff H, König A (1992) Microneme secretion in coccidia: confocal laser scanning and electron microscope study on microneme secretion of Sarcocystis muris using a monoclonal antibody. Eur J Cell Biol 59:405–413

Eschenbacher KH, Klein H, Sommer I, Meyer HE, Entzeroth R, Mehlhorn H, Rüger W (1993) Characterization of cDNA clones encoding a major microneme antigen of Sarcocystis muris (Apicomplexa). Mol Biochem Parasitol 62:27–36

Fine E, Aikawa M, Cochrane AH, Nussenzweig RS (1984) Immuno-electron microscopic observations on Plasmodium knowlesi sporozoites: localization of protective antigen and its precursors. Am J Trop Med Hyg 33:220–226

Kaslow DC, Quakyi IA, Syin C, Raum MG, Keister DB, Coligan JE, McCuchan TF, Miller LH (1988) A vaccine candidate from the sexual stage of human malaria that contains EGF-like domains. Nature 333:74–77

Morrissette NS, Bedian V, Webster P, Roos DS (1994) Characterization of extreme apical antigens from Toxoplasma gondii. Exp Parasitol 79:445–459

Robson KJH, Hall JRS, Jennings MW, Harris TJR, Marsh K, Newbold CI, Tate VE, Weatherhall DJ (1988) A highly conserved sequence in thrombospondin, properdin and in proteins from sporozoites and blood stages of a human malarial parasite. Nature 335:79–82

Robson KJH, Frevert U, Reckman I, Cowan G, Beier J, Scragg SG, Takehara K, Bishop DHL, Pradel G, Sinden RE, Saccheo S, Müller HM, Crisanti A (1995) Thrombospondin related adhesive protein (TRAP) of Plasmodium falciparum: expression during sporozoite ontogeny and binding to human hepatocytes. EMBO J 14:3883–3894

Rogers WO, Malik A, Mellouk S, Nakamura K, Rogers M, Szarfman A, Gordon D, Aikawa M, Hoffman S (1992) Characterization of Plasmodium falciparum sporozoite surface protein 2. Proc Natl Acad Sci 89:9176–9180

Sim BKL, Chitnis CE, Wasniowska K, Hadley TJ, Miller LH (1994) Receptor and ligand domains for invasion of erythrocytes by Plasmodium falciparum. Science 264:1941–1944

Tomley FM, Clarke LE, Kawazoe U, Dijkema R, Kok JJ (1991) Sequence of the gene encoding an immunodominant microneme protein of Eimeria tenella. Mol Biochem Parasitol 49:227–288

Role of Secretory Dense Granule Organelles in the Pathogenesis of Toxoplasmosis

M.-F. Cesbron-Delauw[1], L. Lecordier[1], and C. Mercier[2]

1 Introduction

Multiplication of *Toxoplasma gondii* only occurs intracellularly inside a specialized compartment called the parasitophorous vacuole (PV). Regulated secretory processes are key to the success of the intracellular parasitism of *Toxoplasma* as the parasite extensively modifies the newly formed vacuole using secreted proteins (BECKERS et al. 1994; CHARIF et al. 1990; SIBLEY and KRAHENBUHL 1988). The main structural modification of the PV consists of elaboration of a network of tubular membranes that are continuous with the vacuolar membrane (SIBLEY and KRAHENBUHL 1988; SIBLEY et al. 1986, 1995).

Toxoplasma cells contain three distinct secretory organelles: the rhoptries, the micronemes and the dense granules. Whereas the precise function of these organelles is not known, they are involved in the installation process of the parasite in the host cell. The timing of rhoptry discharge strongly suggests that these organelles participate in the invasion process (SAFFER et al. 1992). Several early morphological observations suggested that the rhoptries secrete molecules which might cause degeneration of the host cell membrane (LYCKE et al. 1975). More recently, the ROP2 protein has been observed to be associated with the PV membrane (PVM), providing evidence that rhoptry contents may participate in the formation of the PVM.

[1] INSERM U415, Institut Pasteur de Lille, 1 rue du Pr. Calmette, BP245, 59019 Lille Cedex, France
[2] Department of Molecular Microbiology, Washington University School of Medicine, St. Louis, MO 63110, USA

In contrast, dense granule release appears to occur continuously during intracellular development of the parasite (LERICHE and DUBREMETZ 1990; DUBREMETZ et al. 1993). A similar granular composition is observed in the dense granules of both the tachyzoites and the bradyzoite encysted forms (TORPIER et al. 1993). The dense granule proteins are associated with both the PV and the cyst wall at different locations (TORPIER et al. 1993), which suggests that the dense granules are more likely to play a major role in the modification of the PV and therefore in the intracellular survival and/or multiplication of the parasite.

2 Biogenesis of Dense Granule Organelles

The dense granules are electron-dense vesicles found in all coccidian parasites. They resemble the secretory vesicles of mammalian cells and, by analogy, are probably formed by budding from the Golgi apparatus. Therefore, it seems likely that the newly synthesized granular proteins are posttranslationally targeted in the endoplasmic reticulum and subsequently accumulate in the Golgi. Indeed, the primary deduced amino acid sequences show that all dense granule proteins bear a NH_2-terminal hydrophobic domain (CESBRON-DELAUW 1994) that fits the characteristic of a signal sequence, which targets proteins in the secretory pathway. Since these polypeptides are thereafter targeted to the dense granules, they presumably contain a specific addressing signal that causes their sorting into the dense granules. In *Toxoplasma* cells, such sequence information is likely required to allow protein sorting into the correct secretory organelles. The recent success in expressing foreign proteins in *T. gondii* allowed us to start a detailed analysis of sequence elements involved in the sorting of proteins to the dense granules.

3 The GRA Proteins

Eight distinct dense granule proteins have been characterized so far; their genes have been cloned and their subcellular locations have been analyzed. In the absence of any particular biological function for most of the dense granule proteins, they have been named by the three letters GRA and identified by a number (SIBLEY et al. 1991): GRA1, a 23 kDa calcium-binding protein (CESBRON-DELAUW et al. 1989), GRA2 (p28) (MERCIER et al. 1993), GRA3 (p30) (BERMUDES et al. 1994a), GRA4 (p40) (MÉVELEC et al. 1992) and the antigenically related GRA5 (p21) (LECORDIER et al. 1993) and GRA6 (p32) (LECORDIER et al. 1995).

Recently, the NTPases NTP1 and NTP3 (Asai et al. 1995; Bermudes et al. 1994b), which were initially described as cytosolic enzymes within *Toxoplasma* cells, have been reported to be located in the dense granules and secreted in the PV (Sibley et al. 1994).

3.1 Differential Targeting of the GRA Proteins Within the Parasitophorous Vacuole

Shortly after invasion, the contents of the secretory granules are released within the PV. Despite the observation that individual dense granules contain a mixture of GRA proteins (Sibley et al. 1995), they appear to be differentially targeted within the vacuole following release (Charif et al. 1990; Achbarou et al. 1991). Several GRA proteins including GRA1, GRA2, GRA4, GRA6 and the NTPases are found in the vacuolar space and are not detected at the delimiting membrane of the PV. The NTPases and GRA1 are soluble proteins found in *T. gondii* cell extracts (Sibley et al. 1994, 1995). GRA1 becomes associated with the network of the PV and possesses several domains characteristic of moieties capable of calcium-regulated interactions (Cesbron-Delauw et al. 1989). GRA2 is found both as a soluble and as a membrane-associated form (Sibley et al. 1995). The predicted amino acid sequence of GRA2 does not contain a typical transmembrane domain but it is characterized by large amphipathic α-helical regions, which might adopt a transmembrane configuration (Mercier et al. 1993). GRA4 and GRA6, which are also detected closely associated with the network of the PV, have a putative classical membrane-spanning domain in their sequence (Achbarou et al. 1991; Lecordier et al. 1995)

In contrast, GRA5 is strictly associated with the delimiting PVM (Lecordier et al. 1993) and GRA3 combines with both the PVM and the intravacuolar network (Achbarou et al. 1991). The deduced amino acid sequence of GRA5 contains a central hydrophobic domain characteristic of a transmembane region, whereas the sequence of GRA3 does not have any typical transmembrane domains. GRA3 contains several short hydrophobic stretches, which are not predicted to span a membrane (Bermudes et al. 1994a).

It has recently been shown that the PVM functions as a molecular sieve allowing diffusion of low molecular mass substances (1300–1900 daltons) from the host cell cytoplasm into the vacuole by an unknown mechanism (Schwab et al. 1994). Parasite proteins that are found within the PVM, such as GRA3 and GRA5, could contribute to the formation of pores or channels across the PVM.

3.2 Dense Granule Secretion

Dense granule secretion shares hallmark features of regulated exocytosis including: (1) packaging in electron-dense vesicles; (2) fusion of these vesicles with the plasma membrane, and (3) triggering by an external signal.

One peculiarity of *Toxoplasma* dense granule exocytosis is that two independent fusion events have been described. First, fusion of dense granules with the plasma membrane from the anterior pole of the parasite and release of an amorphous material into the PV (Leriche and Dubremetz 1990); second, the release of membranous structures, occurring at the posterior end of the parasite (Sibley et al. 1995). Immunoelectron microscopy has recently shown that these multilamellar vesicles contain GRA2 (Sibley et al. 1995). Therefore, it is suggested that these two independent secretory events migth lead to the formation of the intravacuolar network. Similar immunolocalization studies involving GRA4 and GRA6, which are also specifically localized in the network, provide additional evidence for this process.

4 The *GRA* Gene Promoters

The genes encoding the GRA proteins have been cloned; only one (*GRA2*) has an intron (Mercier et al. 1993). Except for the NTPases, the primary amino acid sequences derived from the cDNAs did not provide any clues for the function of the GRA proteins. The GRA polyadenylated mRNAs are very abundant in the tachyzoites and are also expressed in the bradyzoites. Since all GRA proteins are processed in the parasite in a similar manner (storage in the same granule, regulated secretion), their expression might also be coordinately regulated.

This has led us to investigate whether similar elements are involved in the control of the expression of the *GRA* genes. Sequence analysis of the 5' flanking regions of four *GRA* genes that have been cloned in our laboratory (*GRA1, GRA2, GRA5* and *GRA6*) did not reveal either any obvious similarities between them nor any classical eukaryotic promoter sequences. We next used the recently developed, transient DNA transfection system (Soldati and Boothroyd 1993) to define sequence elements that are critical for the expression of each of the four *GRA* genes (Mercier et al. 1996). Sequence comparison of the defined regions revealed the presence of an heptanucleotide motif (A/TGAGACG) found in both orientations. This sequence element is found in the repeated sequence that constitutes the *SAG1* promoter (Soldati and Boothroyd 1995). Mutagenesis confirmed the functionality of the detected element (Mercier et al. 1996).

5 The GRA Antigens

The role of excreted-secreted antigens (ESAs) of *T. gondii* in inducing protective immunity has been shown in several experimental models of toxoplasmosis. ESAs protect highly susceptible nu/nu rats, inducing both antibody--dependent and cellular immune responses (DUQUESNE et al. 1990). In a rat model of congenital toxoplasmosis, ESA have been shown to reduce transplacental infection of the fetus from 70% in the control rats to 3% in ESA-immunized animals (Zenner et al., unpublished). ESA immunization also protects mice against oral infection with a lethal dose of *Toxoplasma* cysts (50%–60% survival rate) (Fig. 1; DARCY et al. 1992).

ESA are obtained upon incubation of extracellular tachyzoites in a cell-free medium containing 10% serum. The majority of these products have been shown to be components of dense granules. Amongst them, GRA2 is probably the most immunogenic. Immunization of mice with high-performance liquid chromatography (HPLC)-purified GRA2 has resulted in a survival rate of 75% (Fig. 1). Protection with GRA2 was also shown by immunizing mice with antigens that had been affinity-purified by the monoclonal antibody F3G3 (SHARMA et al. 1984). Serological studies using human sera and truncated recombinant GRA2 have shown that this antigen contains at least three B cell epitopes (MURRAY et al. 1993). One of them, including the eight COOH-terminal amino acid residues, is recognized by the mouse anti-GRA2 monoclonal antibody TG17-179 (CESBRON-DELAUW et al. 1992).

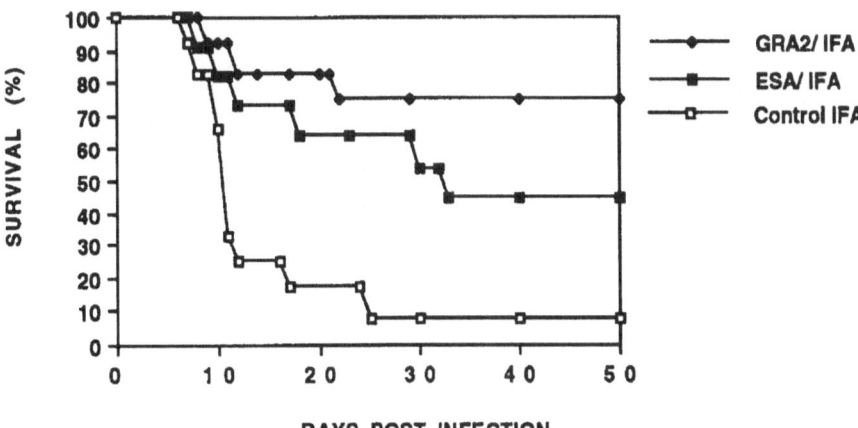

Fig. 1. Survival curves of immunized OF1 mice after oral infection with 1200 cysts of *Toxoplasma. gondii* 76 K strain. Three groups of mice have been immunized subcutaneously with: (1) excreted-secreted antigens (ESAs) in the presence of incomplete Freund adjuvant (IFA; 13 mice), (2) with the 28 kDa antigen (GRA2) purified by HPLC in the presence of IFA (12 mice), and (3) controlmice immunized with IFA alone (15 mice)

Currently, work is in progress to express GRA2 in several cloning systems in order to test its potential as a vaccine component. Recently, successful expression has been obtained in bacillus Calmette-Guerin (BCG) and experiments are also in progress to test the protective activity of such recombinant BCG expressing GRA2 (Abomoelak et al., personal communication).

Acknowledgements. We wish to thank Didier Deslée for excellent technical assistance.

References

Achbarou A, Mercereau-Puijalon O, Sadak A, Fortier B, Leriche MA, Camus D, Dubremetz JF (1991) Differential targeting of dense granule proteins in the parasitophorous vacuole of Toxoplasma gondii. Parasitology 103:321–329

Asai T, Miura S, Sibley LD, Okabayashi H, Takeuchi T (1995) Biochemical and molecular characterization of nucleoside triphosphate hydrolase isozymes from the parasite protozoan Toxoplasma gondii. J Biol Chem 270:11391–11397

Beckers CJM, Dubremetz JF, Mercereau-Puijalon O, Joiner KA (1994) The Toxoplasma gondii protein ROP2 is inserted into the parasitophorous vacuole membrane, surrounding the intracellular parasite, and is exposed to the host cell cytoplasm. J Cell Biol 127:947–961

Bermudes D, Dubremetz JF, Achbarou A, Joiner KA (1994a) Molecular cloning of a complete cDNA encoding the dense granule protein GRA3 from Toxoplasma gondii. Mol Biochem Parasitol 68:247–257

Bermudes D, Peck KR, Afifi MA, Beckers CJM, Joiner KA (1994b) Tandemly repeated genes encode nucleoside triphosphatase hydrolase isoforms secreted into the parasitophorous vacuole of Toxoplasma gondii. J Biol Chem 269:29252–29260

Cesbron-Delauw MF (1994) Dense granule organelles of Toxoplasma gondii: their role in the host-parasite relationship. Parasitol Today 10:293–296

Cesbron-Delauw MF, Guy B, Torpier G, Pierce RJ, Lenzen G, Cesbron JY, Charif H, Lepage P, Darcy F, Lecocq JP, Capron A (1989) Molecular characterization of a 23-kilodalton major antigen secreted by Toxoplasma gondii. Proc Natl Acad Sci USA 86:7537–7541

Cesbron-Delauw MF, Boutillon C, Mercier C, Fourmaux MP, Murray A, Miquey F, Tartar A, Capron A (1992) Amino acid sequence requirements for the epitope recognized by a monoclonal antibody reacting with the secreted antigen GP28.5 of Toxoplasma gondii. Mol Immunol 29:1375–1382

Charif H, Darcy F, Torpier G, Cesbron-Delauw MF, Capron A (1990) Toxoplasma gondii: characterization and localization of antigens secreted from tachyzoites. Exp Parasitol 71:114–124

Darcy F, Maes P, Gras-Masse H, Auriault C, Bossus M, Deslée D, Godard I, Cesbron-Delauw MF, Tartar A, Capron A (1992) Protection of mice and nude rats against toxoplasmosis by a multiple antigenic peptide construction derived from Toxoplasma gondii P30 antigen. J Immunol 149:3636–3641

Dubremetz JF, Achbarou A, Bermudes D, Joiner KA (1993) Kinetics and pattern of organelle exocytosis during Toxoplasma gondii-host cell interaction. Parasitol Res 79:402–408

Duquesne V, Auriault C, Darcy F, Decavel JP, Capron A (1990) Protection of nude rats against Toxoplasma infection by excreted-secreted antigen-specific helper T cells. Infect Immun 58:2120–2126

Lecordier L, Mercier C, Torpier G, Tourvieille B, Darcy F, Liu JL, Maes P, Tartar A, Capron A, Cesbron-Delauw MF (1993) Molecular structure of a Toxoplasma gondii dense granule antigen (GRA5) associated with the parasitophorous vacuole membrane. Mol Biochem Parasitol 59:143–154

Lecordier L, Moleon-Borodowski I, Dubremetz JF, Tourvieille B, Mercier C, Deslée D, Capron A, Cesbron-Delauw MF (1995) Characterization of a dense granule antigen of Toxoplasma gondii (GRA6) associated to the network of the parasitophorous vacuole. Mol Biochem Parasitol 70:85–94

Leriche MA, Dubremetz JF (1990) Exocytosis of Toxoplasma gondii dense granules into the parasitophorous vacuole after host cell invasion. Parasitol Res 76:559–562

Lycke E, Carlberg K, Norrby R (1975) Interaction between Toxoplasma gondii and its host-cells: function of the penetrating enhancing factor of Toxoplasma. Infect Immun 11:853–861

Mercier C, Lecordier L, Darcy F, Deslée D, Murray A, Tourvieille B, Maes P, Capron A, Cesbron-Delauw MF (1993) Molecular characterization of a dense granule antigen (GRA2) associated with the network of the parasitophorous vacuole in Toxoplasma gondii. Mol Biochem Parasitol 58:71–82

Mercier C, Lefebvre-Van-Hende S, Garber G, Lecordier L, Beauchamps P, Capron A, Cesbron-Delauw MF (1996) Common cis-acting elements critical for the expression of several genes of Toxoplasma gondii. Mol Microbiol (In press)

Mévelec M-N, Chardès T, Mercereau-Puijalon O, Bourguin I, Achbarou A, Dubremetz J-F, Bout D (1992) Molecular cloning of GRA4, a Toxoplasma gondii dense granule protein, recognized by mucosal IgA antibodies. Mol Biochem Parasitol 56:227–238

Murray A, Mercier C, Decoster A, Lecordier L, Capron A, Cesbron-Delauw MF (1993) Multiple B-cell epitopes in a recombinant GRA2 secreted antigen of Toxoplasma gondii. Appl Parasitol 34:235–244

Saffer LD, Mercereau-Puijalon O, Dubremetz JF, Schwartzman JD (1992) Localization of Toxoplasma gondii rhoptry protein by immunoelectron microscopy during and after host cell penetration. J Protozool 39:526–530

Schwab JC, Bekers CJM, Joiner KA (1994) The parasitophorous vacuole membrane surrounding intracellular Toxoplasma gondii functions as a molecular sieve. Proc Natl Acad Sci USA 91:509–513

Sharma SD, Araujo FG, Remington JS (1984) Toxoplasma antigen isolated by affinity chromatography with monoclonal antibody protects mice against lethal infection with Toxoplasma gondii. J Immunol 133:2818–2820

Sibley LD, Krahenbuhl JL (1988) Modification of host cell phagosomes by Toxoplasma gondii involves redistribution of surface proteins and secretion of a 32 kDa protein. Eur J Cell Biol 47:81–87

Sibley LD, Krahenbuhl JL, Adams GM, Weidner E (1986) Toxoplasma modifies macrophage phagosomes by secretion of a vesicular network rich in surface proteins. J Cell Biol 103:867–874

Sibley LD, Pfefferkorn ER, Boothroyd JC (1991) Proposal for a uniform genetic nomenclature in Toxoplasma gondii. Parasitol Today 7:327–328

Sibley LD, Niesman IR, Asai T, Takeuchi T (1994) Toxoplasma gondii: secretion of a potent nucleoside triphosphate hydrolase into the parasitophorous vacuole. Exp Parasitol 79:301–311

Sibley LD, Niesman IR, Parmley SF, Cesbron-Delauw MF (1995) Regulated secretion of multi-lamellar vesicles leads to formation of a tubular vesicular network in host-cell vacuoles occupied by Toxoplasma gondii. J Cell Science 108:1669–1677

Soldati D, Boothroyd JC (1993) Transient transfection and expression in the obligate intracellular parasite Toxoplasma gondii. Science 260:349–352

Soldati D, Boothroyd JC (1995) A selector of transcription initiation in the protozoan parasite Toxoplasma gondii. Mol Cell Biol 15:87–93

Torpier G, Charif H, Darcy F, Liu J, Dardé ML, Capron A (1993) Toxoplasma gondii:differential location of antigens secreted from encysted bradyzoites. Exp Parasitol 77:13–22

Serological Recognition
of *Toxoplasma gondii* Cyst Antigens

J.E. Smith[1], G. McNeil[2], Y.W. Zhang[3], S. Dutton[4],
G. Biswas-Hughes[1], and P. Appleford[1]

1   Introduction

Infection with *Toxoplasma gondii* is characterised by an acute phase, associated with the rapid proliferation of tachyzoites, followed by a chronic phase during which the slow-growing bradyzoite stage forms cysts in brain and muscle tissue. Although tissue cysts are often refered to as 'dormant', there is evidence of bradyzoite turnover within the cyst (Pavesio et al. 1992) and of periodic cyst rupture (Ferguson et al. 1989). Bradyzoites released from the cyst may convert into tachyzoites causing recrudesence of acute disease which, in immuno-suppressed individuals, can be fatal (Luft and Remington 1992). Control of dis-ease relies upon recognition and immunoregulation of both stages and it is therefore important to have an understanding of the comparative antigenic structure of tachyzoites and bradyzoites.

In terms of their cellular structure, tachyzoites and bradyzoites are very similar but there are major differences in both the phenotype they produce in the host cell and in their molecular composition. Tachyzoites multiply rapidly to destroy the host cell within 48 h, while bradyzoites divide slowly and secrete proteinaceous material which forms a dense matrix around the bradyzoites and modifies the parasitophorous vacuole membrane to form the cyst wall (Ferguson and Hutchinson 1987). These variations in the growth and morphology of the two stages indicate underlying differences in metabolism which are

---

[1] Department of Biology, Leeds University, Leeds, LS29JT, UK
[2] Department of Paediatrics, University of Oxford, John Radcliffe Hospital, Oxford OX3 9DU, UK
[3] Department of Pathology, Albert Einstein College of Medicine, Yeshiva University, 1300 Morris Park Avenue, Bronx, New York 10461, USA
[4] Department of Life Sciences, Nottingham University, University Park, Nottingham, NG7 2RD, UK

perhaps most clearly reflected by direct comparison of protein profiles. Figure 1a compares the profile of proteins in RH strain tachyzoites and 18691 strain cysts. The dominant protein in tachyzoites is the major Mr 30 000 surface molecule (SAG1) with other prominent bands at Mr 42 000 and 66 000 (ZHANG and SMITH 1995a). In contrast, the cyst has one very abundant protein of Mr 24 000 with other fainter bands including those of Mr 35 000, 42 000, 54 000 and 65 000. The profile of cyst proteins is very similar in two other strains of the parasite, Gleadle and 17025 (Fig. 1b), suggesting that there is no major strain-dependent variation at this level.

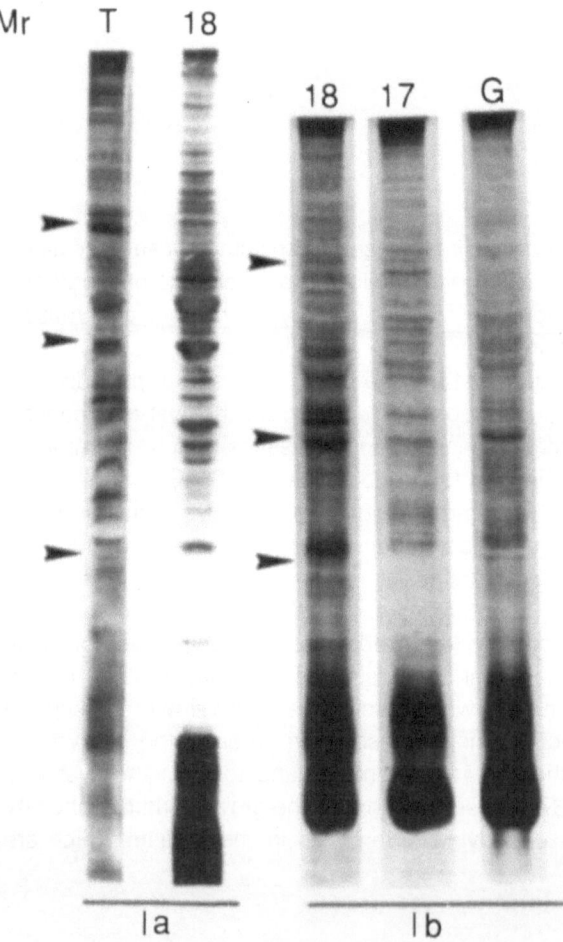

Fig. 1. Comparison of the total protein profile of *Toxoplasma gondii* bradyzoites and tachyzoites. Parasites, 10⁶ RH strain tachyzoites (*7*) or 2000 tissue cysts of the 18691 (*18*), 17025 (*17*) and Gleadle (*G*) strains, were boiled for 5 min in reducing sample buffer, separated on 12% acrylamide gels and stained with silver stain. *Arrowheads* indicate the molecular weight markers BSA (66 000), ovalbumin (45 000) and trypsinogen (24 000)

2 Antigenic Differences Between Tachyzoites and Bradyzoites

The first demonstration that the two stages were antigenically distinct came from LUNDE and JACOBS (1983), who compared the cross-reactivity of tachyzoite- and bradyzoite-specific antisera in an immunofluorescence assay (IFA). The extent of this overlap was first quantified by KASPER (1989), who raised stage-specific antisera and found very low cross-reactivity in a differential ELISA screen. More recently, we have shown that removal of anti-tachyzoite anti-bodies from immune mouse serum via adsorption has no effect on the anti-cyst titer (ZHANG and SMITH 1995a). Taken together, these results imply that there is virtually no antigenic overlap between the two stages and thus immuno-regulation is likely to be highly stage-specific.

Identification of the major antigens of tachyzoites and cysts by western blotting largely confirms this suggestion. The overall profile of antigens in the tachyzoite has been well established by a number of studies using human sera (PARTENEN et al. 1984; POTASMEN et al. 1986; VERHOFSTEDE et al. 1988). The consensus from these studies is that there is a strong humoral response to the tachyzoite with molecules of approximate Mr 30 000, 35 000, 42 000, 54 000, 60 000, 67 000 and 120 000 amongst the most frequently cited antigens. Comparison of the recognition pattern of polyclonal antisera with monoclonal antibodies has enabled identification of some of the major antigens, for example the dominant tachyzoite surface antigen SAG1 (COUVREUR et al. 1988) and the family of dense granule molecules (GRA 1–6, DARCY et al. 1990; CESBRON-DELAUW 1994), originally identified as excreted-secreted antigens.

The overall profile of bradyzoite/cyst antigens has also been analysed by western blotting with immune sera and monoclonal antibodies. The dominant antigen in the cyst has a Mr of approximately 24 000, while molecules of approximate Mr 18 000, 20 000, 28 000, 43 000, 52 000 and 67 000 are also frequently recognised (KASPER 1989; DARCY et al. 1990; WOODISON and SMITH 1990; MAKIOKA et al. 1991). It is clear from the use of monoclonal antibodies that some molecules, such as the dense granule and rhoptry proteins, are common to both stages (VAN GELDER et al. 1993; TORPIER et al. 1993) and the existence of such 'shared' antigens appears to be at odds with the low sero-logical cross-reactivity seen in IFA and ELISA. However, it appears that these molecules are not immunodominant and that the major antigens are stage-specific. A direct comparison of tachyzoite and bradyzoite antigens can be seen in Fig. 2, which shows a two-dimensional western blot of RH strain tachyzoites and RRA strain cysts probed with immune mouse sera. Although common antigens, like the surface protein SAG3 and the microneme molecule MIC3, can be identified, stage-specific antigens can be detected at Mr 30 000 (SAG1) and 35 000 in the tachyzoite and at Mr 18 000, 20 000 and 24 000 in the cyst. One of the major differences between the two stages is in the

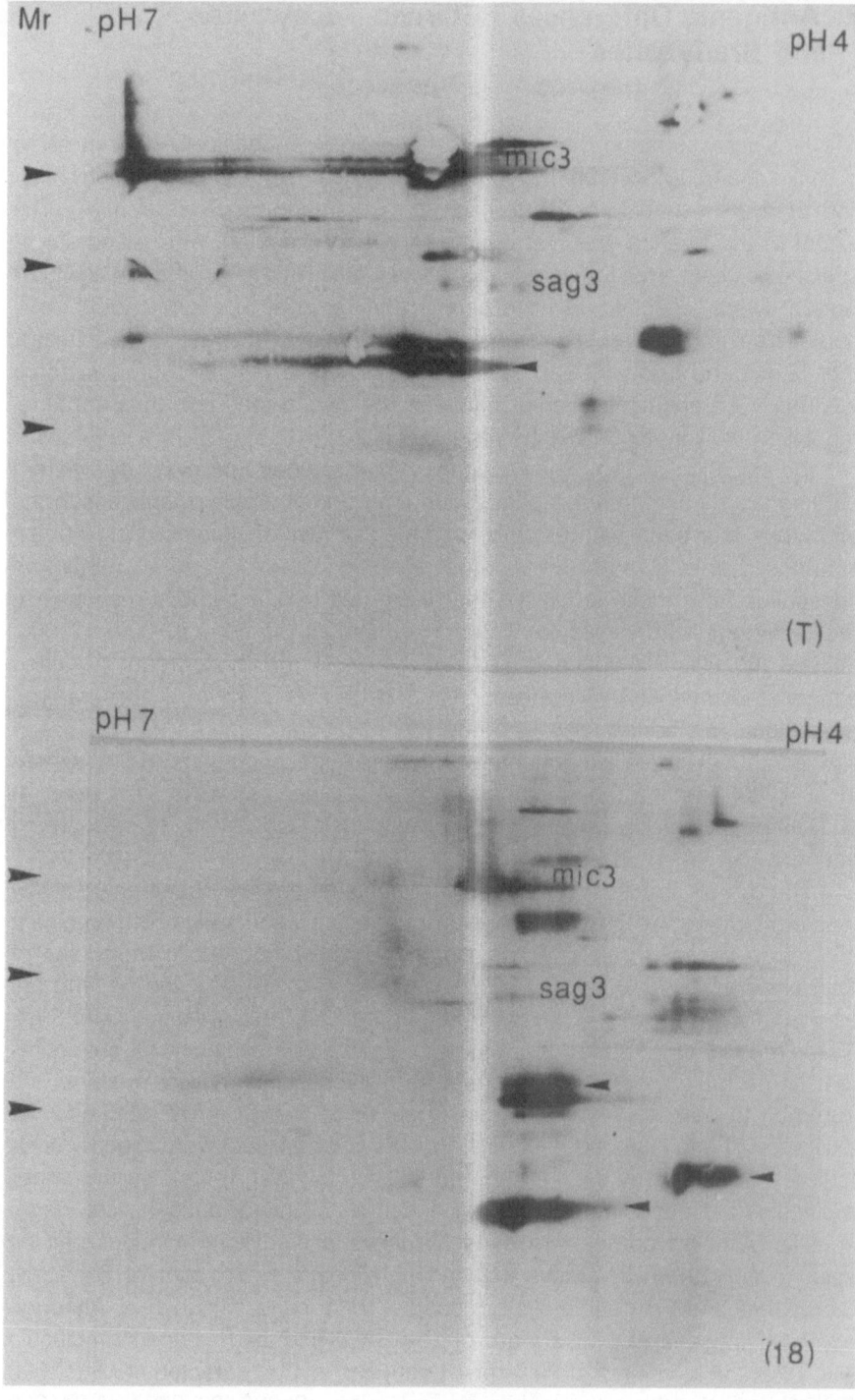

molecular composition of the surface. The bradyzoite possesses at least four specific surface molecules, Pb18 (tentative name SAG4; this volume), Pb21, Pb36 and Pb34 (Tomavo et al. 1991), but lacks two major tachyzoite proteins, SAG1 and SAG2 (Kasper et al. 1985; Woodison and Smith 1990). In addition, specific molecules are found in the cytoplasm of bradyzoites (Weiss et al. 1992; Bohne et al. 1993) and in the matrix (Parmley et al. 1994; Zhang and Smith 1995b) and wall (Weiss et al. 1992) of the cyst. The most immunogenic of these appears to be the cytoplasmic molecule (Bohne et al. 1993), which in our western blotting analysis overlaps with the major Mr 24 000 cyst antigen (Zhang, personal communication).

3 The Serological Response to Bradyzoite/Cyst Antigens During Infection

Western blotting studies have given a good qualitative impression of the serological responses to bradyzoite/cyst antigens during infection. In experimental murine infection, an early antibody response is seen to the tachyzoite, and the first antigen to be recognised in cysts is a shared Mr 42 000 molecule. Antibody recognition of cyst-specific antigens first appears at 21 days postinfection and increases in complexity and strength up to day 90 (Woodison, personal communication). The antibody response in humans is quite similar except that recognition of cyst antigens is even lower. Direct comparison of tachyzoite and cyst antigens probed with randomly selected human sera illustrates this point. Woodison et al. (1993) analysed the sequential reactivity of human sera from heart transplant patients and from an individual accidentally infected with the RH strain. In all cases, the antibody response to the cyst was both lower and occurred later in infection than the response to tachyzoites.

More recently we have made direct quantitative comparisons of the IgG and IgM responses to tachyzoite and cyst antigens during infection using a differential ELISA screen. In experimental murine infection, the IgG response to cysts was much lower than to tachyzoites but the IgM response was preferentially directed at the cyst (Zhang and Smith 1995a). In humans, both IgG and IgM responses to cyst antigens were very low, barely rising above cut-off values (Zhang et al. 1995).

◄─────────────────────

Fig. 2. Two-dimensional analysis of tachyzoite and bradyzoite antigens. Parasites, 10^6 RH strain tachyzoites (*7*) or 18691 strain bradyzoites (*18*), were solubilised in lysis buffer, and separated overnight on 3.5% acrylamide rod gels containing 3.5–10 ampholines. Second dimension separation was performed on 12% acrylamide gels and proteins were transferred to nitrocellulose and probed with pooled sera from 18691 strain infected mice. Spots corresponding to the shared surface antigen SAG3 ,the shared microneme molecule MIC3, the major tachyzoite surface molecule SAG1, and three bradyzoite-specific molecules (Pb 18 000, 20 000 and 24 000) are marked. *Arrowheads* indicate the molecular weight markers BSA (66 000), ovalbumin (45 000) and trypsinogen (24 000)

Both western blotting and ELISA data confirm that the serological response to cyst antigens during natural infection is very low. A number of factors may contribute to this phenomenon: cysts may be inherently less immunogenic; antigen load may be lower; processing and presentation of cyst antigen, predominantly located in brain and muscle tissue, may be inefficient or indeed there may be tolerance to these antigens. We suggest that antigen load may well be important. Ultrastructural studies have shown the cyst wall to be a relatively stable adaptation of the parasitophorous vacuole membrane (FERGUSON and HUTCHINSON 1987) which, unlike the tachyzoite (GRIMWOOD and SMITH 1995), shows no evidence of vacuolar extension or export of antigenic material to the host cell surface. Leakage of antigen from the cyst is therefore likely to be restricted to the rare occasions when cyst rupture occurs (FERGUSON et al 1989).

One point of interest raised by comparative serological studies (ZHANG and SMITH 1995b, ZHANG et al. 1995) is that the response to cysts in human infection is much lower than in experimental murine infection. It is tempting to speculate that either the number or the turnover of cysts is lower in humans. Certainly there is evidence of pathology associated with cystic infection in mouse brain (FERGUSON et al. 1991) and, although no quantitative studies of parasite burden have been completed, reports of cysts in routine human pathological examinations are rare (FERGUSON, personal communication).

In summary, study of the serological response to *Toxoplasma* tachyzoites and bradyzoites has revealed that, although some antigenic molecules are common to both stages, the dominant antigens are stage-specific. The recognition of cyst antigen during *Toxoplasma* infection in humans is very low compared to the anti-tachyzoite response and we suggest that this is partially a reflection of differences in antigen load.

Acknowledgements. We wish to thank Debra Evans for excellent technical assistance and Amanda Lane for reading the manuscript. Peter Appleford is supported on an MRC studentship.

References

Bohne W, Heesemann J, Gross U (1993) Coexistence of heterogeneous populations of Toxoplasma gondii parasites within parasitophorous vacuoles of murine macrophages revealed by a bradyzoite-specific monoclonal antibody. Parasitol Res 79:485–487

Cesbron-Delauw MF (1994) Dense granule organelles of Toxoplasma gondii. Parasitol Today 10:293–296

Couvreur G, Sadak A, Fortier B, Dubremetz JF (1988) Surface antigens of Toxoplasma gondii. Parasitol 97:1–10

Darcy F, Charif H, Caron H, Deslee D, Pierce RJ, Cesbron-Delauw MF, Decoster A, Capron A (1990) Identification and biochemical characterisation of antigens of tachyzoites and bradyzoites of Toxoplasma gondii with cross-reactive epitopes. Parasitol Res 76:473–478

Ferguson DJP, Hutchinson WM (1987) An ultrastructural study of the early development and tissue cyst formation of Toxoplasma gondii in the brains of mice. Parasitol Res 73:483–491

Ferguson DJP, Hutchinson WM, Pettersen E (1989) Tissue cyst rupture in mice chronically infected with Toxoplasma gondii. Parasitol Res 75:599–603

Ferguson DJP, Graham DI, Hutchinson WM (1991) Pathological changes in the brains of mice infected with Toxoplasma gondii: a histological, immunocytochemical and ultrastructural study. Int J Exp Path 72:463–474

van Gelder P, Bosman F, Demeuter F, van Heuverswyn H, Hérion P (1993) Serodiagnosis of Toxoplasma gondii by using a recombinant form of a 54 kDa rhoptry antigen expressed in E. coli. J Clin Microbiol 31:9–15

Grimwood J, Smith JE (1995) Toxoplasma gondii: redistribution of tachyzoite surface protein during host cell invasion and intracellular development. Parasitol Res 81:657–661

Kasper LH (1989) Identification of stage specific antigens of Toxoplasma gondii. Infect Immun 57:668–672

Kasper LH, Currie KM, Bradley MS (1985) An unexpected response to vaccination with a purified major membrane tachyzoite antigen (P30) of Toxoplasma gondii. J Immunol 132:443–449

Luft BJ, Remington JS (1992) Toxoplasmic encephalitis in AIDS. Clin Infect Dis 15:211–222

Lunde MN, Jacobs L (1983) Antigenic differences between endozoites and cystozoites of Toxoplasma gondii. J Parasitol 69:806–811

Makioka AA, Suzuki Y, Kobayashi A (1991) Recognition of tachyzoite and bradyzoite antigens of Toxoplasma gondii by infected hosts. Infect Immun 59:2763–2766

Parmley SF, Yang S, Harth G, Sibley LD, Sucharczuk A, Remington JS (1994) Molecular characterisation of a 65-kilodalton Toxoplasma gondii antigen expressed abundantly in the matrix of tissue cysts. Mol Biochem Parasitol 66:283–296

Partenen P, Turunen HJ, Paasivuo RAT, Leinikki PO (1984) Immunoblot analysis of Toxoplasma gondii antigens by human immunoglobulins G, M and A antibodies at different stages of infection. J Clin Microbiol 20:133–135

Pavesio CEN, Chiappino ML, Setzer PY, Nichols BA (1992) Toxoplasma gondii: differentiation and death of bradyzoites. Parasitol Res 78:1–9

Potasmen I, Araujo FG, Desmonts G, Remington JS (1986) Analysis of Toxoplasma gondii antigens recognised by human sera obtained before and after acute infection. J Infect Dis 154:650–657

Tomavo S, Fortier B, Soête M, Ansel C, Camus D, Dubremetz JF (1991) Characterisation of bradyzoite-specific antigens of Toxoplasma gondii. Infect Immun 59:3750–3753

Torpier G, Charif H, Darcy F, Liu J, Dardé ML, Capron A (1993) Toxoplasma gondii: differential location of antigens secreted from encysted bradyzoites. Exp Parasitol 77:13–22

Verhofstede C, van Gelder P, Rabaey M (1988) The infection stage-related IgG response to Toxoplasma gondii studied by immunoblotting. Parasitol Res 74:516–520

Weiss LM, LaPlace D, Tanowitz HB, Wittner M (1992) Identification of Toxoplasma gondii bradyzoite specific monoclonal antibodies. J Infect Dis 166:213–215

Woodison G, Smith JE (1990) Identification of the dominant cyst antigens of Toxoplasma gondii. Parasitol 100:389–342

Woodison G, Balfour AH, Smith JE (1993) Sequential reactivity of serum against cyst antigens in Toxoplasma infection. J Clin Pathol 46:548–550

Zhang YW, Smith JE (1995a) Toxoplasma gondii: Reactivity of murine sera against tachyzoite and cyst antigens via FAST-ELISA. Int J Parasitol 25:637–640

Zhang YW, Smith JE (1995b) Toxoplasma gondii: Identification and characterisation of a cyst molecule. Exp Parasitol 80:228–233

Zhang YW, Fraser A, Balfour AH, Wreghitt TG, Gray JJ, Smith JE (1995) Serological reactivity against cyst and tachyzoite antigens of Toxoplasma gondii via FAST-ELISA. J Clin Pathol 48:908–911

Toxoplasma gondii: Kinetics of Stage-Specific Protein Expression During Tachyzoite-Bradyzoite Conversion in Vitro

M. Soête and J.F. Dubremetz

1 Introduction

Stage conversion by *Toxoplasma gondii* is a key step in the interaction of the parasite with its host, since it is responsible for the long-term maintenance of the infection in immunocompetent hosts.The ability of T. gondii to persist in the host explains the long-range success of the parasite. Stage conversion is also a major concern in human pathology, as it is the main cause of toxoplasmic reactivation, which is often fatal in AIDS patients. This last aspect has recently stimulated a strong interest in studying the mechanisms of stage conversion in *T. gondii*. Progress has been made possible by identification of stage-specific probes and development of in vitro procedures mimicking the in vivo conversion, facilitating study of the process in culture rather than in the organs of infected hosts. Most of the studies have been concerned with the tachyzoite to bradyzoite switch, and most of this review will deal with this conversion. The reverse phenomenon, i.e., reactivation by cyst rupture, is still very poorly understood, as no in vitro model is yet available for its analysis.

The switch from the acute stage to the cyst stage is actually a way for the parasite to stay hidden within a host cell and therefore out of reach of the host's defenses. Since tachyzoites multiply indefinitely and are eventually released from the host cell, a major change required for cyst formation is a decrease in the multiplication rate, most likely leading to a complete arrest

Unité 42 INSERM, 369 rue J Guesde, 59650 Villeneuve d'Ascq, France

of the mitotic cycle in "mature cysts." Thus, in addition to modifying gene expression, switching leads to an alteration of the parasite's cell cycle, an aspect which remains entirely uninvestigated. Whether a decrease in multiplication is a cause or a consequence of the switch is still unsolved.

The following review will describe the results of in vitro analysis of stage conversion with respect to the characteristics and kinetics of this process.

2 Induction of Differentiation in Vitro

Using newly obtained bradyzoite-specific antibodies, it could be shown that, in parasites grown in vitro in normal culture conditions, a part of the population expressed bradyzoite-specific proteins. This confirmed earlier observations reporting cyst-like structures in vitro. The difference, however, was that, whereas "spontaneous cyst" development had been reported for a limited number of slowly growing *T. gondii* strains, the higher sensitivity and precision of immunodetection showed that this expression could even occur in rapidly growing strains such as RH. The frequency of spontaneous expression was, however, lower in these strains than in the slowly growing ones. Bradyzoite protein expression had never been observed in parasites taken from the mouse peritoneal cavity after routine passage; instead, only tachyzoite-specific molecules were expressed. Thus, it seemed that this low percentage of spontaneous bradyzoite protein expression in vitro was dependent on a trigger related to the cell culture. As adverse culture conditions such as exhaustion of the medium seemed to increase the percentage of parasites expressing bradyzoite proteins, evaluation of the effect of various physical or chemical agents on the cultures led to the design of procedures enhancing the switch from tachyzoites to bradyzoites in vitro (SOÊTE et al. 1993, 1994). The easiest procedure was to use a high pH medium (RPMI adjusted at pH 8); but heat treatment (growing infected cells at 43°C) also induced the switch. Other investigators have used procedures mimicking the immune response in vivo, such as the effect of interferon (IFN)-γ on infected macrophages, which led to the identification of nitric oxide (NO) as a switch inducer (BOHNE et al. 1993, 1994). As NO can act on mitochondrial respiratory enzymes, inhibitors of the respiratory pathway were tested and shown to induce switching in vitro (BOHNE et al. 1994; TOMAVO and BOOTHROYD 1995). Thus, a number of treatments can be used to trigger stage conversion in vitro. However, all these procedures show some toxicity for host cells (especially when using mitochondrial inhibitors), and treatments beyond a few days are often deleterious. It is therefore important to reach a balance between efficiency of switching and survival of the culture. WEISS et al. (1995) even proposed a combination of several factors, e.g. pH, IFN-γ, and interleukin (IL)-6, to obtain continuous production in vitro: their procedure essentially aims at avoiding tachyzoite overgrowth.

What these results suggest in terms of the mechanism of switching is that a stress, which can be brought on for example by affecting the mitochondrial respiration of the parasite, is needed, but the precise pathway leading to bradyzoite gene expression is still under investigation. Experiments using host cells with deficient mitochondria have suggested that the effectors act directly on the parasite and not through alterations in host cell metabolism (BOHNE et al. 1994; TOMAVO and BOOTHROYD 1995).

3 Stage-Specific Proteins and Structures

Differentiation between the tachyzoite and bradyzoite stages involves both the parasite itself and the compartment in which the parasite resides: tachyzoites in the parasitophorous vacuole and bradyzoites in the cyst. During stage conversion in vitro all intermediates between these two compartments can be found. These structures do not seem to be significantly different from those that have been observed in vivo (FERGUSON and HUTCHINSON 1987), although the difficulty of in vivo analysis has impeded description of the process with as much detail as found regarding in vitro analysis.

Most of the molecules described previously in *T. gondii* were identified in the tachyzoite stage, since it is more accessible to experimental studies. Many of these molecules are common to both tachyzoites and bradyzoites. The main tachyzoite-specific proteins are surface proteins: SAG1 (P30) and SAG2 (P22). The bradyzoite stage-specific molecules that have been described so far can be distinguished by their size and localization:

1. Surface molecules: 18 kDa, 21 kDa, 34 kDa, 36 kDa (TOMAVO et al. 1991). The gene encoding the 18 kDa protein has been cloned and sequenced (ODBERG-FERRAGUT et al. unpublished); biochemical data (surface biotinylation, glucosamine incorporation) and sequence data (hydrophobic COOH-terminal) suggest that this protein is GPI anchored, as are the major tachyzoite surface proteins (TOMAVO et al. 1989). The other proteins have not been studied in detail, so far. Preliminary electron microscopy studies have suggested that the 21 kDa molecule might also be located in micronemes (Soête, unpublished).
2. Cytoplasmic proteins: the gene encoding the 29 kDa protein (BAG1) has been cloned and sequenced (see BOHNE et al. 1995; PARMLEY et al. 1995), as has the gene encoding LDH2 (YANG and PARMLEY 1995), the cytoplasmic localization of which has to be confirmed.
3. Cyst wall proteins: the gene encoding a 65 kDa protein (MAG1) has been cloned and sequenced (PARMLEY et al. 1994), while both a 116 kDa (WEISS et al. 1992) and a 29 kDa protein (ZHANG and SMITH 1995) have been studied only at the protein level.

Most of the dense granule proteins (GRA1–6; CESBRON-DELAUW 1994) are common to the tachyzoite and bradyzoite stages and are exocytosed into the vacuolar space or to the cyst wall; therefore, although their abundance and distribution may vary during the two stages, they are not useful markers of differentiation.

The 116 kDa molecule does not accumulate significantly in parasite organelles, but is very abundant in the cyst wall (Soête et al., unpublished).Thus, it may very well be that, in contrast to other material exocytosed by *T. gondii*, the 116kDa molecule might not be stored before secretion, but rather exocytosed immediately upon synthesis. The same findings have been reported for the other cyst wall-specific molecule MAG1 (PARMLEY et al. 1994)

4 Kinetics of Stage Conversion in Vitro

The tachyzoite to bradyzoite switch in vitro is a gradual process, which leads from a parasitophorous vacuole containing typical tachyzoites through a series of intermediate stages to a "true cyst" containing typical bradyzoites. There are still some concerns about the maturity of the latter, as discussed below.

When using peritoneal tachyzoites and pH switching on strain PLK (cloned ME49), it is necessary to wait for 24 h after invasion to be able to detect the early expression of surface bradyzoite proteins (36 kDa, 18 kDa). At this stage, a significant deposit of the 116 kDa cyst wall molecule is already found by immunofluorescence staining. SAG1 is still present (whether it just persists in situ or is coexpressed is unknown, as mRNA data are not available yet). Parasites completely lacking SAG1 are first seen at day 4. The last known molecule to appear during transformation is the 21 kDa protein, which is found in part of the population (and often only some of the parasites within a single vacuole) at 7 days. Although it remains to be confirmed, this molecule may appear only when the mitotic cycle is arrested.

Electron microscopy of the switch shows a gradual increase in micronemes and amylopectin granule content in the parasites. However, these characteristics are only clearly found after 7 days, whereas alteration of the vacuolar space is obvious as early as 24 h after invasion. At this stage, the vacuolar network is already disorganized, and a significant accumulation of dense material is already found at 48 h. This means that the cyst wall forms very early, long before total differentiation of the bradyzoite is completed. As differences in the organization of the compartment are likely to reflect a dramatic change in the metabolic exchanges with the host cell, one of the first events in the switch is likely to be restriction of the trafficking of metabolites between the parasite and the host cell.

In the RH strain, the characteristics of switching are somehow different, except for cyst wall deposition which is rather similar to what happens with other strains. The main differences concerning the RH strain in vitro are: (1) it never expresses the 21 kDa protein, (2) the increases in micronemes and amylopectin are less pronounced, and (3) dividing parasites are still common at day 7, whereas they are almost absent with strain PLK at the same stage. Whether this means that RH differentiates at a slower rate or whether it cannot reach the same level of differentiation as the other strains is not yet known, but the second proposal seems to be more likely, as suggested by the almost complete inability of RH to produce cysts in vivo.

All in all, it remains to be determined when in vitro-obtained cysts are fully mature and, indeed, what is the definition of maturity (for example: complete mitotic arrest). In addition, the timing and level of maturity might be different among strains.

Reactivation has not been studied in vitro, although we know that releasing the stress usually leads to a rapid resumption in tachyzoite multiplication. It is not yet known whether any reactivation occurs before parasite release, i.e., within the cysts. In addition, it is also unknown if the tachyzoite to bradyzoite switch can be reversed within the same vacuole and, if so, up to what stage. Since alteration of the vacuolar space occurs very early during the switch, one might think that the interaction with the host cell (and especially through the parasitophorous vacuole pores) may be irreversibly modified early on, unless the parasite could reorganize the compartment from the cyst to the vacuole, which seems unlikely.

This issue of reactivation is rather difficult to study, since all cultures undergoing switching in vitro usually include a large variety of stages of switching and it is not possible to follow the fate of a single vacuole/cyst while simultaneously analysing specific protein expression.

5 Perspectives

Many biological questions remain to be solved concerning the respective properties of tachyzoites and bradyzoites. The operational definition of the bradyzoite, i.e., pepsin resistance and the ability to produce oocysts in cats in 72 h, has not been tested yet with culture-switched parasites. This would reveal which of the developmental changes, among those we follow during in vitro conversion, are needed to form functional bradyzoites. The function of stage-specific surface molecules may be related to stage-specific features of the cycle, such as the need for the bradyzoite to resist the pepsin and trypsin in the gut. These proteins may also be involved in host cell specificity. It is also important to understand the function of cyst wall deposition with respect to metabolic exchanges with the host cell, and possibly with respect to the survival of the infected host cell, to ensure long-lasting protection for the parasite.

The in vitro system will also serve many more experimental purposes, such as the search for switch mutants that would either never switch, would develop exclusively as bradyzoites, or would express only part of the conversion phenotype (such as lacking the cyst wall). In vitro switching could also be used to study stage-specific gene knockout mutants. By dissecting the switching process, these investigations will tell us what molecular mechanisms are involved in differentiation.

What is still needed is to develop a reactivation model in vitro, which would be a direct continuation of the bradyzoite induction model and would complete the in vitro model of disease. There is no doubt that the progress made so far has paved the way for such a model, which would greatly facilitate an understanding of the effect of this important aspect of the disease on the immunocompromised host.

References

Bohne W, Heesemann J, Gross U (1993) Induction of bradyzoite-specific Toxoplasma gondii antigens in gamma interferon-treated mouse macrophages. Infect Immun 61:1141–1145

Bohne W, Heesemann J, Gross U (1994) Reduced replication of Toxoplasma gondii is necessary for induction of bradyzoite-specific antigens: a possible role for nitric oxide in triggering stage conversion. Infect Immun 62:1761–1767

Bohne W, Gross U, Ferguson DJP, Heesemann J (1995) Cloning and characterization of a bradyzoite-specifically expressed gene (hsp 30/bag1) of Toxoplasma gondii related to genes encoding small heat shock proteins of plants. Mol Microbiol 16:1221–1230

Cesbron-Delauw MF (1994) Dense-granule organelles of Toxoplasma gondii: their role in the host-parasite relationship. Parasitol Today 10:293–296

Ferguson DJP, Hutchison WM (1987) An ultrastructural study of the early development and tissue cyst formation of Toxoplasma gondii in the brains of mice. Parasitol Res 73:483–491

Parmley SF, Weiss LM, Yang S (1995) Cloning of a bradyzoite-specific gene of Toxoplasma gondii encoding a cytoplasmic antigen. Mol Biochem Parasit 73:253–257

Parmley SF, Yang SM, Harth G, Sibley LD, Sucharczuk A, Remington JS (1994) Molecular characterization of a 65-kilodalton Toxoplasma gondii antigen expressed abundantly in the matrix of tissue cysts. Mol Biochem Parasitol 66:283–296

Soête M, Fortier B, Camus D, Dubremetz JF (1993) Toxoplasma gondii: kinetics of bradyzoite-tachyzoite interconversion in vitro. Exp Parasitol 76:259–264

Soête M, Camus D, Dubremetz JF (1994) Experimental induction of bradyzoite-specific antigen expression and cyst formation by the RH strain of Toxoplasma gondii in vitro. Exp Parasitol 78:361–370

Tomavo S, Boothroyd JC (1995) Interconnection between organellar functions, development and drug resistance in the protozoan parasite, Toxoplasma gondii. Int J Parasitol 25:1293–1299

Tomavo S, Fortier B, Soête M, Ansel C, Camus D, Dubremetz JF (1991) Characterization of bradyzoite-specific antigens of Toxoplasma gondii. Infect Immun 59:3750–3753

Tomavo S, Schwarz RT, Dubremetz JF (1989) Evidence for glycosyl-phosphatidyl inositol anchoring of Toxoplasma gondii major surface antigens. Mol Cell Biol 9:4576–4580

Weiss LM, Laplace D, Takvorian PM, Tanowitz HB, Cali A, Wittner M (1995) A cell culture system for study of the development of Toxoplasma gondii bradyzoites. J Euk Microbiol 42:150–157

Weiss LM, Laplace D, Tanowitz HB, Wittner M (1992) Identification of Toxoplasma gondii bradyzoite-specific monoclonal antibodies. J Infect Dis 166:213–215

Yang S, Parmley SF (1995) A bradyzoite-specifically expressed gene of Toxoplasma gondii encodes a polypeptide homologous to lactate dehydrogenase. Mol Biochem Parasitol 73:291–294.

Zhang YW, Smith JE (1995) Toxoplasma gondii: Identification and characterization of a cyst molecule. Exp Parasitol 80:228–233

Bradyzoite-Specific Genes

W. Bohne[1], S.F. Parmley[2], S. Yang[2], and U. Gross[1]

1 Introduction

Stage conversion between tachyzoites and bradyzoites is a central event for pathogenicity and persistence of *Toxoplasma gondii*. The acute form of the infection, which is associated with tachyzoites, is normally overcome by the onset of the specific immune response. However, the parasite is able to persist by forming intracellularily located tissue cysts which contain metabolically and replicatively dormant bradyzoites. These cysts are found predominantly in the brain of its host. The differentiation process is reversible; reactivation of brain cysts from chronically infected patients is thought to be the major cause of toxoplasmic encephalitis in acquired immunodeficiency syndrome (AIDS) patients (LUFT et al. 1984; WONG et al. 1984). Since to date no drug is available that is able to cure patients from a chronic infection by killing bradyzoites, the tissue cyst stage represents a lifelong risk for reactivation in immunocompromised patients.

During the interconversion between tachyzoites and bradyzoites, the expression of stage-specific components is induced. Several stage-specific antigens have been characterized by using monoclonal antibodies (MAbs) or polyclonal antisera (KASPER 1989; OMATA et al. 1989; TOMAVO et al. 1991; WEISS et al. 1992; BOHNE et al. 1993; GROSS et al. 1995). SAG1 and SAG2 are major

[1] Institute of Hygiene and Microbiology, University of Würzburg, Josef-Schneider-Strasse 2, 97080 Würzburg, Germany.
[2] Department of Immunology and Infectious Diseases, Research Institute, Palo Alto Medical Foundation; 860 Bryant Street, Palo Alto, CA 94301, USA

surface proteins of tachyzoites that are not expressed in bradyzoites (BURG et al. 1988; PRINCE et al. 1990). Transcription of *SAG1* is under the control of multiple 27-bp repetitive elements in the 5' untranslated region. It has been shown in studies utilizing reporter genes that the transcription efficiency depends on the number and location of these elements, and that they probably serve as a selector for transcription (SOLDATI and BOOTHROYD 1995). It is not known whether these elements are also responsible for stage-specific transcription of *SAG1*.

Besides structural changes in the parasite itself, the differentiation from tachyzoites to bradyzoites is accompanied by the formation of the cyst wall, probably by a modification of the parasitophorous vacuole (FERGUSON and HUT-CHISON 1987). In addition, the parasite reduces its metabolism and replication rate in the cyst stage as an adaptation for persistence in the host. The interconversion requires an initial decision to start the differentiation, followed by a coordinated regulation of gene expression. Triggers for these steps in vivo are unknown, but recent progress in analyzing stage conversion has come from in vitro studies, in which it was demonstrated that stage conversion is inducible (BOHNE et al. 1993, 1994; SOÊTE et al. 1994). A detailed analysis of stage conversion might lead to therapeutic concepts for influencing the differentiation process of *T. gondii* with the aim of preventing reactivation of a chronic infection and cyst formation after an acute infection. Cloning and analyzing those genes that are differentially expressed in tachyzoites and bradyzoites is important in this context to investigate the differentiation on a molecular level. In this review, we will give an update on the current knowledge of bradyzoite-specific genes.

Several strategies exist for the detection of stage-specifically expressed genes including differential display polymerase chain reaction (PCR) and subtractive cDNA cloning. The generation of bradyzoite-specific MAbs and polyclonal antisera has, however, made immunoscreening of cDNA expression libraries the most promising strategy for identifying genes expressed only in bradyzoites. A major problem in the construction of bradyzoite-specific cDNA libraries is the limited amount of bradyzoites that can either be obtained by isolating cysts from brains of infected mice or by in vitro induction of bradyzoite development. This problem has been overcome by using random PCR, resulting in amplification of the entire cDNA population (FROUSSARD 1992). As a disadvantage, this method results in truncated cDNA fragments which lack the 5' and the 3' ends. In addition, shorter fragments might be preferentially amplified, and clones containing longer cDNA fragments might therefore be underrepresented in the obtained cDNA libraries. Nevertheless, expression libraries constructed by this method were useful for identifying several bradyzoite-specific genes (PARMLEY et al. 1994; BOHNE et al. 1995).

Bradyzoites were isolated either from cysts derived from mouse brains or were induced in vitro. Starting from a tachyzoite culture, the induction of bradyzoite formation is possible by exposing the parasites to stress conditions (BOHNE et al. 1994; SOÊTE et al. 1994). Cultivation of parasites at alkaline pH

Table 1. Bradyzoite-specific expressed genes

Gene	Molekular mass (kDa)	Homology	mRNA Tachyzoites	Bradyzoites	Localization
MAG1	65	–	+	+	cyst matrix
BAG1	28–30	sHSP	–	+	cytosol
LDH2	35	LDH	–	+	cytosol

(pH 8.0–8.5) for several days has turned out to be one of the least laborious methods of obtaining acceptable yields of bradyzoites (Soête et al. 1994). However, stage conversion is not completed by this method and only 30%–70% of parasites (strain NTE) express bradyzoite-specific antigens, making a purification step necessary in order to eliminate tachyzoites. By utilizing the bradyzoite-specific MAb 4F8, in vitro induced bradyzoites could be separated from concomitant tachyzoites by magnetic cell sorting (MACS). A fraction of these in vitro induced bradyzoites still co-expressed tachyzoite-specific proteins (e.g., SAG1). They probably differ in several aspects from mature bradyzoites in cysts; in vitro induced bradyzoites represent an early form of a bradyzoite that has not completed the entire differentiation process. However, for isolating mRNA from genes that are expressed early during bradyzoite development, these in vitro induced bradyzoites might be even more useful than mature cysts. It is possible that transcription rates are significantly reduced in dormant mature bradyzoites, and transcription rates of early induced genes might be maximal at the beginning of the differentiation. Some bradyzoite-specific expressed genes have recently been successfully cloned and characterized (Table 1).

2 BAG1

BAG1 is a 28–30 kDa antigen that is specifically expressed in bradyzoites, but not in tachyzoites or oocysts. The corresponding gene, designated BAG1, was isolated from an expression library by using MAb 7E5 (Bohne et al. 1995). The same gene was isolated independently by using MAb 74.1.8. and was named BAG5 (Parmley et al. 1995). Both MAbs were described originally in two different publications as bradyzoite-specific antibodies (Weiss et al. 1992; Bohne et al. 1993). To avoid confusion with the designation in the future, we propose that gene and gene product should be referred to as BAG1 and BAG1, respectively.

   Both MAbs react with recombinant BAG1 expressed in *Escherichia coli*. However, these MAbs are directed to different epitopes. MAb 74.1.8. but not MAb 7E5 reacts with recombinant BAG1 that is lacking the first 28 N-terminal amino acids, suggesting that the reactive epitope of MAb 7E5 is located within this region. Polyclonal antisera that were generated against purified recombinant BAG1 react like MAb 7E5 with a 28–30 kDa antigen expressed only in bradyzoites. This anti-BAG1 antiserum was used in immunoelectron microscopy to localize the reactive antigen within bradyzoites. BAG1 is exclusively located in the cytoplasm (BOHNE et al. 1995). Labeling was found neither in the organelles of the parasite nor in the matrix ground substance and the cyst wall. Tachyzoites were completely nonreactive, confirming stage-specific expression of this antigen.

The *BAG1* gene contains three introns; its transcript has an open reading frame of 687 bp that codes for 229 amino acids. The carboxyl-terminal region of the predicted protein has similarities to the family of small heat shock proteins (sHSP), especially those from plants (LINDQUIST and CRAIG 1988). The GVL motif, a typical characteristic for sHSPs, is completely conserved in BAG1. Further homologies of BAG1 to other proteins are found in the amino-terminal region, where a 50 amino acid domain shares homology to the B domain of synapsin Ia, a peripheral membrane protein that is involved in binding of synaptic vesicles to the cytoskeleton (SÜDHOF 1990).

The homology of BAG1 to sHSPs might reflect a possible function of the protein as a stress protein. Differences in expression of sHSP during differentiation and development is a common phenomenon found in many species (LINDQUIST and CRAIG 1988). Moreover, some sHSPs (especially HSP27) were found to accumulate in cells with a reduced replication rate (PAULI et al. 1990). It has been shown recently that beside HSPs of higher molecular weight, also some sHSPs have chaperone function. At least in vitro, bradyzoite formation is inducible by exposing the parasite to stress conditions (pH 8 or nitric oxide). In vivo, similar stress situations might trigger the conversion to bradyzoites in the brain, for example by nitric oxide release of activated astroglia or microglia cells (CHAO et al. 1992; PETERSON et al. 1995). In this environment, inducible stress proteins might have a protective function at a point where the parasite has to initiate the conversion process accompanied with major changes in gene expression and metabolism. However, BAG1 is not only expressed during the interconversion but also in mature cysts, suggesting that it is also important for mature bradyzoites.

Expression of BAG1 seems to be regulated at the mRNA level. *BAG1* transcripts were detectable by reverse transcriptase-polymerase chain reaction (RT-PCR) only in cysts but not in tachyzoites. In cell cultures infected with tachyzoites, *BAG1* transcripts could be detected after induction of bradyzoite differentiation by pH 8.0 shift of the cell culture medium, indicating that the steady-state level of *BAG1* mRNA is upregulated during bradyzoite differentiation. Developmentally expressed sHSPs are not necessarily induced by heat-shock treatment (ZIMMERMANN et al. 1983; BOND and SCHLESINGER 1987; GYORGYEY et al. 1991). Indeed, the heat-shock response elements (HRE) normally found in the promoter region of heat shock-inducible genes (AMIN et al. 1988) are absent from the promoter region of *BAG1*. Furthermore, no enhanced level of *BAG1* transcripts could be found in reverse PCR analysis after heat-shock treatment, indicating that *BAG1* is not upregulated after heat-induced stress.

As for other genes of *T. gondii*, no consensus sequence such as a TATA box, is found in the promoter region of *BAG1*. However, an interesting feature has been identified in the 5′ untranslated region. A DNA stretch of 9 bp (TGCTGTGTC) which is located 204 bp upstream of the transcription start of *BAG1* is also present in the 5′ untranslated region of the gene encoding the matrix antigen MAG1 (PARMLEY et al. 1994; BOHNE et al 1995). A potential role of this region for transcriptional regulation is under investigation. However, transcriptional regulation of *MAG1* seems to differ from those of *BAG1*, since *MAG1* transcripts are also detectable in tachyzoites (see below).

3 *MAG1*

MAG1 is a 65 kDa antigen that is expressed only in the cyst stage, but not in the tachyzoite stage. The *MAG1* gene was isolated from a mature cyst cDNA expression library by immunoscreening with polyclonal anticyst sera (PARMLEY et al. 1994). Antisera generated against purified recombinant MAG1 reacted with an abundant 65 kDa antigen in immunoblots of lysates from mature cysts, but not significantly with lysates from tachyzoites isolated from mouse peritoneal fluid. The anti-MAG1 antiserum was also used in immuno-electron microscopy and found to localize the native MAG1 antigen in abundance in the cyst matrix and at a lower level in the cyst wall. No significant labeling was detected on the surface of bradyzoites, in the cytoplasm, or in organelles within bradyzoites, suggesting that the antigen is rapidly secreted. The recombinant MAG1 antigen is recognized strongly by sera from chronically infected animals. In addition, the major band in immunoblots of purified cysts that is recognized by sera from infected animals co-migrates with the 65 kDa band detected with anti-MAG1 antisera. These results suggest that MAG1 is an immunodominant cyst antigen.

The *MAG1* transcript is 2 kb and has an open reading frame of 1356 bp that codes for 452 amino acids. The 25 residue amino-terminal region of the polypeptide is predicted to function as a signal sequence, which is consistent with the secretion of the antigen into the cyst matrix. The *MAG1* gene contains two introns. One intron is 110 bp and is located 95 bp downstream of the first ATG codon. The other intron is 503 bp and is located 4 bp upstream of the first ATG. This is the first *T. gondii* gene that has been found to contain an intron in the 5′ untranslated region. The functional significance of this intron is unknown, but could be important for the developmental regulation of the MAG1 protein. Comparison of the nucleic acid sequence and the predicted amino acid sequence with databases did not reveal any homology to known genes or proteins. Apart from the 9 bp sequence in the 5′ untranslated sequence of *MAG1* that has identity with the *BAG1* promoter region (mentioned above), no consensus sequences, such as a TATA box was found in the upstream region of *MAG1*.

Preliminary RT-PCR experiments performed on mouse-derived tachyzoites and bradyzoites suggest that the *MAG1* transcript is expressed in similar levels in mature bradyzoites and rapidly dividing tachyzoites (S.F. PARMLEY and S. YANG, unpublished). The detection of the *MAG1* transcript in tachyzoites was unexpected since there is no detectable MAG1 protein at this stage. One explanation is that the tachyzoite preparation contained parasites that had converted (or were in the process of converting) to bradyzoites. This explanation seems to be ruled out by the absence of detectable bradyzoite-specific *BAG1* transcript in the same tachyzoite preparation. However, if certain bradyzoite genes are turned on at earlier times in the intermediate steps leading towards conversion to bradyzoites, then transcripts such as *MAG1* might be detectable before others such as *BAG1*. Indeed, when immunoblots were probed with bradyzoite MAb 4A12 (TOMAVO et al., 1991), a low level of expression of the 4A12-reactive 36 kDa bradyzoite-specific antigen was detected in lysates from the same tachyzoite preparation (S.F. PARMLEY and S. YANG, unpublished). An alternative explanation is that *MAG1* is developmentally regulated after transcription. In this case the transcript might carry regulatory sequences affecting mRNA stability or translation in the tachyzoite stage.

4 *LDH2*

LDH2 is a 35 kDa antigen that is also expressed in the bradyzoite stage, but not in the tachyzoite stage. A partial *LDH2* cDNA was isolated from a mature cyst cDNA expression library by immunoscreening with polyclonal anticyst sera (YANG and PARMLEY 1995). When the sequence of the cDNA insert was compared with sequences in the databases, a strong homology to lactate dehydrogenase enzymes (LDH; L-lactate: NAD+ oxidoreductase, E.C. 1.1.1.27) from other organisms was discovered. PCR primers derived from the cDNA sequence were found to specifically amplify a DNA fragment from bradyzoite cDNA, but not from tachyzoite cDNA. Antisera generated against purified recombinant LDH2 reacted strongly with a 35 kDa antigen in immunoblots of lysates from mature cysts. However, this antiserum also reacted weakly with a 33 kDa antigen in lysates from tachyzoites isolated from mouse peritoneal fluid. By two-dimensional (2-D) gel electrophoresis and immunoblot, the approximate isoelectric point (pI) and molecular weight of the native bradyzoite LDH2 antigen were found to be 7.0 and 35 kDa, respectively (S. YANG and S.F. PARMLEY, unpublished). The antigen that was detected with the anti-LDH2 serum in tachyzoite lysates had an approximate pI of 6.0 and molecular weight of 33 kDa, suggesting that another isoform of LDH is expressed by tachyzoites (named LDH1). These data are consistent with previous isoenzyme studies which demonstrated that tachyzoites grown in tissue culture express a second LDH isoform not seen in tachyzoites harvested from mouse peritoneum (DARDÉ et al. 1990). In light of recent discoveries that tachyzoites grown in vitro contain a small population of organisms that have undergone spontaneous stage conversion to bradyzoites (BOHNE et al. 1993; SOÊTE et al. 1993), the second LDH isoform seen in these isoenzyme studies most likely is identical to the bradyzoite LDH2 protein. Furthermore, expression of *LDH2* mRNA was induced when tachyzoites were treated under conditions that have been shown to induce conversion to bradyzoites such as incubation in media with alkaline pH (pH 8) or in media containing mitochondrial inhibitors, antimycin A, and oligomycin (S. YANG and S.F. PARMLEY, unpublished).

The entire *LDH2* gene was cloned and characterized. In addition, the gene corresponding to the tachyzoite *LDH1* was cloned and characterized (S. YANG and S.F. PARMLEY, unpublished). Preliminary RT-PCR studies of *LDH1* suggest that the transcript is present in both tachyzoites and bradyzoites, although the LDH1 polypeptide is only expressed in tachyzoites. The *LDH1* mRNA is 1822 bp long with 143 bp of 5′ nontranslated sequence and 692 bp of 3′ nontranslated sequence. The *LDH2* mRNA is 2593 bp with 268 bp of 5′ nontranslated sequence and 1347 bp of 3′ nontranslated sequence. The *LDH1* transcript has an open reading frame of 987 bp, while the *LDH2* transcript has an open reading frame of 978 bp. Both *LDH1* and *LDH2* have introns of similar size (588 and 538 bp, respectively) at the same relative position in their coding regions (approximately 125 bp downstream of the first ATG). Very little homology is found in the sequences of the introns and regions flanking the coding sequence between the two LDH genes of *T. gondii*. However, the two *T. gondii* LDH genes share 64% of identical nucleotides in the coding region. Most of the nucleotide difference occurs at the third position of the codons and does not result in amino acid differences. Consequently, LDH1 and LDH2 share 71.4% of identical amino acids, and an additional 7.3% are conservative substitutions. *LDH1* encodes three more amino acids than *LDH2*. The predicted molecular masses and pIs of LDH1 (35 550 daltons and 5.96, respectively) and LDH2 (35 342 daltons and 7.08, respectively) were similar to the values estimated for the corresponding native antigens. When compared with sequences in the protein databases, the *T. gondii* LDHs were found to share the most identical amino acids with an LDH from *Plasmodium falciparum* (BZIK et al. 1993; YANG and PARMLEY 1995), which is also an Apicomplexan protozoa. *T. gondii* LDH1 and LDH2 share 46.5% and 48.5% identical amino acids with the *P. falciparum* LDH, respectively. When the amino acid sequence of LDH2 was aligned with LDHs from a variety of species, a pentapeptide insertion (KSDKE) was found, which was also found in LDH of *P. falciparum* (BZIK et al. 1993). A similar insertion was found in LDH1 (KPDSE).

The preliminary studies of these two developmentally regulated LDHs suggest a possible correlation between alteration in carbohydrate or energy metabolism and stage conversion in this parasite. For example, LDH2 may have unique properties which are required for the bradyzoite to adapt to the particular microenvironment and metabolic status of the cyst stage.

5 *SAG4*

The bradyzoite-specific 18 kDa protein (tentative name SAG4) has been purified by affinity chromatography with MAb T8 2B1 using parasites grown in vitro under the switching conditions described by SoÊTE and coworkers (1994). Amino acid sequence was obtained from the N-terminus and from internal proteolytic fragments by Edman degradation. The sequence information was used to design degenerate oligonucleotide primers, which were used for PCR ampli- fication of *T. gondii* genomic DNA. Amplification products could be obtained from strains PLK, RH, and 76 K. The amplified fragment (190 bp) was se- quenced, and its translation product agreed perfectly with the polypeptide sequence obtained from the purified protein. This fragment was then used as a probe to clone the complete gene (F.C. ODBERG et al., manuscript in preparation). Genomic DNA was subjected to simple or double restriction enzyme digests, blotted onto filters, and hybridized. A preliminary restriction map was obtained and a 4 kb *Bam*HI fragment was chosen to prepare a size-selected library to be inserted into a λZAP expression vector. Positive clones were obtained from strains PLK and RH and were then sequenced. No difference was found in the promoter region between these two strains. The complete cDNA could be obtained using 5′ and 3′ rapid amplification of cDNA ends (RACE) from total RNA of the PLK strain grown in vitro under switching conditions. The coding sequence is 516 bp long, with no intron present. A putative polyadenylation site is found 709 bp downstream of the stop codon. The 5′ untranslated mRNA is 257 bp long. Genomic DNA has been sequenced 699 bp upstream of the initiation site. No sequence homology has been found in this part with *BAG1* or *MAG1*. The deduced protein sequence is 172 amino acids long and contains a putative signal sequence, a hydrophobic C-terminal tail likely for a glycosylphosphatidylinositol (GPI) anchor, and no transmembrane domain. Comparison with sequence data bases did not show any significant homology to published protein sequences.

6 Bradyzoite Promoter Studies

Plasmids harboring the chloramphenicol acetyltransferase (CAT) gene con- tained within 5′- and 3′-flanking sequences from *T. gondii* genes have been used successfully to monitor promoter activity and to define minimal promoter sequences (KIM et al. 1993; SOLDATI et al. 1995). These studies were performed in tachyzoites using *T. gondii* genes that are expressed in the tachyzoite stage, such as *SAG1* (a tachyzoite-specific gene) and *TUB1* (a constitutively expressed house-keeping gene). To date, no bradyzoite-specific genes have been tested since they have only recently been reported, but bradyzoite genes are expected to have little or no promoter activity in tachyzoites. However, treatments that induce bradyzoite gene expression in vitro might be used in combination with transfection to analyze bradyzoite promoters. To test this system, an LDH2/CAT plasmid was constructed with 1 kb of 5'-flanking sequences (670 bp nontranscribed region and 330 bp transcribed but not translated sequence) and 2 kb of 3'-flanking sequences (400 bp nontranscribed and 1.6 kb transcribed but not translated sequence) from the bradyzoite *LDH2* gene flanking the CAT reporter gene. P strain (cloned from the ME49 strain) tachyzoites were transfected with LDH2/CAT or TUB1/CAT (KIM et al. 1993) as a positive control. Transient transfectants were treated in parallel in neutral media (pH 7.4) or in alkaline media (pH 8.0). There was a fivefold reduction in parasite numbers in transfectants grown for 48 h in alkaline media compared with neutral media; therefore, the CAT values were normalized by parasite number. The normalized CAT activities from the TUB1/CAT plasmid were similar in transfectants grown in neutral or alkaline media. The CAT activities from LDH2/CAT-transfected tachyzoites were near baseline and were increased 15-fold by treatment with alkaline media (S.F. PARMLEY and S. YANG, unpublished).

The putative BAG1 promoter was also investigated in transient transfections utilizing CAT as a reporter. No CAT activity was detectable in tachyzoites, whereas an induction of CAT activity could be observed during bradyzoite differentiation after pH shift of the cell culture medium (W. BOHNE et al., unpublished).

These results confirm that the in vitro pH shift induces differentiation of tachyzoites to bradyzoites at the level of promoter activation as predicted from the shift in expression of stage-specific transcripts observed previously.

7 Outlook

The generation of knockout mutants might be especially powerful for analyzing gene function of bradyzoite-specific genes (KIM et al. 1993; DONALD and ROOS 1994). In contrast to essential tachyzoite-specific genes, whose deletion would be lethal for the parasite, it should be feasible to delete bradyzoite genes in the tachyzoite stage regardless of whether they are essential for bradyzoite development. The capability of knockout mutants to differentiate into bradyzoites could then be investigated in vitro and in vivo. An alternative approach to study gene function would be the permanent expression of bradyzoite genes in tachyzoites, by generating transfectants with the bradyzoite genes under control of a tachyzoite promoter. An altered phenotype might give hints to a possible function of the gene product. Analysis of bradyzoite-specific promoters might lead to the delineation of stage-specific regulatory sequences and the corresponding regulatory factors in order to define the mechanisms of stage conversion at the molecular level.

Acknowledgments. Studies performed by the authors have been supported by grants from the Bundesministerium für Bildung und Forschung (01 KI 9454), the Deutsche Forschungsgemeinschaft (Gr 906/5–2) and the United States Department of Agriculture (91–37204–6878).

References

Amin J, Ananthan J, Voellmy R (1988) Key features of heat shock regulatory elements. Mol Cell Biol 8:3761–3769

Bohne W, Heesemann J, Gross U (1993) Induction of bradyzoite-specific Toxoplasma gondii antigens in gamma interferon-treated mouse macrophages. Infect Immun 61:1141–1145

Bohne W, Heesemann J, Gross U (1994) Reduced replication of Toxoplasma gondii is necessary for induction of bradyzoite-specific antigens:a possible role for nitric oxide in triggering stage conversion. Infect Immun 62:1761–1767

Bohne W, Gross U, Ferguson DJP, Heesemann J (1995) Cloning and characterization of a brady-zoite-specifically expressed gene (hsp30/bag1) of Toxoplasma gondii related to genes encoding small heat-shock proteins of plants. Mol Microbiol 16:1221–1230

Bond U, Schlesinger MJ (1987) Heat-shock proteins and development. Adv Genet 24:1–29

Burg JL, Perelman D, Kasper LH, Ware PL, Boothroyd JC (1988) Molecular analysis of the gene encoding the major surface antigen of Toxoplasma gondii. J Immunol 141:3584–3591

Bzik DJ, Fox BA, Gonyer K (1993) Expression of Plasmodium falciparum lactate dehydrogenase in Escherichia coli. Mol Biochem Parasitol 59:155–166

Chao C, Hu CS, Molitor TW, Shaskan EG, Peterson PK (1992) Activated microglia mediate neuronal cell injury via a nitric oxide mechanism. J Immunol 149:2736–2741

Dardé ML, Bouteille B. Pestre-Alexandre M (1990) Comparisons of isoenzyme profiles of Toxoplasma gondii tachyzoites produced under different culture conditions. Parasitol Res 76:367–371

Donald RGK, Roos DS (1994) Homologous recombination and gene replacement at the dihydrofolate reductase-thymidylate synthase locus in Toxoplasma gondii. Mol Biochem Parasitol 63:243–253

Ferguson DJP, Hutchison WM (1987) An ultrastructural study of the early development and tissue cyst formation of Toxoplasma gondii in the brains of mice. Parasitol Res 73:483–491

Froussard P (1992) A random-PCR method (rPCR) to construct whole cDNA library from low amounts of RNA. Nucleic Acids Res 20:2900

Gross U, Bormuth H, Gaissmaier C, Dittrich C, Krenn V, Bohne W, Ferguson DJP (1995) Monoclonal rat antibodies directed against Toxoplasma gondii suitable for studying tachyzoite-bradyzoite interconversion in vivo. Clin Diag Lab Immunol 2:542–548

Gyorgyey J, Gartner A, Nemeth K, Magyar Z, Hirt H, Heberle Bors E, Dudits D (1991) Alfalfa heat shock genes are differentially expressed during somatic embryogenesis. Plant Mol Biol 16:999–1007

Kasper LH (1989) Identification of stage-specific antigens of Toxoplasma gondii. Infect Immun 57:668–672

Kim K, Soldati D, Boothroyd JC (1993) Gene replacement in Toxoplasma gondii with chloramphenicol acetyltransferase as selectable marker. Science 262:911–914

Lindquist S, Craig EA (1988) The heat-shock proteins. Annu Rev Genet 22:631–677

Luft BJ, Brooks RG, Conley FK, McCabe RE, Remington JS (1984) Toxoplasmic encephalitis in patients with acquired immune deficiency syndrome. JAMA 252:913–917

Omata Y, Igarashi M, Ramos MI, Nakabayashi I (1989) Toxoplasma gondii: antigenic differences between endozoites and cystozoites defined by monoclonal antibodies. Parasitol Res.75:189–193

Parmley SF, Yang S, Harth G, Sibley LD, Sucharczuk A, Remington JS (1994) Molecular characterisation of a 65-kilodalton Toxoplasma gondii antigen expressed abundantly in the matrix of tissue cysts. Mol Biochem Parasitol 66:283–296

Parmley SF, Weiss LM, Yang S (1995) Cloning of a bradyzoite-specific gene of Toxoplasma gondii encoding a cytoplasmic antigen. Mol Biochem Parasitol 73:253–257

Pauli D, Tonka CH, Tissieres A, Arrigo AP (1990) Tissue-specific expression of the heat shock protein HSP27 during Drosophila melanogaster development. J Cell Biol 111:817–828.

Peterson PK, Gekker G, Hu S, Chao CC (1995) Human astrocytes inhibit intracellular multiplication of Toxoplasma gondii by a nitric oxide-mediated mechanism. J Infect Dis 171:516–518

Prince JB, Auer KL, Huskinson J, Parmley SF, Araujo FG, Remington JS (1990) Cloning, expression, and cDNA sequence of surface antigen P22 from Toxoplasma gondii. Mol Biochem Parasitol 43:97–106

Soête M, Fortier B, Camus D, Dubremetz JF (1993) Toxoplasma gondii: kinetics of bradyzoite-tachyzoite interconversion in vitro. Exp Parasitol 76:259–264

Soête M, Camus D, Dubremetz JF (1994) Experimental induction of bradyzoite-specific antigen expression and cyst formation by the RH strain of Toxoplasma gondii in vitro. Exp Parasitol 78:361–370

Soldati D, Boothroyd JC (1995) A selector of transcription initiation in the protozoan parasite Toxoplasma gondii. Mol Cell Biol 15:87–93

Südhof TC (1990) The structure of the human synapsin I gene and protein. J Biol Chem 265:7849–7852

Tomavo S, Fortier B, Soête M, Ansel C, Camus D, Dubremetz JF (1991) Characterization of bradyzoite-specific antigens of Toxoplasma gondii. Infect Immun 59:3750–3753

Weiss LM, LaPlace D, Tanowitz HB, Wittner M (1992) Identification of Toxoplasma gondii bradyzoite-specific monoclonal antibodies. J Infect Dis166:213–215

Wong J, Gold WM, Brown AE, Lange M, Fried R, Grieco M, Mildvan D, Giron J, Tapper ML, Lerner CW, Armstrong D (1984) Central-nervous-system toxoplasmosis in homosexual men and parenteral drug abusers. Ann Intern Med 100:36–42

Yang S, Parmley SF (1995) A bradyzoite-specifically expressed gene of T. gondii encodes a polypeptide homologous to lactate dehydrogenase. Mol Biochem Parasitol 73:291–294

Zimmermann JL, Petri WL, Meselson M (1983) Accumulation of specific subsets of D. melanogaster heat shock mRNA in normal development without heat shock. Cell 32:1161–1170

C

The Host: Immunogenetics and Immune Response

Immunogenetics in Pathogenesis of and Protection Against Toxoplasmosis

R. McLeod[1,2,3], J. Johnson[1], R. Estes,[1] and D. Mack[1,2]

1 Human Genes and Varied Clinical Manifestations and Transmission Rates of Congenital Toxoplasmosis

Clinical manifestations and congenital transmission rates vary markedly among individuals with *Toxoplasma* infection (reviewed in BOYER and McLEOD 1996; REMINGTON et al. 1995). An influence of host immune response genes on the outcome of congenital infection is supported by observations of concordance

[1]Michael Reese Hospital and Medical Center, 31st Street and S. Lake Shore Drive, Chicago, IL 60616, USA
[2]University of Illinois, 1740 West Taylor, Chicago, IL 60612, USA
[3]University of Chicago, 5940 S. Maryland Avenue, Chicago, IL 60637, USA

of manifestations in monozygotic versus discordance of manifestations in dizygotic twins (Couvreur et al. 1976; reviewed in Remington et al. 1995), ethnic variations in incidence and severity of manifestations of disease (McAuley et al. 1994) and recent HLA typing studies (Mack et al. 1996b). Data from studies of murine models indicate that variables such as parasite strain (McLeod et al. 1984, 1988; Suzuki et al. 1989; Sibley and Boothroyd 1992; Suzuki and Joh 1994), inoculum size (Johnson et al., unpublished), host age (Johnson et al. unpublished), sex hormones (Roberts et al. 1995), and immune status (Gazzinelli et al. 1993a,b) also clearly modulate and effect outcome of infection. Studies are under way to define precisely how host genetics, as well as certain of the variables mentioned above, determine the small proportion of immunologically competent older children and adults who develop various disease manifestations. These studies are focused on the role of genetics in determining disease manifestations such as those described in systemic (reviewed in Boyer and McLeod 1996), retinal (Couvreur and Thulliez 1996), neurologic disease (Townsend et al. 1975), or lymphadenopathic toxoplasmosis (McCabe et al. 1987; Montoya and Remington 1995). In addition, these studies address whether immunogenetics influence the small proportion of acutely infected women who transmit infection to their fetuses early in gestation (Hohfeld et al. 1994), the 40% of *T. gondii* seropositive patients with AIDS who develop toxoplasmic encephalitis (Luft and Remington 1992), the severity of manifestations of congenital infection (McAuley et al. 1994; Mack et al. 1996b) or the occurrence or frequency of episodes of reactivation in recurrent retinochoroiditis in individuals with toxoplasmic retinochoroiditis (Mets et al. 1992; Mets et al. 1996) .

2 Endpoints in Murine Models

Endpoints that have been studied include survival (McLeod et al. 1984, 1989b; Blackwell et al. 1993, 1994; Suzuki and Remington 1993), parasite burden in the brain during the acute and/or subacute infection (Jones and Erb 1985; McLeod et al. 1989b; Brown and McLeod 1990; Brown et al. 1995; Gazzinelli et al. 1993a,b, 1994; Deckert-Schlüter et al. 1995), parasite burden in the brain during chronic infection with recrudescence (Suzuki et al. 1989; Suzuki and Remington 1993; Beaman et al. 1994; Decker-Schlüter et al. 1995), encephalitis during acute, subacute and chronic infections (i.e., patterns of inflammation in brain and other tissues in response to the parasite and cytokine response) (Suzuki et al. 1991, 1994; Burke et al. 1994; Brown et al. 1995; Deckert-Schlüter et al. 1995), congenital transmission (Johnson 1994), and the capacity for protection by immunization (McLeod et al. 1984).

These studies have used different murine models (Williams et al. 1978; McLeod et al. 1989b; Suzuki et al. 1989, 1991) and different endpoints (McLeod et al. 1984, 1989a,b; Suzuki et al. 1989, 1991, 1994). More recently, models which are more like human infections have been developed and characterized (McLeod et al. 1984, 1989a,b). Routes of infection which have included subcutaneous, peritoneal, peroral and congenital are important, as in certain strains of mice mortality is dependent on route of infection (McLeod et al. 1989a; Brown and McLeod 1994).

Different numbers of host genes (McLeod et al. 1989b) and in some instances different genes (McLeod et al. 1989b; Brown et al. 1995) uniquely influence various endpoints. A series of studies (Williams et al. 1978; McLeod et al. 1984, 1989a,b; Brown and McLeod 1990; Suzuki et al. 1991; Brown et al. 1995) have demonstrated that immunogenetics provide a powerful and rational tool (Paigen 1995; Nadeau et al. 1995) to characterize pathogenesis and protection in toxoplasmosis.

3 Chromosome Locations and Numbers of Alleles of Human and Mouse Genes Likely or Proven to Be Important in Resistance to Toxoplasmosis

Certain mouse and human major histocompatibility complex (MHC) genes (mouse chromosome 17, and human chromosome 6), Nramp (mouse chromosome 1, human chromosome 2); cytokines (e.g., IL-4, mouse chromosome 11, human chromosome 5; TNF-α, mouse chromosome 17, human chromosome 6; IFN-γ, mouse chromosome 10, human chromosome 12), and cytokine receptors have a profound influence on the outcome of toxoplasmosis. Allelic variations have been demonstrated to be important in resistance phenotype for only certain of the MHC and Nramp genes. As human MHC alleles are highly polymorphic (Fig. 1), analysis is complex and requires studies of substantial numbers of individuals. In spite of this polymorphism, the paradigm

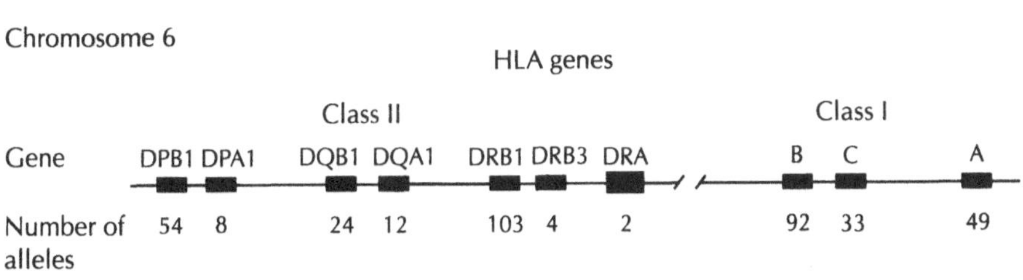

Chromosome 6

HLA genes

	Class II						Class I		
Gene	DPB1 DPA1	DQB1 DQA1	DRB1 DRB3 DRA				B	C	A
Number of alleles	54 8	24 12	103 4	2			92 33		49

Fig. 1. Map of the location of polymorphic HLA class I and class II genes in the human MHC on chromosome 6. The number of known alleles of each gene, whether identified serologically or by sequence analysis, is indicated. (From Hill et al. 1992 and McLeod et al. 1995, with modifications based on data from Bodner et al. 1994)

provided by identification of human malaria resistance and susceptibility genes, and the peptides their gene products present (HILL et al. 1992), provides an elegant example of the feasibility and implications of such an approach. Although there are likely to be important differences in the pathogenesis of human and murine infections, the identification of homologies between human genes and mouse genes demonstrates the importance of using a candidate gene approach (McLEOD et al. 1995; NADEAU et al. 1995) in studying pathogenesis of and protection against toxoplasmosis. These homologies also promise to be useful for understanding the pathogenesis of *T. gondii* infection and toxoplasmosis.

4 Genetic Influences on Outcomes in Murine Models

4.1 Gene Mapping

Gene mapping has proved to be a very powerful tool to define resistance mechanisms (McLEOD et al. 1995; NADEAU et al. 1995; PAIGEN 1995). The approach has been to begin with studies utilizing resistant and susceptible inbred strains of mice, followed by recombinant inbred strains of mice, on the initially defined resistant and susceptible backgrounds in conjunction with a well-characterized linkage map (McLEOD et al. 1989b) and then to use congenic, mutant and transgenic mice to identify the genes important for key protective effector functions which determine those endpoints discussed above (BROWN and McLEOD 1990; BROWN et al. 1995). For example, utilizing A/J (H-2^d, resistant) and C57/BL6/J (H-2^b, susceptible) mice, and then AXB/BXA recombinant inbred strains of mice, genes influencing resistance to survival and cyst formation have been identified (McLEOD et al. 1989a, 1993a,b; BROWN and McLEOD 1990; BROWN et al. 1995). Thus, genetics can be extremely useful for the identification of potentially critical, initiating effector mechanisms important in pathogenesis or for protection.

4.2 Survival in Acute Infection

The first studies on the influence of genetics on survival during acute infection with tachyzoites of a moderately virulent (type III) *T. gondii* strain were performed in 1978 by WILLIAMS et al. Following intraperitoneal infection, linkage between certain H-2 alleles and greater susceptibility, measured as diminished survival, were demonstrated. A second susceptibility gene appeared to be linked to the H-13 locus. These studies suggested that susceptibility to intraperitoneal infection by *T. gondii* in mice, measured as survival, was effected by at least two genes, one linked to the H-2 locus and one to the H-13 locus. Furthermore, these studies demonstrated that more than a single mechanism of resistance should be considered to explain the observed genetic controls of susceptibility (WILLIAMS et al. 1978). In a series of studies, first using inbred and then AXB/BXA recombinant inbred strains of mice (McLEOD et al. 1989a), following peroral infection with cysts of the avirulent ME49 strain of *T. gondii*, it was demonstrated that at least five genes influenced survival and that there were linkages to the H-2 complex, to the *Int-1* (now called *Wnt1*) locus, and to a gene conferring resistance to *Ectromelia*. Perhaps the latter is the H-13 locus on mouse chromosome 2, as separate studies have indicated that this gene governs resistance to *Ectromelia* (McLEOD et al. 1993a,b). Survival is clearly a polygenic trait (McLEOD et al. 1989b, 1993a,b). The mechanism and significance of the linkage of a gene in the region of the *Wnt1* gene with survival remains to be defined. Since *Wnt* loci are contiguous to mouse mammary tumor virus (MMTV) insertions and MMTVs encode superantigens which result in clonal deletion of T cell receptors (TCRs) with particular Vb chains (MARRACK et al. 1991; MARRACK and KAPPLER 1994), the observation concerning linkage of *Wnt1* and survival following peroral infection suggested a possible relationship between survival and TCR usage influenced by MMTV (McLEOD et al. 1993a). This observation prompted studies which have identified *T. gondii* superantigen(s) that effect human (MACK et al. 1996a) and mouse T cells (DENKERS et al. 1994). One hypothesis to explain the *Wnt1* linkage and survival is that a lack of response to *T. gondii* superantigen(s) caused by deletion of T cells by infection with certain MMTVs could confer protection for mice of certain strains which have TCRs that bear selected Vb chains. Mice lacking these MMTV(s) would possess superantigen reactive T cells which would be bound by the *T. gondii* superantigen and elicit a harmful immune response. An alternative hypothesis is that T cells with TCRs with Vβ chains important for response to nominal protective *T. gondii* antigens could be clonally deleted as occurred for MMTV7 and subsequent polyoma virus infection (LUKACHAR et al. 1995). Identification of *T. gondii* superantigen(s) could have important implications for vaccine design, as in certain mouse strains or individuals they could elicit detrimental responses.

4.3 Parasite Burden in Brain in Acute and Subacute Infection

JONES and ERB (1985) noted that the numbers of cysts varied in different mouse strains infected with *T. gondii*. In successive studies utilizing inbred (McLEOD et al. 1984, 1989a), congenic (BROWN and McLEOD 1990), recombinant inbred (McLEOD et al. 1989a), and ultimately mutant and transgenic mice (BROWN and McLEOD 1990; BROWN et al. 1995), parasite burden, measured as brain cyst number, was found to be controlled largely by the MHC class I gene, L^d, (Figs. 2–4) with a small influence of the *Nramp* gene (Fig. 3). MHC class II

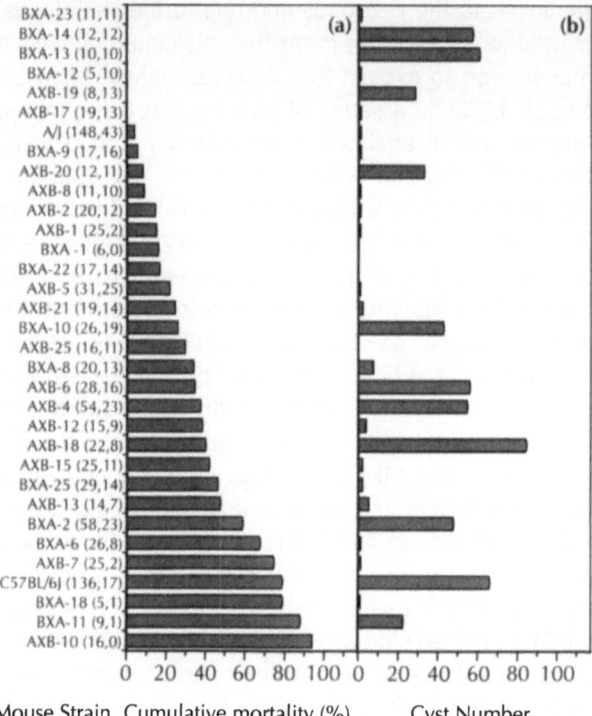

Mouse Strain   Cumulative mortality (%)   Cyst Number
(number*)

Fig. 2a,b. Survival and brain cyst burden in AXB/BXA recombinant inbred strains of mice. Cumulative mortality (**a**) and cyst number (**b**) 30 days after peroral infection of AXB and BXA recombinant inbred strains of mice. *The first number within *parentheses* represents the number of mice studied for mortality, the second number within *parentheses* represents the number of mice studied for brain cyst number. In this study, genetics of two traits, survival and brain cyst number after peroral *Toxoplasma gondii* infection, were studied by using recombinant inbred strains of mice derived from resistant A/J (indicated by the letter *A*) and susceptible C57BL/6J (indicated by the letter *B*) progenitors, F_1 progeny of crosses between A/J and C57BL/6J mice, and congenic mice (B10 background). The continuous variation in the percentage survival indicated that control of this trait involved multiple genes. Analysis of strain distribution pattern of survival of AXB/BXA recombinant mice indicated that survival is regulated by a minimum of five genes. One of these genes appears to be linked to the H-2 complex, another is related to an as yet unmapped gene controlling resistance to *Ectromelia* virus and another the *Wnt1* locus. The large versus low magnitude phenotypes indicated that cyst formation is regulated by one or only a few genes. Associations of defined traits with resistance or susceptibility to *Toxoplasma* cyst formation were also analyzed. Cyst number is regulated by a locus on chromosome 17 within 0–4 cM of the H-2 complex ($p=0.001$). Mice with the H-2a haplotype are resistant and those with the H-2b haplotype are susceptible. This analysis also indicated that the *Bcg* (*Nramp*) locus on chromosome 1 may effect cyst number (map distance=12 cM, $p=0.05$). Resistance to cyst formation is a dominant trait. (From MCLEOD et al. 1989b)

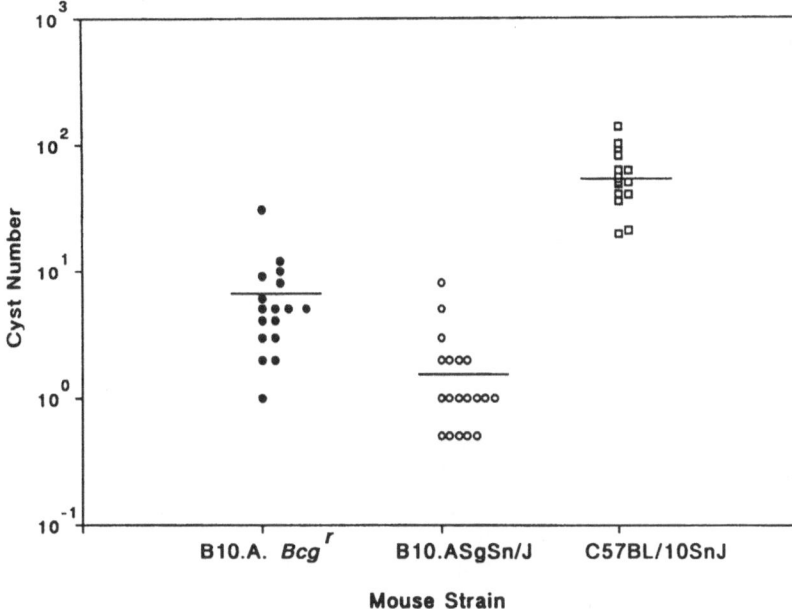

Fig. 3. Studies with congenic mice demonstrate a major influence of the H-2 complex and minor influence of the *Bcg* (*Nramp*) locus on the number of brain cysts following peroral infection. Numbers of cysts (per 10 µl) in brains of B10 congenic mice 30 days after peroral infection with *Toxoplasma gondii*. *Circles* represent mice which have the H-2a haplotype. *Squares* represent mice that have the H-2b haplotype. *Solid symbols* represent mice that are *Bcg* resistant and *open symbols* represent mice that are *Bcg* susceptible. Data are from two replicate experiments with similar results. Differences between H-2a and H-2b mice were highly significant regardless of their *Bcg* type ($p < 0.001$). The smaller differences between B10.A.*Bcg*[r] and B10.ASgSn/J mice also were significant ($p < 0.01$). In these experiments, control A/J mice had low cyst numbers and C57BL/6J mice had high cyst numbers, as in all other experiments. (From MCLEOD et al. 1989b)

genes were also found to influence cyst numbers (BROWN and MCLEOD 1990; MACK et al. 1996b). Resistance is a dominant trait (MCLEOD et al. 1989b).

4.4 Encephalitis in Acute and Subacute Infection

Resistance to encephalitis has also been found to be regulated by the *L* gene, with the L^d allele conferring resistance (Fig. 4; BROWN et al. 1995). This con- clusion was supported by SUZUKI et al. (1995), who found that the 140 kb H2–D region, which includes the L^d gene but not the *TNF* gene, influences the development of toxoplasmic encephalitis. Genes other than MHC genes have also been shown to influence brain parasite burden, encephalitis and cytokine patterns (DECKERT-SCHLÜTER et al. 1995). Cytokine patterns have differed

(a)

(b) Mouse chromosome 17

in reports using different strains of mice studied at varying times after infection (HUNTER et al. 1994; BROWN et al. 1995; DECKERT-SCHLÜTER et al. 1995).

4.5 Chronic Infection and Recrudescent Disease

SUZUKI and REMINGTON (1993) have demonstrated that susceptibility to acute infection in mice does not always correlate with susceptibility to chronic infection. For example, resistance against acute and chronic infection with *T. gondii* was compared between BALB/c and CBA/Ca mice. Doses of ME49 strain *T. gondii* cysts which were lethal for the BALB/c mice did not cause mortality early after infection in CBA/Ca mice. In marked contrast, during chronic infection CBA/Ca mice died but there was no mortality in the BALB/c mice. These results indicate that susceptibility to chronic infection did not always correlate with susceptibility to acute infection.

Progression of chronic encephalitis has also been found to vary with the strain of mouse (SUZUKI et al. 1991; DECKERT-SCHLÜTER et al. 1995). Genetically deficient mice (e.g., SCID mice) have been used to demonstrate that CD4$^+$ T cells are of critical importance in containment of chronic infection (BEAMAN et al. 1994).

4.6 Congenital Infection

Host genetics also appear to influence the outcome of congenitally transmitted infection (JOHNSON 1994). These studies showed that approximately one half of mice born to mothers fed *T. gondii* cysts at 11 days of gestation survived

---

Fig. 4. Mapping studies which indicate that resistance to brain parasite burden and encephalitis are regulated by the L^d gene. *Top,* MHC haplotypes of mouse strains used to determine the controlling locus for cyst formation following peroral *Toxoplasma gondii* infection. *Bottom,* schematic diagram indicating the location of these MHC loci on mouse chromosome 17. Control of resistance to cyst burden following peroral infection with *T. gondii* had been mapped previously to a region of mouse chromosome 17 of approximately 140 kb (MCLEOD et al. 1989b; BROWN and MCLEOD 1990). This region is contiguous with and contains the class 1 gene L^d. Resistance to development of toxoplasmic encephalitis had also been reported to be controlled by genes in this region of H-2 by SUZUKI et al. (1991). *TNF-α, D* and *L* genes as well as unidentified genes are in this region. Studies were performed to identify the gene(s) in the 140 kb region that confers resistance to cysts and encephalitis (BROWN et al. 1995). In this study relative resistance to *T. gondii* organisms and cyst burden in brain, and toxoplasmic encephalitis 30 days following peroral *T. gondii* infection were correlated with presence of the L^d gene in inbred, recombinant, mutant and C3H.L^d transgenic mice. Mice that were resistant to cysts and encephalitis had little detectable brain cytokine mRNA expression, whereas mice that were susceptible had elevated levels of mRNA for a wide range of cytokines, consistent with their greater amounts of inflammation. This work definitively demonstrates that an L^d-restricted response decreases the number of organisms and cysts within the brain and thereby limits toxoplasmic encephalitis and levels of IFN-γ, TNF-α, IL-2, IL-6, IL-10, TGF-β, IL-1a, IL-1b and macrophage inhibitory protein mRNA in the brain 30 days after peroral infection. *Boxed X* indicates that the haplotype at this locus has not been determined. *Hatched square* indicates deletion of the L^d gene. (From BROWN et al. 1995)

until or after they were weaned. In contrast to long-term survival of congenitally infected neonates, no effect of MHC haplotype on early survival was observed in a group of backcross progeny. The ability of mice infected as neonates to survive until weaning was dependent upon IFN-γ and on Thy-1$^+$ cells but not on CD4$^+$ or CD8$^+$ cells. Mice that survived to maturity after infection as neonates were slightly more resistant to challenge with virulent *T. gondii* parasites than were sham-infected controls. However, these mice were less resistant than mice that were infected as adults.

4.7 Factors Which Modulate Genetic Susceptibility

In addition, there are differences in genetic susceptibility to different strains of parasites that appear to be influenced by the strain of mouse that was infected (SUZUKI et al. 1991). Route of infection also may lead to differences in susceptibility of varying strains of mice (McLEOD et al. 1989a; BROWN and McLEOD 1994; Suzuki et al., unpublished).

Studies of the *Nramp* gene and *T. gondii* infection again demonstrated the major influence of background genes other than the MHC (BLACKWELL et al. 1994). The work of DECKERT-SCHLÜTER et al. (1995) also demonstrates the importance of genes other than the MHC genes, on outcome measured as survival, parasite burden, encephalitis and cytokines produced in the spleen and brain 11 and 100 days after infection. In these studies, BALB.K (H-2^k) mice are resistant and BALB.G (H-2^q) mice are susceptible. In marked contrast, BIO.BR (H-2^k) mice are susceptible and BIO.G (H-2^q) mice are resistant. The greater quantity of mRNA for IFN-γ, and less consistently TNF-α, IL-6 and IL-2 in the brain and lesser amounts of IFN-γ in the spleen, appeared to correlate with greater resistance. In contrast, in studies of brain cytokines 30 days after peroral infection, BROWN et al. (1995) found that cytokine production was greater in susceptible mice without the L^d gene. It is not possible to directly compare results from these studies because, in the strains of mice used, certain of the MHC haplotypes studied and methods differed considerably in these studies.

5 Genetic Regulation of Effector Functions

5.1 Use of Genetics to Define Likely Effector Functions

The identification of the critical relevant genes (e.g., L^d, Ia^b, $DQ3$, etc.), has guided a search for the key effector functions (BROWN et al. 1995; BROWN and McLEOD 1990; MACK et al. 1996b) and peptide(s), recognized. For example, studies which link class I MHC-mediated resistance to CD8$^+$ T lymphocytes which produce IFN-γ and/or are cytolytic provide paradigms of possible mechanisms (HILL et al. 1992; BROWN et al. 1995; MACK et al. 1996b; MIYAHIRA et al. 1995; Johnson et al., unpublished), whereby L^d-restricted immune responses may influence the outcome of toxoplasmosis. Studies of malaria and rheumatoid arthritis provide such representative examples (HILL et al. 1992; ZANELLI et al. 1995). In addition, in studies of rheumatoid arthritis, identification of the key protective roles of certain MHC class II DR loci and the inflammation-promoting roles of certain MHC class II DQ loci and the peptides bound by them support a role for self-reactive and inflammation-promoting DQ-restricted Th1 CD4$^+$ T cells which are down-modulated by Th2 cytokine-producing CD4$^+$ T cells which, in turn, are DR-restricted (ZANELLI et al. 1995). Interestingly, the same disease-associated haplotypes (*DQ3* susceptible and possibly *DR17* resistant) have been noted in susceptibility to hydrocephalus in human congenital toxoplasmosis (MACK et al. 1996b).

5.2 Linking Genes and the Effector Functions They Specify

Use of mutant mice or transgenic mice can provide a critical means for defining relevant effector functions. For example, the differences between L^d transgenic mice and their nontransgenic controls following peroral infection have been used to elucidate the importance of cytotoxic T lymphocytes (CTLs) and IFN-γ and inducible nitric oxide synthase (iNOS) production in response to peroral *T. gondii* infection (Johnson et al., unpublished). As effector mechanisms regulated by specific genes are established, it should be possible to identify the peptides bound by the gene products which elicit such responses and the proteins from which they are derived (HUNT et al. 1992). The association of the L^d gene with resistance to *Toxoplasma* parasite burden in the brain as well as toxoplasmic encephalitis (BROWN et al. 1995) suggested that there should be means of egress for peptides from the parasitophorous vacuole into a cytoplasmic, class I MHC presentation and processing pathway (Fig. 5). Recently, however, it has been determined that certain types of macrophages and dendritic cells can ingest proteins and traffic the peptides derived from them to the classical MHC class I pathway.

5.3 Implications for Identification of Protective Mechanisms and Parasite Antigens

Identification of genes important in resistance and susceptibility has enabled the identification of key protective proteins in malaria (HILL et al. 1992) and also the analysis of *T. gondii* peptides and the proteins from which they are derived which are recognized by CTLs in human *Toxoplasma* infection (AOSAI et al. 1994). Such information is useful in determining proteins that have potential efficacy as vaccines. The ability of peptides from such proteins to bind

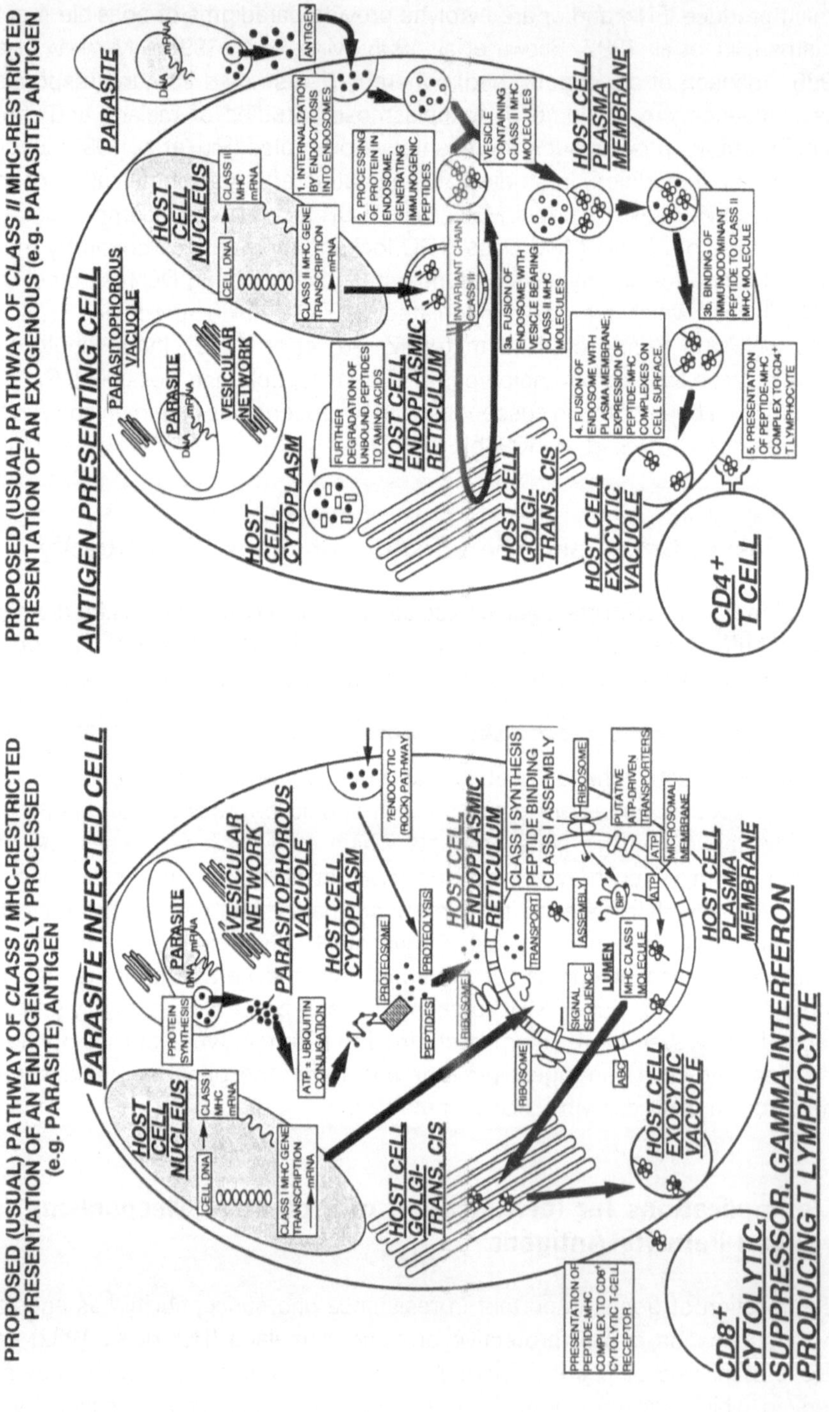

to polymorphic human MHC gene products will be important in the development of vaccines to protect humans.

5.4 Use of Gene Knockout Mice to Study Pathogenesis

The use of gene knockout mice provides a powerful tool to characterize critical effector functions (reviewed in KAUFMANN 1994). MHC class I, MHC class II, cytokine, and effector mechanism knockout mice all have been utilized. In some instances alternative effector functions which substitute for the effector function that has been eliminated have been key in protection. Thus, while studies have demonstrated that the effector function which has been eliminated is not necessary they have not proven that the knocked out effector function would not also be sufficient in the absence of the alternate effector function. The studies of DENKERS et al. (1993) of class I (i.e., β2-microglobulin) knockout mice illustrate the importance of additional mechanisms of protection in addition to the effector function which was knocked out. These studies have demonstrated the importance of IFN-γ produced by NK cells in establishing early protection in class I knockout mice. (DENKERS et al. 1993). Studies with knockout mice also have demonstrated that MHC class II genes may enhance susceptibility measured as diminished survival following peroral *T. gondii* infection. Conversely, class II genes may play an important role later in limiting cyst formation in the brain (Johnson et al., unpublished).

Studies using cytokine knockout mice have demonstrated the importance of IL-4 (ROBERTS et al. 1996) and IL-10 (GAZZINELLI et al. 1996) in the pathogenesis of and protection against toxoplasmosis. Again, important caveats in these studies of knockout mice are that, in the absence of a particular effector function, compensatory immune functions have developed and have proven to be critical in protection. Furthermore, the disruption of the cytokine gene could also have effected other critical immune functions during development in addition to the immediate effect of the cytokine on the outcome being studied. Data from studies of the gene knockout mice, as well as other work in which cytokines have been ablated, indicate that a substantial part of protection against toxoplasmosis is established by early cytokine production, (IL-12 produced by macrophages and then IFN-γ produced by NK cells), which leads T cells to produce IL-2 and IFN-γ and macrophages to produce iNOS. IL-4 and IL-10 may then down-modulate these immune responses to protect the host from excessive inflammation (GAZZINELLI et al. 1992, 1993a,b, 1994; SHER

Fig. 5. Usual pathways of MHC class I and class II MHC restricted processing and presentation of peptides to CD8[+] or CD4[+] T cells. Class I cytoplasmic processing usually results in stimulation of CD8[+] T lymphocytes that are cytolytic and/or produce IFN-γ or are suppressor T lymyphocytes. Class II processing usually results in stimulation of CD4[+] T lymphocytes. Development of a Th1 (i.e., IFN-γ and IL-2 producing) versus Th2 (i.e., IL-4, IL-5, and IL-10 producing) phenotype depends in part on the interaction with costimulatory molecules such as B7-1 and B7-2 during sensitization. (Adapted from ABBAS et al. 1991 with modifications)

et al. 1993; BURKE et al. 1994; KHAN et al. 1994). TNF-α and IFN-γ activate antimicrobial pathways (e.g., NO) in macrophages (SIBLEY et al. 1991) and NO may also drive tachyzoite to bradyzoite interconversion (BOHNE et al. 1994).

6 Human Resistance and Susceptibility Genes: Use of Transgenic Mouse Models, Disease Associations and Transmission

Human MHC class I and class II transgenes have been introduced into mice and the effect of these genes on outcome of *Toxoplasma* infection studied (BROWN et al. 1994; Johnson et al., unpublished). The first of these studies using *HLA-B27* and *Cw3* transgenes demonstrated that the class I gene products require the presence of additional human molecules (possibly human β2-microglobulin) in order to function effectively and may even worsen outcome without such additional molecules. Introduction of *HLA-B27* into B10 mice made them even more susceptible to cyst formation than B-10 mice without the gene, whereas introduction of the *Cw3* gene did not alter susceptibility (Fig. 6). Elimination of class II genes enhanced resistance, measured as survival, and reintroduction of other class II molecules (e.g., Ie) as transgenes enhanced susceptibility early (Johnson et al., unpublished). Implications of this latter work are that harmful cytokines may be produced by Th1 or Th2 type CD4$^+$ T lymphocytes leading to increased mortality. As has been shown with the *HLA-B27* transgenic mouse model, there appear to be genetic influences on susceptibility and resistance in humans (BROWN et al. 1994). Class II gene products, however, also are important for eliciting immune responses which restrict later parasite burden. For example, the human MHC class II gene, *DQ3*, appears to confer a susceptibility to hydrocephalus in congenital toxoplasmosis, and *DQ3* and DR transgenic mice are being used to further char-

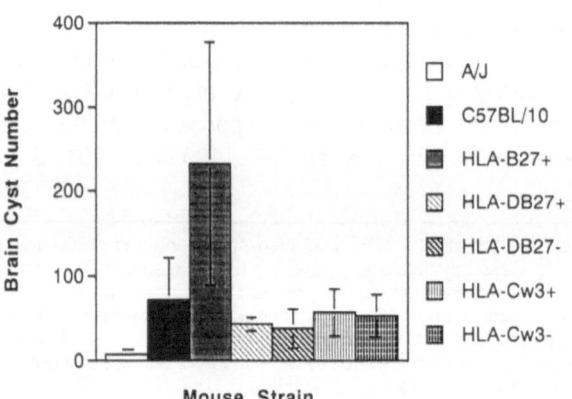

Fig. 6. Effect of human MHC transgenes on parasite burden in brain. Cyst numbers in brains of control and HLA-transgenic mice. Presence (+) or absence (–) of transgene expression in littermates is indicated. (From BROWN et al. 1994)

acterize the roles of *DQ3* and DR gene products in susceptibility versus resistance (Mack et al. 1996b).

As discussed above, the fact that only a small percentage of women transmit the infection in the first trimester suggests that genetic factors may influence magnitude of parasitemia and therefore congenital transmission of *T. gondii*. In addition, the marked differences between dizygotic twins and presence of infection at all or manifestations of infection as contrasted with the similarities in monozygotic twins suggest that genetics also influence these findings (Couvreur et al. 1976). Interestingly, there appear to be an excess of children of Asian background and a diminished number of children of African-American background in a recent study of children with congenital toxoplasmosis (McAuley et al. 1994). Whether demographic and/or genetic and/or other factors influence these differences are being studied at present (McLeod et al. 1995).

7 Conclusions

Identification of murine and human resistance genes have provided insights into the pathogenesis of and protection against toxoplasmosis. The murine L^d gene has been definitively identified as a resistance gene (Brown et al. 1995) and the Ia^b and Ie^b genes as susceptibility genes for diminished survival, but Ia^b and Ie^b also contribute later to resistance against brain cyst burden (Brown and McLeod 1990; Johnson et al., unpublished) . Identification of such genes is useful in determining the effector functions that these genes specify or regulate and the parasite antigens that elicit them (Pamer et al.1991; Hill et al.1992). Gene knockout mice also provide a useful means to identify critical effector mechanisms. Human MHC transgenic mouse models have proven to be helpful in definitively identifying the key roles of these genes and in characterizing the roles of such human gene products in pathogenesis and protection (Brown et al. 1994) and promise to be extremely beneficial in elucidating pathogenic and protective mechanisms (McLeod et al. 1989b). Such immunogenetic findings and tools will be key to the development of protective preparations for humans by establishing the antigens that confer protection as well as identifying those that may cause harm by eliciting detrimental immune responses (McLeod et al. 1993a,b).

Acknowledgements. This work was supported by NIH-NIAID-TMP R01 AI 16945 and 27530 and the Research to Prevent Blindness Foundation. The assistance of Vicki Aitchison and Ellen Holfels in the preparation of this manuscript is greatly appreciated. J. Alexander's, C. Roberts', M. Kirisits', and E. Mui's review of and insightful comments concerning this manuscript and C. Robert's and J. Alexander's permission to discuss their as yet unpublished data are gratefully acknowledged. Helpful discussions with E. Skamene, C. David, C. Grumet, J. Miller and M. Roizen during the evolution of the work described herein are also gratefully acknowledged.

References

Abbas AK, Lichtman AH, Pober JS (1991) Cellular and molecular immunology. Saunders, Philadelphia

Aosai F, Yang TH, Ueda M, Yano A (1994): Isolation of naturally processed peptides from a Toxoplasma gondii-infected human B lymphoma cell line that are recognized by cytotoxic T lymphocytes. J Parasitol 80:260–266

Beaman MH, Araujo FG, Remington JS (1994) Protective reconstitution of the SCID mouse against reactivation of toxoplasmic encephalitis. J Infect Dis 160:375–383

Blackwell J, Roberts CW, Alexander J (1993) Influence of genes within the MHC on mortality and brain cyst development in mice infected with Toxoplasma gondii: kinetics of immune regulation in BALB H-2 congenic mice. Parasite Immun 15:317–324

Blackwell JM, Roberts CW, Roach TI, Alexander J (1994) Influence of macrophage resistance gene LSH/ITy/Bcg (candidate NRamp on Toxoplasma gondii infection in mice). Clin Exp Immunol:97:107–112

Bodner JG, Marsh SGE, Albert ED, Bodmer WF, Dupont B, Erlich HD, Mach B, Mayr WR, Patnam P, Sasazuld T (1994) Nomenclature report: nomenclature for factors of the HLA system. Tissue Antigens 44:1–18

Bohne W, Heesemann J, Gross U (1994) Reduced replication of Toxoplasma gondii is necessary for induction of bradyzoite-specific antigens: a possible role for nitric oxide in triggering stage conversion. Infect Immun 62:1761–1767

Boyer K, McLeod R (1996) Toxoplasma gondii and Toxoplasmosis. In: Long S, Proeber C, Pickering, L (eds) Pediatric infectious diseases. Churchill Livingstone (in press)

Brown CR, McLeod R (1990) Class I MHC genes and CD8[+] T cells determine cyst number in Toxoplasma gondii infection. J Immunol 145:3438–3441

Brown CR, McLeod R (1994) Mechanisms of survival of mice during acute and chronic T. gondii infection. Parasitol Today 10:290–292

Brown C, David C, Kahn SJ, McLeod R (1994) Effect of human MHC class I genes on outcome of T. gondii cyst formation. J Immunol 152:4537–4541

Brown CR, Hunter C, Estes R, Beckmann E, Forman J, David C, Remington JS, McLeod R (1995) Definitive identification of a gene that confers resistance against toxoplasmosis. Immunology 85:419–429

Burke JM, Roberts CW, Hunter CA, Murray M, Alexander J (1994) Temporal differences in the expression of mRNA for IL-10 and IFN-γ in the brains and spleens of C57BL/10 mice infected with Toxoplasma gondii. Parasite Immunol 15:305–314

Couvreur J, Desmonts G, Girre JY (1976) Congenital toxoplasmosis in twins. J Ped 69:235–240

Couvreur J, Thulliez P (1996) Toxoplasmose acquise à localisation oculaire ou neurologique – 49 cas. Presse Med 25:438–442

Deckert-Schlüter, Albrecht S, Hof H, Wiestler OD, Schlüter D (1995) Dynamics of the intracerebral and splenic mRNA production in Toxoplasma gondii -resistant and -susceptible congenic strains of mice. Immunology 85:408–419

Denkers E, Gazzinelli R, Martin D, Sher A (1993) Emergence of NK1,1[+] cells as effectors of IFN-γ dependent immunity to Toxoplasma gondii in MHC class I deficient mice. J Exp Med 175:1465–1472

Denkers EY, Casper P, Sher A (1994) Toxoplasma gondii possesses a superantigen activity that selectively expands murine T cell receptor V β 5 bearing CD8[+] lymphocytes. J Exp Med 180:985–994

Gazzinelli RT, Oswald IP, James SL, Sher A (1992) IL-10 inhibits parasite killing and nitrogen oxide production by IFN-γ activated macrophages. J Immunol 148:1792–1796

Gazzinelli RT, Denkers EY, Sher A (1993a) Host resistance to Toxoplasma gondii, a model for studying the selective induction of cell-mediated immunity by intracellular parasites. Infect Agents Dis 2:139–149

Gazzinelli RT, Eltoum I, Wynn TA, Sher A (1993b) Acute cerebral toxoplasmosis is induced by in vivo neutralization of TNF-α and correlates with the down-regulated expression of inducible nitric oxide synthase and other markers of macrophage activation. J Immunol 151:3672–3681

Gazzinelli RT, Wysocka M, Hayashi S, Denkers EY, Hieny S, Casper P, Trinchieri G, Sher A (1994) Parasite-induced IL-12 stimulates early IFN-γ synthesis and resistance during acute infection with Toxoplasma gondii. J Immunol 153:2533–2543

Gazzinelli RT, Wysocka M, Hieny S, Kersten T, Carrera L, Cheever A, Kühn R, Müller W, Trinchieri G, Sher A (1996) In the absence of endogenous IL-10 mice acutely infected with Toxoplasma gondii succumb to a lethal immune response associated with increased synthesis of IL-12, IFN-γ and TNF-α. J. Immunol. (in press)

Hill AVS, Elvin J, Willis AC, Aidoo M, Allsopp CE, Gotch FM, Gai XM, Takiguchi M, Greenwood BM, Townsend AR (1992) Molecular analysis of the association of HLA-B53 and resistance to severe malaria. Nature 360:434–439

Hohfeld P. Daffos F, Costa JM, Thulliez P, Forestier F, Vidaud M (1994) Prenatal diagnosis of congenital toxoplasmosis with a polymerase-chain reaction test on amniotic fluid. N Engl J Med 331:695–699

Hunt DF, Henderson RA, Shabanowitz J, Sakaguchi K, Michel H, Sevilir N, Cox AL, Apella E, Engelhard VH (1992) Characterization of peptides bound to the class I MHC molecule HLA-A2 by mass spectrometry. Science 255:1261–1263

Hunter CA, Subauste CS, Van Cleave VH, Remington JS (1994) Production of IFN-γ by natural killer cells from Toxoplasma gondii-infected SCID mice: regulation by interleukin-10, interleukin-12, and tumor necrosis factor-α. Infect Immun 62:2818–2824

Jones TC, Erb P (1985) H-2 Complex-linked resistance in murine toxoplasmosis. J Inf Dis 151:739–740

Johnson LL (1994) Resistance to Toxoplasma gondii in mice infected as neonates or exposed in utero. Infect Immun 62:3075–3079

Kaufmann SHE (1994) Bacterial and protozoal infections in genetically disrupted mice. Curr Opin Immunol 5:518–525

Khan IA, Matsuura T, Kasper LH (1994) Interleukin-12 enhances murine survival against acute toxoplasmosis. Infect Immun 62:1639–1642

Luft BJ, Remington JS (1992) Toxoplasmic encephalitis in AIDS. Clin Inf Dis 15:211–222

Lukachar AE, Yupo M, Carroll JP, Abromson-Leeman SR, Laning C, Darf ME, Benjamin T (1995) Susceptibility to tumors induced by polyoma virus is conferred by an endogenous mouse mammary tumor virus superantigen. J Exp Med 151:1683–1692

Mack D, Estes RG, McLeod R (1996a) Toxoplasma gondii superantigen (manuscript in preparation)

Mack D, Johnson J, Estes R, David C, Grumet C, Terasaki P, McLeod R (1996b) Toxoplasmosis susceptibility genes (manuscript in preparation)

Marrack P, Kappler J (1994) Subversion of the immune system by pathogens. Cell 76:323–332

Marrack P, Kushnir E, Kappler J (1991) A naturally inherited superantigen encoded by a mammary tumour virus. Nature 349:524–526

McAuley J, Boyer K, Patel D, Mets M, Swisher C, Roizen N, Wolters C, Stein L, Stein M, Schey W, McLeod R (1994) Early and longitudinal evaluations of treated infants and children and untreated historical patients with congenital toxoplasmosis: the Chicago collaborative treatment trial. Clin Inf Dis 18:38–72

McCabe RE, Brooks RG, Dorfman RF, Remington JS (1987) Clinical spectrum in 107 cases of toxoplasmic lymphadenopathy. Rev Infect Dis 9:754–774

McLeod R, Cohen H, Estes R (1984) Immune response to ingested Toxoplasma: description of a mouse model of Toxoplasma acquired by ingestion. J Infect Dis 149:234–244

McLeod R, Frenkel JK, Estes RG, Mack DG, Eisenhauer P, Gibori G (1988) Subcutaneous and intestinal vaccination with tachyzoites of Toxoplasma gondii and acquisition of immunity to peroral and congenital Toxoplasma challenge. J Immunol 140:1632–1637

McLeod R, Eisenhauer P, Mack DG, Filice G, Spitalny G (1989a) Immune responses associated with early survival after peroral infection with Toxoplasma gondii. J. Immunol 142:3247–3255

McLeod R, Skamene E, Brown CR, Eisenhauer P, Mack D (1989b) Genetic regulation of early survival and cyst number after peroral Toxoplasma gondii infection of AXB/BXA recombinant inbred and B10 congenic mice. J Immunol 143:3031–3034

McLeod R, Mack DG, Brown C, Skamene E (1993a) Secretory IgA, antibody to SAG1, H-2 class I restricted CD8[+] T-lymphocytes and the INT-1 locus in protection against Toxoplasma gondii. In: Smith J (ed) Toxoplasmosis. NATO ASI Series H. Cell Biology; vol. 78. Springer, Berlin,Heidelberg, New York, pp 131–151

McLeod R, Brown C, Mack D (1993b) Immunogenetics influence outcome of T. gondii infection. Res Immunol Ann Inst Pasteur 144:61–66

McLeod R, Bushman E, Arbuckle D, Skamene E (1995) Immunogenetics in the analysis of resistance to intracellular pathogens. Curr Opin Immunol 7:539–552

Mets M, Holfels E, McLeod R, and the Toxoplasmosis Study Group (1992) Ophthalmologic findings in congenital toxoplasmosis. Invest Ophthal Vis Sci 33:4

Mets M, Holfels E, McLeod R and the Toxoplasmosis Study Group (1996) The eye manifestations of toxoplasmosis. Am J Ophthalmol (in press)

Miyahira Y, Murata K, Rodriguez D, Rodriguez JR, Esteban M, Rodrigues M, Zavala F (1995) Quantification of antigen specific CD8$^+$ T cells using an ELLISPOT assay. J Immunol Method 181:45–54

Montoya G, Remington J (1995) Studies on the serodiagnosis of toxoplasmic lymphadenitis. Clin Inf Dis 20:781–789

Nadeau JH, Arbuckle LD, Skamene E (1995) Genetic dissection of inflammatory responses. J Inflamm 45:27–48

Paigen K (1995) A miracle enough: the power of mice. Nature Med 1:215–220

Pamer EG, Harty JT, Bevan PNJ (1991) Precise prediction of a dominant class I MHC-restricted epitope of Listeria monocytogenes. Nature 353:852–855

Remington JS, McLeod R, Desmonts G (1995) Congenital Toxoplasmosis. In: Remington JS, Klein J (eds) Infections of the fetus and newborn infant. Saunders, Philadelphia, 140–268

Roberts CW, Cruickshank SM, Alexander J (1995) Sex determined resistance to Toxoplasma gondii is associated with temporal differences in cytokine production. Infect Immun 63:2549–2555

Roberts CW, Ferguson DJP, Jebbari H, Satoskar A, Bluethmann H, Alexander J (1996) Different roles for IL4 during the course of Toxoplasma gondii infection. Infect Immun 64:897–904

Sher A, Oswald IP, Hieny S, Gazzinelli RT (1993) Toxoplasma gondii induces a T-independent IFN-γ response in natural killer cells that requires both adherent accessory cells and tumor necrosis factor-α. J Immunol 150:3982–3989

Sibley DL, Adams LB, Fukutomi Y, Krahenbuhl JL (1991) Tumor necrosis factor-α triggers antitoxoplasmal activity of IFN-γ primed macrophages. J Immunol 147:2340–2345

Sibley DL, Boothroyd J (1992) Virulent strains of Toxoplasma gondii comprise a single clonal lineage. Nature 359:82–85

Suzuki Y, Joh K (1994) Effect of the strain of Toxoplasma gondii on the development of toxoplasmic encephalitis in mice treated with antibody to IFN-γ. Parasitol Res 80:125–130

Suzuki Y, Remington JS (1993) Susceptibility to chronic infection with Toxoplasma gondii does not correlate with susceptibility to acute infection. Infect Immun 61:2264–2288.

Suzuki Y, Conley F, Remington JS (1989) Differences in virulence and development of encephalitis during chronic infection vary with the strain of Toxoplasma gondii. J Infec Dis 159:790–793

Suzuki Y, John K, Orellana MA, Conley FK, Remington JS (1991) A gene(s) within the H-2D region determines the development of toxoplasmic encephalitis in mice. Immunology 74:732–739

Suzuki Y, Joh K, Kwon OC, Yang Q, Conley FK, Remington JS (1994) MHC class I gene(s) in the D/L region but not the TNF-α gene determines development of toxoplasmic encephalitis in mice. J Immun 97:4649–4654

Townsend JJ, Wolinsky JS, Barringer JR, Johnson PC (1975) Acquired toxoplasmosis. Arch Neurol 32:335–343

Williams DM, Grumet FC, Remington JS (1978) Genetic control of murine resistance to Toxoplasma gondii. Infect Immun 9:416–420

Zanelli E, Gonzalez-Gay MA, David CS (1995) Could HLA-DRI be the protective locus in rheumatoid arthritis? Immunol Today 16:274–278

Cells and Cytokines in Resistance to *Toxoplasma gondii*

C.A. Hunter[1,2], Y. Suzuki[1], C.S. Subauste[1,2], and J.S. Remington[1,2]

1 Introduction

With the advent of the acquired immunodeficiency syndrome (AIDS) epidemic the importance of toxoplasmosis as an opportunistic infection has markedly increased. This parasite has long been recognized as an important cause of disease in other patients with deficiencies in T cell functions including those with Hodgkin's and non-Hodgkin's lymphoma, those with acute lymphocytic leukemia, and those patients undergoing bone marrow transplantation (ISRAELSKI and REMINGTON 1992, 1993). The fact that life-threatening toxoplasmosis occurs in these patients, in contrast to immunologically normal individuals, indicates the importance of T cells in resistance to this infection. However, other aspects of the immune system are also involved in resistance or susceptibility to this disease. In this review we will present data from recent studies from our laboratory on the role of a major histocompatibility complex (MHC) class I gene in resistance to toxoplasmic encephalitis (TE) in murine models, the role of interleukin (IL)-6 and tumor necrosis factor (TNF)-α in the pathogenesis of

1 Department of Immunology and Infectious Diseases, Research Institute, Palo Alto Medical Foundation, Palo Alto, CA 94301, USA

2 Division of Infectious Diseases and Geographic Medicine, Department of Medicine, Stanford University School of Medicine, Stanford, CA 94305, USA

TE, the response of human γδ T cells to *T. gondii*, and the role of cytokines in the regulation of T cell-independent resistance to the parasite.

2 Critical Role of the Ld Gene of MHC Class I Antigens in Preventing TE

We first reported that a gene(s) within the H-2D region of the MHC regulates development of TE in chronically infected mice; mice with the b or k allele are susceptible and those with the d allele are resistant (SUZUKI et al. 1991). This genetic regulation is consistent with those regulating formation of cysts in brains of mice as reported by BROWN and McLEOD (1990). We examined whether the *D* gene or the *L* gene of the MHC class I antigens of the H-2D region is most critical for resistance against development of TE using B10.D2-H-2^{dm1} (dm1) and BALB/c-H-2^{dm2} (dm2) mice with a mutation in the D/L region (SUZUKI et al. 1994a). Both dm1 and dm2 mice were infected with ten cysts of the ME49 strain. B10, BALB/c, B10.D2 and B10.A(18R) mice were infected at the same time and used as controls. The susceptible B10 mice developed remarkable inflammatory changes in their brains, whereas B10.A(18R) mice which have the d haplotype only in the H-2D region did not (Table 1). These results confirm that gene(s) within the H-2D region determine development of TE. The dm1 mice, which have the mutant *D/L* hybrid gene formed by fusion of the 5′ part of the D^d gene and the 3′ part of the L^d gene, developed TE in contrast to their background B10.D2 mice (Table 1) (SUZUKI et al. 1994a). The dm2 mice, which have a complete deletion of the L^d gene, had significantly more *T. gondii* cysts in their brains than did dm1 mice and developed large

Table 1. H-2 haplotypes of mice and presence of acute inflammation in the brains of mice infected with *T. gondii* for 8 weeks[a]

Strain of mice[b]	K	A	H-2 Haplotype E	S	D D	L	Acute focal inflammation	Necrosis of brain tissue
B10	b	b	b	b	b	b	+	−
BALB/c	d	d	d	d	d	d	−	−
B10.D2	d	d	d	d	d	d	−	−
B10.A(18R)	b	b	b	b	d	d	−	−
dm1	d	d	d	d	d*[c]	d	+	−
dm2	d	d	d	d	d	−	+	+

Asterisks used for emphasis.
[a]Adapted from SUZUKI et al. (1994a).
[b]Mice were infected i.p. with ten cysts of the ME49 strain of *T. gondii*. Eight weeks after infection, histological studies were performed. Three mice were used for each experimental group.
[c]A fusion of the 5′ part of the D^d gene and the 3′ part of the L^d gene.

areas of necrosis in their brains that were not observed in dm1 mice (Table 1). These results indicate that a gene(s) in the D/L region determines whether TE will occur and that the L^d gene plays a critical role in resistance against development of TE (Suzuki et al. 1994a). In regard to the importance of the L^d gene for prevention of TE in mice, McLeod and her colleagues recently reported that relative resistance to *T. gondii* organisms and cyst burden in the brain and TE correlates with the presence of the L^d gene in inbred, recombinant mutant and transgenic mice (Brown et al. 1995).

The critical role of the L^d gene, a MHC class I gene, in preventing TE in genetically resistant mice strongly suggests an important role for CD8$^+$ T cells in the protective immune response against development of TE since MHC class I antigens play a critical role in antigen recognition by CD8$^+$ T cells. It was previously reported that CD8$^+$ T cells are important for prevention of cyst formation in the brain (Brown and McLeod1990) and for resistance against TE in genetically susceptible mice (Gazzinelli et al. 1992a). It may be that *T. gondii* antigen(s) which bind to the L^d molecule are important for activation of CD8$^+$ T cells which play a critical role in prevention of TE in genetically resistant mice.

3 Importance of TNF-α for Preventing Progression of TE

We have previously reported that polymorphisms in the TNF-α gene correlated with susceptibility of mice to development of TE and with elevated levels of TNF-α mRNA in brains of infected mice (Freund et al. 1992). However, it was not clear in that study whether the polymorphisms in the TNF-α gene result in differences in induction of mRNA for this cytokine and whether the difference predisposes to development of TE. The dm2 mice are a suitable model to attempt to answer this question since they have the same TNF-α gene as BALB/c mice, but the dm2 mice develop TE whereas BALB/c mice do not (Suzuki et al. 1994a). Six weeks after infection, total RNA was obtained from brains of BALB/c and dm2 mice. Histological studies performed at the same time revealed remarkable inflammatory changes in the brains of the dm2 but not the BALB/c mice. Messenger RNA for TNF-α were specifically detected in the total RNA of infected dm2 but not BALB/c mice by reverse-transcribing the RNA followed by amplification of TNF-α-specific cDNA using polymerase chain reaction (PCR) (Suzuki et al. 1994a). These results clearly indicate that polymorphisms in the TNF-α gene are neither a cause of induction of mRNA of TNF-α in the brains of infected mice nor the cause of development of TE. Injection of neutralizing antibodies against TNF-α resulted in worsening of the TE in infected dm2 mice, but did not induce TE in infected BALB/c mice (Suzuki et al. 1994a). Thus, TNF-α appears to be produced in the brain after

TE has developed and is responsible for preventing progression of TE. In support of this conclusion, GAZZINELLI et al. (1993a) have previously reported that anti-TNF-α antibodies increased the severity of TE in B6 mice, which are genetically susceptible to development of TE.

4 Importance of IL-6 in the Immunopathogenesis of TE

In order to examine whether IL-6 is involved in the development or prevention of TE, genetically susceptible B6 mice were infected with the ME49 strain of *T. gondii* and treated with a monoclonal antibody (mAb) against IL-6 for 4 weeks, beginning 4 weeks after infection (SUZUKI et al. 1994b). Treatment with anti-IL-6 mAb resulted in a remarkable decrease in the numbers of *T. gondii* tachyzoites and cysts and numbers of foci of acute focal inflammation associated with tachyzoites (SUZUKI et al. 1994b). Paradoxically, mice treated with anti-IL-6 mAb had higher serum levels of IL-6 than controls (SUZUKI et al. 1994 b). These results reveal the importance of IL-6 in the immunopathogenesis of TE, although it is not clear whether IL-6 plays a pathogenic or protective role in development of TE.

5 γδ T Cell Responses to *T. gondii*

Studies of T cell-mediated immunity against *T. gondii* have demonstrated the role of both MHC-restricted CD8$^+$ and CD4$^+$ T cells (YANO et al. 1989; SUBAUSTE et al. 1991). In these studies, the role of T cells presumably bearing the αβ T cell receptor (TCR) has been addressed, but they have not determined whether T cells bearing the γδ TCR play a role in the immune response. Increasing evidence indicates that γδ T cells participate in the immune response against certain intracellular organisms (BORN et al. 1991). Indeed, it has been proposed that γδ T cells may represent a more primitive arm of the T cell immune response that recognize a limited range of antigens and may act as a first line of defense against certain pathogens and tumors (JANEWAY et al. 1993).

    Preliminary experiments conducted in our laboratory indicated that purified peripheral blood T cells from either *T. gondii*-seronegative or seropositive individuals proliferated when incubated with autologous peripheral blood mononuclear cells (PBMC) infected with *T. gondii*. No significant difference was observed between the *T. gondii*-induced proliferation of T cells from *T. gondii*-seronegative individuals and that of T cells from seropositive individuals

(SUBAUSTE et al. 1995). Cytofluorometric and cell count analyses demonstrated that this *T. gondii*-mediated proliferation of T cells was accompanied by an increase in the percentage and number of γδ T cells, regardless of the donors' serological status for antibodies to *T. gondii* (Table 2). There was no significant difference observed in the increase in the number of γδ T cells between seronegative and seropositive individuals. The stimulatory effect of *T. gondii*-infected PBMC on γδ T cells was confirmed in proliferation assays using preparations of purified resting γδ T cells. It appears unlikely that this response was mediated by prior exposure to cross-reactive antigens, since apparently unprimed γδ T cells from *T. gondii*-seronegative umbilical cord blood also responded to cells with intracellular *T. gondii* (SUBAUSTE et al. 1995). Therefore, our data suggest that human γδ T cells have an inherent reactivity to *T. gondii*. These observations support the possibility that γδ T cells act as a first line of defense against *T. gondii*.

Data obtained in our laboratory using different preparations of *T. gondii* suggest that intracellular localization of *T. gondii* tachyzoites is required for a preferential γδ T cell response. Whereas incubation of T cells with PBMC infected with UV-attenuated *T. gondii* or PBMC that had internalized formalin-killed *T. gondii* resulted in preferential activation (assessed by expression of CD25 and HLA-DR molecules) and expansion of the γδ T cell population, incubation with *T. gondii* lysate antigens (TLA) did not (SUBAUSTE et al. 1995). The fact that cells which contained intracellular parasites, but not those treated with TLA, induced a preferential γδ T cell response may be relevant to in vivo infection with *T. gondii* since the parasite resides within cells of multiple tissues in infected persons. The absence of a preferential γδ T cell response when TLA was used would suggest that in order for *T. gondii* antigens to induce this response, they either need to originate from intracellular parasites or that

Table 2. γδ T cell receptor expression on T cells before and after 7 days of incubation with either uninfected or *T. gondii*-infected peripheral blood mononuclear cells (PBMC)

| | Seronegative (n=19) | | Seropositive (n=5) | | All (n=24) | |
	% γδ	Fold increase[a]	% γδ	Fold increase[a]	% γδ	Fold increase[a]
Day 0	3.0 (0.4–11.9)[b]	N.A.	4.0 (1.9–7.1)	N.A.	3.3 (0.4–11.9)	N.A.
Day 7						
PBMC	0.8 (0.1–2.2)	0	2.2 (1.8–2.6)	0	1.1 (0.1–2.6)	0
PBMC-UVTg	31.0 (11.2–74.2)	25.7 (2.4–127.0)	28.1 (12.6–57.0)	18.7 (5.2–70.9)	31.4 (11.2–74.2)	24.0 (2.4–127.0)

Data are given as means with the range given in parentheses.
N.A. not applicable; UVTg, UV-attenuated *T. gondii*
[a]Fold increase in the numbers of γδ T cells
[b]Outside parentheses, mean values; inside parentheses, range

the γδ T cells recognize a modification of the cells in response to the presence of intracellular tachyzoites. In this regard, it has been reported that γδ T cells respond to the evolutionary highly conserved heat shock proteins (HSP) expressed on stressed cells or present in certain microorganisms (O'BRIEN et al. 1989). Pertinent to the immune response to *T. gondii* are studies that detected a 65-kDa HSP in peritoneal macrophages from mice infected with *T. gondii* (NAGASAWA et al. 1992; HISAEDA et al. 1995). In addition, γδ T cells appeared to be required for expression of HSP65 during infection with *T. gondii* (HISAEDA et al. 1995). Although γδ T cells can recognize HSP, it appears that these molecules are not the major ligand for γδ T cells reactive to certain pathogens such as *Mycobacterium tuberculosis* (KABELITZ et al. 1990). Molecules related to a thymidine 5'-triphosphoryl-X nucleotide conjugate, isopentenyl pyrophosphate and related prenyl pyrophosphate derivatives isolated from *M. tuberculosis,* have recently been reported to be recognized by human γδ T cells (CONSTANT et al. 1994; TANAKA et al. 1995). These ligands may also originate from cells infected with intracellular pathogens and thus function as primitive antigens that indicate stress of the infected cells (CONSTANT et al. 1994; TANAKA et al. 1995).

Microorganisms such as *M. tuberculosis, Francisella tularensis*, and *Plasmodium falciparum* induce expansion of the subset of γδ T cells bearing Vγ9 chains (GOERLICH et al. 1991; KABELITZ et al. 1991; SUMIDA et al. 1992). Similarly, experiments conducted in our laboratory revealed that *T. gondii*-reactive γδ T cells from either umbilical cord blood or from peripheral blood of adults expressed the Vγ9 and Vδ2 chains (SUBAUSTE et al. 1995). Given the relatively limited genetic diversity of the γδ TCR, it is possible that Vγ9$^+$ γδ T cells recognize similar ligands present in each of these stimulants. Whether the γδ T cell response to *T. gondii* is mediated by a superantigen as proposed for other microorganisms (PFEFFER et al. 1992; SUMIDA et al. 1992), or whether this response is a specific, TCR-mediated event that requires a particular V gene pair for antigen recognition remains to be determined. Relevant to our observations of the in vitro activation and expansion of human γδ T cells are the reports of an increase in γδ T cells in peripheral blood from patients with symptomatic acute toxoplasmosis (DE PAOLI et al. 1992; SCALISE et al. 1992). This increase was mainly due to an increase in Vδ2$^+$ cells and was not observed when retesting was performed during convalescence, suggesting that the γδ T cell response was confined to the early phase of the infection.

Whereas several studies have reported that the response of γδ T cells is not MHC-restricted (HOLOSHITZ et al. 1989; BOOM et al. 1992), γδ T cells can recognize antigen in association with MHC molecules (KOZBOR et al. 1989). Results of our experiments in which *T. gondii*-infected MHC incompatible Epstein-Barr virus-transformed B lymphblastoid B cell lines (EBV-LCL) were used in T cell proliferation experiments as well as in cellular cytotoxicity assays indicate that recognition of *T. gondii*-infected cells by γδ T cells is not restricted by polymorphic MHC-I or MHC-II molecules (SUBAUSTE et al. 1995). However, we cannot exclude the possibility that nonclassical MHC or MHC-like molecules such as CD1 (Porcelli et al. 1992) are involved in the γδ T cell response to *T. gondii*.

Recent reports indicate that γδ T cells play a protective role against pathogens including *T. gondii* (Mombaerts et al. 1993; Rosat et al. 1993; Skeen and Ziegler 1993; Hisaeda et al. 1995). Intraperitoneal infection with *T. gondii* results in an increase in the percentage of γδ TCR$^+$ T cells obtained from the peritoneal cavities and spleens of infected mice (Hisaeda et al. 1995). In addition, in vivo depletion of γδ T cells results in shortened survival of these mice after *T. gondii* infection (Hisaeda et al. 1995). The mechanism(s) by which γδ T cells confer protective immunity remain to be determined. We have demonstrated that human γδ T cells produce interferon (IFN)-γ, IL-2, and TNF-α upon stimulation with *T. gondii*-infected cells (Subauste et al. 1995). In addition, human γδ T cells are cytotoxic for *T. gondii*-infected cells (Subauste et al. 1995). The relevance of these observations lies in the fact that production of these cytokines (Sharma et al. 1985; Suzuki et al. 1988; Johnson 1992) and perhaps lysis of *T. gondii*-infected cells (Subauste et al. 1991) confer protection against the parasite. Finally, the early γδ T cell response with resulting production of IFN-γ and lack of production of IL-4 may play an important role in promoting differentiation of a Th1 type response (O'Garra and Murphy 1994) associated with protective immunity to *T. gondii*.

The expansion of the γδ T cell population following infection with T. gondii (De Paoli et al. 1992; Hisaeda et al. 1995) and the protective role of these cells in vivo (Hisaeda et al. 1995) correlate with our results obtained in vitro and suggest that the rapid induction of a remarkable primary γδ T cell response we observed appears to be an important component of the early immune response to *T. gondii*. A γδ T cell response may be particularly relevant to *T. gondii*, since the peroral route is the most common route by which *T. gondii* infection is acquired, and γδ T cells present in the intestinal mucosa (Goodman and Lefrancois 1988) may be one of the first cell types of the immune system to interact with the parasite.

6 T Cell-Independent Resistance to *T. gondii*

Although T cells are important in protective immunity against *T. gondii* there is a T cell-independent mechanism of resistance to this parasite. Mice with the severe combined immune deficiency (SCID) lack T and B lymphocytes, yet possess a mechanism of resistance to *T. gondii*. This mechanism of resistance is dependent on IL-12, TNF-α, and IFN-γ (Gazzinelli et al. 1993a; Sher et al. 1993; Hunter et al. 1993, 1994). The importance of these three cytokines was demonstrated in studies in which SCID mice were treated with neutralizing antibodies specific for IL-12, TNF-α, or IFN-γ. Each of these treatments resulted in significantly earlier mortality than in control mice. Furthermore, when spleno-

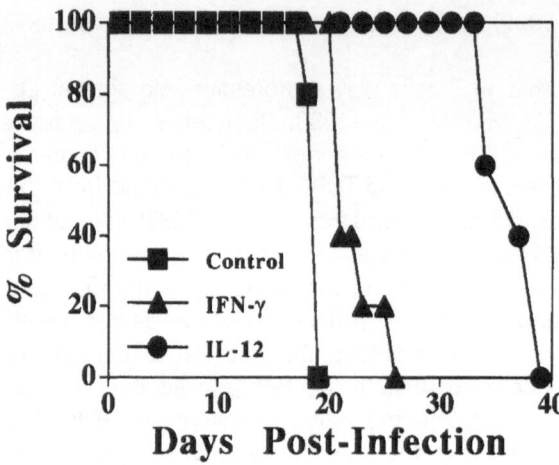

Fig. 1. Administration of IL-12 or IFN-γ delays time to death of SCID mice infected with *T. gondii*. SCID mice were treated with IL-12 (100 ng i.p.) or IFN-γ (5×10⁵ U i.p.) 24 h prior to infection and every day thereafter. Groups of five mice were infected i.p. with 20 cysts of the ME49 strain of *T. gondii* and mortality monitored on a daily basis

cytes from SCID mice were incubated with live or heat killed tachyzoites of *T. gondii*, this stimulus resulted in production of IFN-γ by natural killer (NK) cells. Addition of neutralizing antibodies specific for IL-12 or TNF-α inhibited this parasite-induced production of IFN-γ.

In studies in which infected SCID mice were treated on a daily basis with IL-12, beginning 24 h prior to infection, there was a remarkable delay in time to death (Gazzinelli et al. 1993b). Further studies revealed that a similar treatment regimen with IFN-γ ($5×10^5$ U/day) resulted in a significant delay in time to death of SCID mice infected with *T. gondii*. However, the protective effect of exogenous IFN-γ was not as remarkable as the results with IL-12 (Fig. 1). The difference in time to death observed in mice treated with IL-12 or IFN-γ suggests that the protective effects of IL-12 may not be due solely to stimulation of NK cell production of IFN-γ. IL-12 has other effects on the immune response including enhancement of NK cell proliferation and cytotoxicity. These additional effects of IL-12 could also contribute to the observed protective activity of IL-12.

Since endogenous TNF-α is important in resistance to *T. gondii* [likely through its role to enhance IL-12 induced production of IFN-γ and/or activate macrophages (Johnson 1992; Gazzinelli et al. 1993b; Hunter et al. 1993)], we treated infected SCID mice with TNF-α (1 µg/day). Daily administration of TNF-α resulted in earlier mortality than in control mice (data not shown). Thus, although endogenous TNF-α is important for resistance to *T. gondii*, too much TNF-α can be detrimental to the host (Beutler and Cerami 1987).

7 IL-1β Is Required for IL-12-Induced Production of IFN-γ by NK Cells

As a result of our findings in which we compared the production of IFN-γ by splenocytes from infected and uninfected SCID mice, we proposed the possibility that another factor may be important in the ability of IL-12 plus TNF-α to stimulate production of IFN-γ by NK cells (HUNTER et al. 1994). Studies to explore this possibility revealed a role for IL-1β in the IL-12 induced production of IFN-γ by NK cells. Our results are similar to those of TRINCHIERI and colleagues, who demonstrated that IL-1β was important for the ability of IL-12 to stimulate the production of IFN-γ by human PBMC (D'ANDREA et al. 1993). We observed that anti-IL-1β, but not anti-IL-1α, inhibited completely the ability of *T. gondii* to stimulate production of IFN-γ by SCID mouse splenocytes. Anti-IL-1β also inhibited the ability of IL-12 plus TNF-α to stimulate SCID mouse splenocytes to produce IFN-γ. Further studies revealed that recombinant murine IL-1β enhanced production of IFN-γ by NK cells stimulated with IL-12 (HUNTER et al. 1995b). Although TNF-α enhances IL-12-induced production of IFN-γ by NK cells, in our experiments, addition of anti-TNF-α to cultures of SCID-mouse splenocytes stimulated with IL-12 plus IL-1β did not result in a marked reduction in production of IFN-γ. Moreover, IL-12-induced production of IFN-γ by purified IL-2 activated NK cells was not affected by anti-TNF-α but was completely ablated by anti-IL-1β. These data suggest that the requirement for IL-1β in NK cell production of IFN-γ can be independent of TNF-α and indicate an important role for IL-1β in the NK cell mediated resistance to *T. gondii* (HUNTER et al. 1995b). The significance of these findings in vitro to the in vivo situation was illustrated by the ability of an antibody specific for the type I IL-1 receptor to inhibit the protective effect of administration of IL-12 in SCID mice infected with *T. gondii* (HUNTER et al. 1995b).

8 TGF-β Is an Antagonist of T Cell-Independent Resistance to *T. gondii*

IL-10 is an important antagonist of cell-mediated immunity. This cytokine can inhibit the production of IFN-γ by NK cells as well as the production of proinflammatory cytokines by macrophages. Furthermore, preincubation of macrophages with IL-10 inhibits the ability of IFN-γ plus TNF-α to activate these cells to inhibit parasite replication (GAZZINELLI et al. 1992b; TRIPP et al. 1993). Since IL-10 and TGF-β have similar effects on macrophage function, we investigated the role of TGF-β in NK-mediated resistance to *T. gondii*. Stimulation of splenocytes from SCID mice with tachyzoites of *T. gondii* resulted in production of low levels of IFN-γ with a concomitant increase in levels of TGF-β.

Addition of anti-TGF-β to these cultures enhanced the parasite-induced production of IFN-γ (HUNTER et al. 1995a). Interestingly, work from this laboratory has previously demonstrated that TGF-β is produced by murine macrophages infected with *T. gondii* in vitro (BERMUDEZ et al. 1993). Together, these data suggest that parasite-induced production of TGF-β by macrophages inhibits the IL-12 induced production of IFN-γ by NK cells. Further studies revealed that TGF-β antagonized the ability of IL-12 in combination with TNF-α or IL-1β to stimulate production of IFN-γ by splenocytes from SCID mice. These latter results indicate that the effect of TGF-β may be directly on NK cells (HUNTER et al. 1995b). In agreement with our results, a recent study demonstrated that TGF-β inhibited the production of IFN-γ by human NK cells (BELLONE et al. 1995).

We used several approaches to determine the relevance of our in vitro studies to the in vivo situation. Immunohistochemistry demonstrated enhanced production of TGF-β in the spleens and brains of SCID mice infected with *T. gondii*. To characterize the role of the endogenous TGF-β, we administered high doses of anti-TGF-β to SCID mice infected with *T. gondii*. This treatment resulted in a small, but significant, delay in time to death (HUNTER et al. 1995a). In further studies, we tested the effect of exogenous TGF-β on the course of the infection in SCID mice. Treatment with TGF-β resulted in earlier mortality and antagonized the protective effect of treatment with IL-12. Together, these data indicate that TGF-β is a potent antagonist of the T cell-independent mechanism of resistance to *T. gondii* in vivo. In addition to this antagonistic effect of TGF-β on production of IFN-γ, other activities of this cytokine may also be relevant to the mechanism whereby TGF-β antagonizes resistance against *T. gondii*. Other workers (SILVA et al. 1991; OSWALD et al. 1992; BARRAL et al. 1993) have proposed that the ability of exogenous TGF-β to increase the severity of disease in *Trypanosoma cruzi* and *Leishmania major* infected mice is through its ability to inhibit activation of macrophages by IFN-γ to kill these parasites. We have found that pretreatment of macrophages with TGF-β does not result in enhanced multiplication of intracellular *T. gondii*. Moreover, TGF-β does not affect the ability of IFN-γ plus TNF-α to activate macrophages to inhibit repli-

Table 3. Effect of TGF-β on replication of *T. gondii* in murine macrophages

| Culture conditions | Parasites/100 infected cells | | |
	2 h	20 h	Fold increase
Control	1.25	5.32	4.25
IFN-γ + TNF-α	1.29	1.22	0.94
TGF-β	1.20	5.24	4.37
IFN-γ + TNF-α + TGF-β	1.49	1.19	0.80

Peritoneal macrophages were preincubated with TGF-β (3 ng/ml for 18 h) ± IFN-γ (40 U/ml) + TNF-α (400 U/ml) for 18 h prior to infection with tachyzoites of *T. gondii*. Similar results were observed with higher concentrations of TGF-β (10 ng/ml) used prior to and/or after activation of macrophages by IFN-γ + TNF-α.

cation of intracellular *T. gondii* (Table 3). Thus, the mechanism(s) through which TGF-β interferes with resistance against *T. gondii* may in some respects be mechanistically different from that observed with *T. cruzi* and *Leishmania* species.

Acknowledgements. This work was supported by Public Health Service Grants AI 047017, AI 30230, and AI 35956 from the National Institutes of Health.

References

Barral A, Barral-Netto M, Yong EC, Brownwell CE, Twardzik DR, Reed SG (1993) Transforming growth factor β as a virulence mechanism for Leishmania braziliensis. Proc Natl Acad Sci 90:3442–3446

Bellone G, Aste-Amezaga M, Trinchieri G, Rodeck U (1995) Regulation of NK cell functions by TGF-β1. J Immunol 155:1066–1073

Bermudez LE, Covaro G, Remington JS (1993) Infection of murine macrophages with Toxoplasma gondii is associated with release of transforming growth factor β and downregulation of expression of tumor necrosis factor receptors. Infect Immun 61:4126–4130

Beutler B, Cerami A (1987) Cachectin-tumor necrosis factor: a cytokine that mediates injury initiated by invasive parasites. Parasitol Today 3:345–346

Boom WH, Chervenak KA, Mincek MA, Ellner JJ (1992) Role of the mononuclear phagocyte as an antigen-presenting cell for human γδ T cells activated by live Mycobacterium tuberculosis. Infect Immun 60:3480–3488

Born WK, Harshan K, Modlin RL, O'Brien R (1991) The role of γδ T lymphocytes in infection. Curr Opin Immunol 3:455–459

Brown CR, McLeod R (1990) Class I MHC genes and CD8 T cells determine cyst number in Toxoplasma gondii infection. J Immunol 145:3438–3441

Brown CR, Hunter CA, Estes RG, Beckmann E, Forman J, David C, Remington JS, McLeod R (1995) Definitive identification of a gene that confers resistance against toxoplasmosis. Immunology 85:419–428

Constant P, Davodeau F, Peyrat MA, Poquet Y, Puzo G, Bonneville M, Fournie JJ (1994) Stimulation of human γδ T cells by nonpeptidic mycobacterial ligands. Science 264:267–270

D'Andrea A, Aste-Amezaga M, Valiante NM, Ma X, Kubin M, Trinchieri G (1993) Interleukin 10 inhibits human lymphocyte interferon γ-production by suppressing natural killer cell stimulatory factor/IL-12 synthesis in accessory cells. J Exp Med 178:1041–1048

De Paoli P, Basaglia G, Gennari D, Crovatto M, Modolo ML, Santini G (1992) Phenotypic profile and functional characteristics of human gamma delta T cells during acute toxoplasmosis. J Clin Microbiol 30:729–731

Freund YR, Sgarlato G, Jacob CO, Suzuki Y, Remington JS (1992) Polymorphisms in the tumor necrosis factor α (TNF-α) gene correlates with murine resistance to development of toxoplasmic encephalitis and with levels of TNF-α mRNA in infected brain tissue. J Exp Med 175:683–688

Gazzinelli R, Xu Y, Hieny S, Cheever A, Sher A (1992a) Simultaneous depletion of CD4+ and CD8+ T lymphocytes is required to reactivate chronic infections with Toxoplasma gondii. J Immunol 149:175–180

Gazzinelli RT, Oswald IP, James SL, Sher A (1992b) IL-10 inhibits parasite killing and nitrogen oxide production by IFN-γ-activated macrophages. J Immunol 148:1792–1796

Gazzinelli RT, Eltoum I, Wynn TA, Sher A (1993a) Acute cerebral toxoplasmosis is induced by in vivo neutralization of TNF-α and correlates with the down-regulated expression of inducible nitric oxide synthase and other markers of macrophage activation. J Immunol 151:3672–3681

Gazzinelli RT, Hieny S, Wynn TA, Wolf S, Sher A (1993b) Interleukin-12 is required for the T-lymphocyte-independent induction of interferon-γ by an intracellular parasite and induces resistance in T-cell deficient hosts. Proc Natl Acad Sci 90:6115–6119

Goerlich R, Hacker G, Pfeffer K, Heeg K, Wagner H (1991) Plasmodium falciparum merozoites primarily stimulate the Vγ9 subset of human γ/δ T cells. Eur J Immunol 21:2613–2616

Goodman T, Lefrancois L (1988) Expression of the γδ TCR on intestinal CD8$^+$ intraepithelial lymphocytes. Nature 333:855–858

Hisaeda H, Nagasawa H, Maeda K, Maekawa Y, Ishikawa H, Ito Y, Good RA, Himeno K (1995) γδ T cells play an important role in hsp65 expression and in acquiring protective immune responses against infection with Toxoplasma gondii. J Immunol 154:244–251

Holoshitz J, Koning F, Coligan JE, De Bruyn J, Strober S (1989) Isolation of CD4⁻ CD8⁻ mycobacteria-reactive clones from rheumatoid arthritis synovial fluid. Nature 339:226–229

Hunter CA, Abrams JS, Beaman MH, Remington JS (1993) Cytokine mRNA in the central nervous system of SCID mice infected with Toxoplasma gondii: importance of T-cell-independent regulation of resistance to T. gondii. Infect Immun 61:4038–4044

Hunter CA, Subauste CS, Van Cleave VH, Remington JS (1994) Production of gamma-interferon by natural killer cells from Toxoplasma gondii-infected SCID mice: regulation by interleukin-10, interleukin-12, and tumor necrosis factor alpha. Infect Immun 62:2818–2824

Hunter CA, Bermudez L, Beernink H, Waegell W, Remington JS (1995a) Transforming growth factor-β inhibits interleukin-12-induced production of interferon-γ by natural killer cells: a role for transforming growth factor-β in the regulation of T-cell independent resistance to Toxoplasma gondii. Eur J Immunol 25:994–1000

Hunter CA, Chizzonite R, Remington JS (1995b) Interleukin-1β is required for the ability of IL-12 to induce production of IFN-γ by NK cells: a role for IL-1β in the T cell independent mechanism of resistance against intracellular pathogens. J Immunol 155:4347–4354

Israelski DM, Remington JS (1992) AIDS associated toxoplasmosis. In: Sande MA, Volderding PA (eds) The medical management of AIDS. Saunders, Philadelphia, pp 319–345

Israelski DM, Remington JS (1993) Toxoplasmosis in patients with cancer. Clin Infect Dis 17 (Suppl): S423-S435

Janeway CA, Jones B, Hayday A (1993) Specificity and function of T cells bearing γδ receptors. Immunol Today 9:73–76

Johnson LL (1992) A protective role for endogenous tumor necrosis factor in Toxoplasma gondii infection. Infect Immun 60:1979–1983

Kabelitz D, Bender A, Schondelmaier S, Schoel B, Kaufmann SHE (1990) A large fraction of human peripheral blood γδ$^+$ T cells is activated by Mycobacterium tuberculosis but not by its 65-kD heat shock protein. J Exp Med 171:667–679

Kabelitz D, Bender A, Prospero T, Wesselborg S, Janssen O, Pechhold K (1991) The primary response of human γδ$^+$ T cells to Mycobacterium tuberculosis is restricted to Vγ9-bearing cells. J Exp Med 173:1331–1338

Kozbor D, Trinchieri G, Monos DS, Isobe M, Russo G, Haney JA, Zmijewski C, Croce CM (1989) Human TCR-γ$^+$/δ$^+$, CD8$^+$ T lymphocytes recognize tetanus toxoid in an MHC restricted fashion. J Exp Med 169:1847–1851

Mombaerts P, Arnoldi J, Russ F, Tonegawa S, Kaufmann SHE (1993) Different roles of αβ and γδ T cells in immunity against an intracellular bacterial pathogen. Nature 365:53–56

Nagasawa H, Oka M, Maeda K, Jian-Guo C, Hisaeda H, Ito Y, Good RA, Himeno K (1992) Induction of heat shock protein closely correlates with protection against Toxoplasma gondii infection. Proc Natl Acad Sci 89:3155–3158

O'Brien RL, Happ MP, Dallas A, Palmer E, Kubo R, Born WK (1989) Stimulation of a major subset of lymphocytes expressing T cell receptor γδ by an antigen derived from Mycobacterium tuberculosis. Cell 57:667–674

O'Garra A, Murphy K (1994) Role of cytokines in determining T-lymphocyte function. Curr Opin Immunol 6:458–466

Oswald IP, Gazzinelli RT, Sher A, James S (1992) IL-10 synergises with IL-4 and transforming growth factor-β to inhibit macrophage cytotoxic activity. J Immunol 148:3578–3582

Pfeffer K, Schoel B, Plesnila, N, Lipford GB, Kromer S, Deusch K, Wagner H (1992) A lectin binding, protease-resistant mycobacterial ligand specifically activates Vγ9$^+$ human γδ T cells. J Immunol 148:575–583

Porcelli S, Morita CT, Brenner MB (1992) CD1b restricts the response of human CD4⁻8⁻ T lymphocytes to a microbial antigen. Nature 360:593–597

Rosat JP, MacDonald HR, Louis JA (1993) A role for γδ$^+$ T cells during experimental infection of mice with Leishmania major. J Immunol 150:550–555

Scalise F, Gerli R, Castellucci G, Spinozzi F, Fabietti GM, Crupi S, Sensi L, Britta R, Vaccaro R, Bertotto A (1992) Lymphocytes bearing the γδ T-cell receptor in acute toxoplasmosis. Immunol 76:668–670

Sharma SD, Hofflin JM, Remington JS (1985) In vivo recombinant interleukin 2 administration enhances survival against a lethal challenge with Toxoplasma gondii. J Immunol 135:4160–4163

Sher A, Oswald IP, Hieny S, Gazzinelli R (1993) Toxoplasma gondii induces a T-independent IFN-γ response in natural killer cells that requires both adherent accessory cells and tumor necrosis factor-α. J Immunol 150:3982–3989

Silva JS, Twardzik DR, Reed SG (1991) Regulation of Trypanosoma cruzi infections in vitro and in vivo by transforming growth factor β (TGF-β). J Exp Med 174:539–545

Skeen MJ, Ziegler HK (1993) Induction of murine peritoneal γ/δ T cells and their role in resistance to bacterial infection. J Exp Med 178:971–984

Subauste CS, Koniaris AH, Remington JS (1991) Murine CD8⁺ cytotoxic T lymphocytes lyse Toxoplasma gondii-infected cells. J Immuol 147:3955–3959

Subauste CS, Chung JY, Do D, Koniaris AH, Hunter CA, Montoya JG, Porcelli S, Remington JS (1995) Preferential activation and expansion of human peripheral blood γδ T cells in response to Toxoplasma gondii in vitro and their cytokine production and cytotoxic activity against T. gondii-infected cells. J Clin Invest 96:610–619

Sumida T, Maeda T, Takahashi H, Yoshida S, Yonaha F, Sakamoto A, Tomioka H, Koike T, Yoshida S (1992) Predominant expression of Vγ9/Vδ2 T cells in a tularemia patient. Infect Immun 60:2554–2558

Suzuki Y, Orellana MA, Schreiber RD, Remington JS (1988) Interferon-γ: the major mediator of resistance against Toxoplasma gondii. Science 240:516–518

Suzuki Y, Joh K, Orellana MA, Conley FK, Remington JS (1991) A gene(s) within the H-2D region determines the development of toxoplasmic encephalitis in mice. Immunol 74:732–739

Suzuki Y, Joh K, Yang Q, Conley FK, Remington JS (1994a) MHC class I gene(s) in the D/L region but not the TNF-α gene determines development of toxoplasmic encephalitis. J Immunol 153:4649–4654

Suzuki Y, Yang Q, Conley FK, Abrams JS, Remington JS (1994b) Antibody against interleukin-6 reduces inflammation and numbers of cysts in brains of mice with toxoplasmic encephalitis. Infect Immun 62:2773–2778

Tanaka Y, Morita CT, Tanaka Y, Nieves E, Brenner MB, Bloom BR (1995) Natural and synthetic non-peptide antigens recognized by human γδ T cells. Nature 373:155–158

Tripp CS, Wolf SF, Unanue ER (1993) Interleukin 12 and tumor necrosis factor α are costimulators of interferon-γ production by natural killer cells in severe combined immunodeficiency mice with listeriosis, and interleukin-10 is a physiological antagonist. Proc Natl Acad Sci 90:3725–3729

Yano A, Aosai F, Ohta M, Hasekura H, Sugane K, Hayashi S (1989) Antigen presentation by Toxoplasma gondii-infected cells to CD4⁺ proliferative T cells and CD8⁺ cytotoxic cells. J Parasitol 75:411–416

Role of Macrophage-Derived Cytokines in the Induction and Regulation of Cell-Mediated Immunity to *Toxoplasma gondii*

R.T. Gazzinelli[1,2,3], D. Amichay[4], T. Sharton-Kersten[2], E. Grunwald[2], J.M. Farber[4], and A. Sher[2]

1 Introduction

In its intermediate hosts, infection with *Toxoplasma gondii* is characterized by an early phase when the tachyzoites (the rapidly multiplying stage of the parasite) can be found in different tissues accompanied by a mononuclear inflammatory reaction in small necrotic foci. Parasite multiplication during this stage of disease is most rapid in the liver, lymphoid tissues, lung, and brain. With development of immunity, the tachyzoites are cleared from the host tissues, the necrotic foci regenerate and bradyzoites (the dormant stage of *T. gondii*) form inside cysts mainly in the central nervous system without causing inflammatory reactions and necrosis (FRENKEL 1988). This strong cell-

[1]Biochemistry and Immunology Department, Federal University of Minas Gerais, Av. Antônio Carlos 6627, Belo Horizonte – 30161–970, MG, Brazil
[2]Laboratory of Parasitic Diseases, NIAID, National Institutes of Health, Bethesda, MD 20892, USA
[3]Centro de Pesquisas Rene Rachou, Fundação Oswaldo Cruz, Belo Horizonte, MG – 30190–002, Brazil
[4]Laboratory of Clinical Investigation, NIAID, National Institutes of Health, Bethesda, MD 20892, USA

mediated immunity (CMI) induced by *T. gondii* is maintained by Type 1 $CD4^+CD8^-$ (Th1) and $CD4^-CD8^+$ T lymphocytes, as well as interferon (IFN)-γ and protects the host against rapid parasite growth and consequent pathology (GAZZINELLI et al. 1993a). However, immunity is not able to eliminate the infection since the bradyzoites inside cysts can resist protective cell-mediated response. Nevertheless, spontaneous release of parasites from cysts occurs and may boost immunologic memory, thereby explaining the long-lasting immunity induced by this infection. In immunocompetent hosts, these released bradyzoites elicit a strong inflammatory response and are eventually destroyed by the immune system. However, in immunocompromised hosts, the bradyzoites released from cysts can transform into tachyzoites, resulting in rapid parasite proliferation, severe tissue damage, and a disease, toxoplasmic encephalitis, which, if left untreated, is usually fatal. In fact, *T. gondii* has emerged as the major opportunistic infection of parasitic origin in the current acquired immunodeficiency syndrome (AIDS) epidemic (CANNING 1990; KREISS and CASTRO 1990).

Given the ease with which it can be manipulated, the broad range of vertebrate cells it infects, and the appropriateness of the mouse to study both acute and chronic infections, *T. gondii* would seem to be an ideal opportunistic pathogen to study the induction of CMI as well as regulation of parasite growth by the immune system. The studies reviewed here focus in the role of macrophage as pivotal cell for both the induction and regulation of CMI by *T. gondii*.

2 Induction of Cytokine Synthesis by Macrophages Exposed to *T. gondii* Tachyzoites

It is clear from many studies that, in common with other pathogens (MELTZER et al. 1990; VOGEL 1992; ROACH et al. 1993), *T. gondii* has the capacity to nonspecifically trigger cytokine production by macrophages. We have now repeatedly shown that exposure to either live tachyzoites or tachyzoite extracts induces cytokine synthesis by inflammatory macrophages (GAZZINELLI et al. 1993c, 1994; GRUNVALD et al. 1996). As shown in Fig. 1, interleukin (IL)-1β, IL-10, IL-12(p40) and tumor necrosis factor (TNF)-α mRNAs are all expressed in high levels by inflammatory macrophages exposed to *T. gondii* extracts obtained from tachyzoites of the RH strain. The production of the same cytokines also occurs in vivo both during acute and chronic infection with *T. gondii* (HUNTER et al. 1993; GAZZINELLI et al. 1993b, 1994). Recent studies performed in our laboratory indicate that the parasite molecule(s) which trigger monokine induction are heat stable at 100°C but differ in their sensitivity to protease digestion or periodate treatment. Thus, the *T. gondii* factor(s) that stimulate TNF-α and IL-1β are resistant to proteinase k digestion, while the IL-12(p40) and IL-10 inducing molecules are partially sensitive. In contrast, the factors responsible for the induction of all four molecules were found to be sensitive to periodate oxidation, although at different levels. Together, these findings indicate that *T. gondii*-derived molecules, which trigger cytokine release by macrophages, are glycoconjugates which belong to distinct groups in terms of their biochemical properties (GRUNVALD et al. 1996).

The regulatory effects of IFN-γ on the profile of monokine synthesis by macrophages exposed to *T. gondii* products are of particular interest. Whereas IFN-γ enhances the synthesis of IL-12(p40) and TNF-α, it has a potent inhibitory activity on expression of IL-1β and IL–10 mRNAs by the same macrophage population (Fig. 1). We assume that simultaneous enhancement of IL-12 and TNF-α as well as diminishment of IL-10 synthesis will favor induction of CMI. However, it is not clear how inhibition of IL-1β by macrophages will affect establishment and/or maintenance of CMI during acute toxoplasmosis. Be-

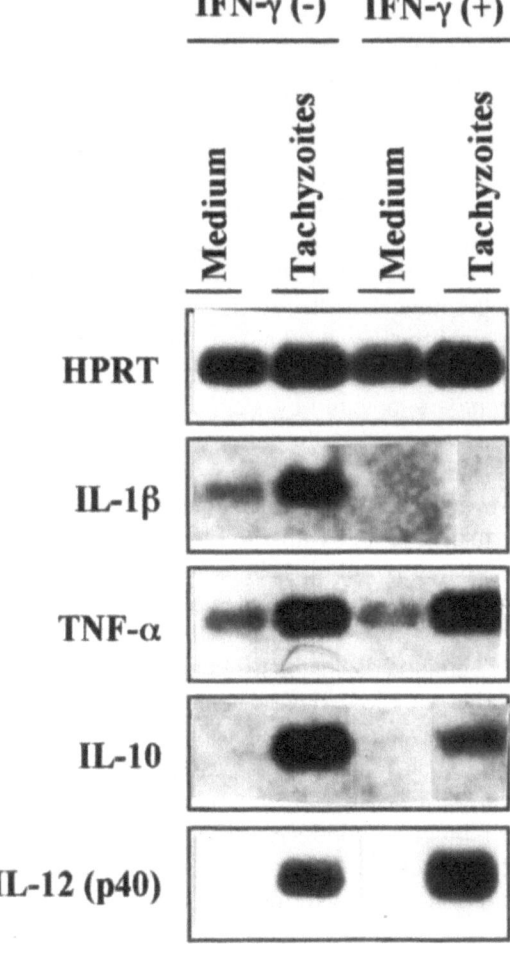

Fig. 1. Tachyzoite extract induces murine inflammatory macrophages to express genes encoding IL-1β, TNF-α, IL-10 and IL-12(p40). Simultaneous exposure with tachyzoite extract and IFN-γ leads to an enhancement of TNF-α and IL-12(p40), contrasting with inhibition of IL-1β and IL-10 gene expression. Murine peritoneal macrophages were harvested 5 days after intraperitoneal thioglycolate injection and were cultured for 6 h in the presence or absence of tachyzoite extract (5 µg/ml) and/or rIFN-γ (100 units/ml). Total RNA was extracted from macrophages, reverse transcribed, and amplified by polymerase chain reaction (PCR) for Southern blot analysis of gene expression. As a control for equal amplification and gel loading, the constitutively expressed hypoxanthine-guanine phosphoribosyl transferase gene (*HPRT*) message was assayed simultaneously (GAZZINELLI et al. 1993c, 1994)

cause IL-1β, in addition to synergizing with IL-12, has an important role in inducing different chemokines, we are now investigating the role of IL-1β in the recruitment of specific leukocyte populations to the site of infection and in parasite clearance.

3 Induction of T Cell-Independent IFN-γ Synthesis: Role in Resistance to Infection

During acute infection with *T. gondii*, natural killer (NK) cells are likely to be the major source of IFN-γ before the establishment of specific T CMI, and are therefore a major barrier for tachyzoite proliferation during early stages of infection. Our studies have defined IL-12 as the key cytokine produced by tachyzoite-activated macrophages, responsible for induction of T cell-independent IFN-γ synthesis by NK cells (Gazzinelli et al. 1993c). While TNF-α and IL-1β (Gazzinelli et al. 1993c; Hunter and Remington 1994; Hunter et al. 1994) synergize with IL-12 on its ability to induce IFN-γ synthesis by NK cells, IFN-γ itself also potentiates this pathway by augmenting the synthesis of IL-12 by macrophages previously exposed to parasite products (Gazzinelli et al. 1993c, 1994) . In contrast, IL-10 and transforming growth factor (TGF)-β (Sher et al. 1993; Hunter et al. 1995a) are potent inhibitors of microbial stimulation of IFN-γ by a T-cell-independent pathway. Whereas IL-10 is a major regulator of IL-12 synthesis by macrophages (D'Andrea et al. 1993; Gazzinelli et al. 1996), the mechanisms by which TGF-β antagonizes IL-12 effects is not completely understood. However, the experiments showing that the beneficial effects of in vivo treatment with rIL-12 on survival of severe combined immunodeficiency (SCID) mice infected with *T. gondii* are inhibited by a simultaneous treatment with TGF-β (Hunter et al. 1995a) suggest that the latter cytokine inhibits the action of IL-12 rather than its synthesis.

The importance of this pathway in resistance to in vivo infection was also revealed in studies using SCID mice infected with *T. gondii* and treated with anti-IL-12 or anti-IFN-γ antibodies (Gazzinelli et al. 1993c; Hunter et al. 1995b). The mortality of infected SCID mice is clearly accelerated after treatment with these anti-cytokines antibodies. Finally, treatment of immunocompetent mice with either anti-IL-12 or anti-NK antibodies increases their susceptibility to infection with *T. gondii* (Gazzinelli et al. 1994; Hunter et al. 1994). In addition, treatment with rIL-12 starting at 1 day before infection promotes resistance of immunocompetent mice to a lethal dose of *T. gondii*, further suggesting the importance of the stimulation of IFN-γ by the T-cell-independent pathway in resistance of animals with an intact immune system (Khan et al. 1994).

The possibility that this pathway can be potentiated in immunodeficient animals was explored by treating SCID mice with recombinant IL-12 (Gazzinelli et al. 1993c; Hunter et al. 1994). A delay in mortality of these animals infected with *T. gondii* was observed when they were simultaneously treated with rIL-12. Together, the above observations suggest that IL-12 may be useful in controlling reactivated toxoplasmosis in AIDS patients and in particular as an adjunct to chemotherapy.

4 Induction of Chemokine Synthesis During Acute Infection with *T. gondii*

It is likely that, after initiation of monokine synthesis by macrophages and IFN-γ by NK cells during acute toxoplasmosis, the synthesis of specific chemokines is of crucial importance in determining the cell types that migrate to sites of infection and control parasite growth. The chemokines are a group of cytokines that are produced by a wide range of cells (e.g., macrophages, fibroblasts, endothelial cells, and smooth muscle cells) and are known to function as chemotactic and activating factors for a variety of inflammatory cells (MILLER and KRANGEL 1992; KELVIN et al. 1993). In contrast to the classical chemotactic agents such as C5a and platelet-activating factor (PAF), the chemokines selectively attract and activate distinct populations of leukocytes.

The chemokines are a family of more than 20 8- to 12-kDa cytokines whose most prominent structural feature is in the arrangement of four invariant cysteine residues, and their most prominent biological activity is their ability to act as chemotactic factors. With one or two exceptions, the chemokines can be assigned to one of two subfamilies: (i) the C-X-C subfamily, where invariant cysteines 1 and 2 are separated by one residue, and (ii) the C-C subfamily, where the invariant cysteines 1 and 2 are adjacent. Members of the C-X-C chemokines can be divided further based on a sequence motif (E-L-R) that is present in IL-8 and related C-X-C chemokines such as macrophage inflammatory protein (MIP)-2, KC, epithelial neutrophil activating peptide (ENA)-78, and growth-related gene (GRO)-α/β/γ that attract neutrophils (MILLER and KRANGEL 1992; KELVIN et al. 1993), but absent from the C-X-C chemokines interferon-γ-inducible protein (IP)-10, macrophage gene induced by interferon-γ (MIG), and platelet factor (PF)-4. IP-10 and MIG are inactive on neutrophils, but instead target T lymphocytes (TAUB et al. 1993; LIAO et al. 1995). C-C chemokines, e.g., regulated upon activation, normal T cell expressed and secreted (RANTES), MIP-1α, MIP-1β, TCA3, JE, and monocyte chemotactic protein (MCP)-1/2/3, target a variety of cells, predominantly monocytes and lymphocytes (MILLER and KRANGEL 1992; KELVIN et al. 1993), but also eosinophils, basophils and mast cells (BAGGIOLINI and DAHINDEN 1994). Some members of the chemokine family received different names depending on the species (human versus mouse) they have been isolated from. In order to clarify this point, a list of names follows with the distinctions of human and mouse chemokines: IL-8, PF-4 and ENA-78 are human chemokines with no mouse homologues; MIG, RANTES and MIP have the same name for both human and mouse homologues. IP-10, GRO, MCP and I-309 are human chemokines which are known as CRG-2, KC, JE, and TCA-3, respectively, in the murine system.

Besides microbial products (e.g., LPS, virus dsRNA, and *Staphylococcus* enterotoxin A), the monokines IL-1 and TNF-α are potent inducers of chemokine synthesis in different cell types (VAN DAMME 1995). In fact, many of the original chemotactic effects attributed to IL-1 and TNF-α were shown later to be mediated by chemokines (VAN DAMME 1995). Interestingly, the cytokine IFN-γ has a dual role in controlling chemokine synthesis. IFN-γ is a potent inducer of MIG and IP-10 (LUSTER et al. 1985; FARBER 1990; VANGURI and FARBER 1990). In contrast, IFN-γ appears to inhibit the production of different C-X-C chemokines containing the E-L-R motif such as KC (OHMORI and HAMILTON 1994), IL-8 and ENA-78 (SCHNYDER-CANDRIAN et al. 1995). If these activities of IFN-γ are important in vivo, they could explain the character of the cellular infiltrates in CMI responses.

Preliminary studies illustrate that, during infection with *T. gondii*, the composition of inflammatory exudate in the peritoneal cavity of IFN-γ knockout (GKO) mice is enriched with neutrophils, eosinophils, and macrophages with relatively fewer lymphocytes than in wild type animals (data not shown). Similarly, when mice chronically infected with *T. gondii* are treated with anti-IFN-γ, an intense neutrophil infiltrate is observed in the central nervous system (CNS) in areas of inflammation and necrosis caused by parasite replication (GAZZINELLI et al. 1992). We have demonstrated that, in marked contrast to wild-type mice, the GKO mice do not show activation of MIG and IP-10 genes during infection with *T. gondii* (D. AMICHAY, R.T. GAZZINELLI, G. KARUPIAH, T. MOENCH, A. SHER, and J. FARBER, manuscript in preparation). Since MIG and IP-10 have been suggested to be important attractants for lymphocytes (TAUB et al. 1993; LIAO et al. 1995), the changes in the composition of the inflammatory infiltrate in the GKO mice may be in part due to the failure to produce these chemokines. Therefore, it is possible that IFN-γ determines the composition of the inflammatory infiltrate during infection with *T. gondii* by inducing chemokines to attract lymphocytes and by inhibiting those chemokines that attract neutrophils. We are currently investigating this hypothesis.

5 Role of IL-12 in Biasing T Cell Differentiation Towards the Th1 Phenotype During Acute Toxoplasmosis

Because IFN-γ is known to play a major role in the differentiation of Th1 cells (GAJEWSKI and FITCH 1988; SCOTT 1991), the possibility that IL-12 could be a potent inducer of Th1 lymphocyte generation has been examined in numerous systems. The overall conclusion is that IL-12 plays an important role in driving the differentiation of Th precursor lymphocytes towards the Th1 phenotype

(HSIEH et al. 1993; SEDER et al. 1993). In addition, it has been shown in many of these systems that, although the effects of IL-12 on T cell differentiation are dependent on IFN-γ, the latter cytokine by itself is not sufficient to drive T cell differentiation towards the Th1 phenotype (SEDER et al. 1993). Moreover, we believe that IFN-γ, induced by IL-12, is produced by different types of cells (i.e., NK cells and T cells), can inhibit expansion of Th2 cells, and therefore favors differentiation and expansion of lymphocytes possessing the Th1 phenotype.

The above studies suggest that, to exert its effects on T cell differentiation, IL-12 is needed during the initial encounter of T cells with antigen. To further analyze the requirement of IL-12 in the maintenance of an already established Th1 response, we studied the role of IL-12 on cytokine synthesis as well as resistance in mice either acutely or chronically infected with *T. gondii*. Our results demonstrate that IL-12 synthesis is required for IFN-γ production and resistance to *T. gondii* during acute (first week) but not chronic (after 30 days) toxoplasmosis. In contrast, treatment with a monoclonal antibody against anti-IFN-γ enhances susceptibility to the parasite in either stage of infection. Together, these results indicate that IL-12 is required to initiate IFN-γ synthesis by lymphocytes, but once the Th1 response is established, IL-12 is no longer required for its maintenance (GAZZINELLI et al. 1994).

6 IL-10 is a Physiological Regulator of Cytokine Synthesis by *T. gondii*-Stimulated Host Cells

IL-10 was first identified by its ability to inhibit IFN-γ synthesis by fully differentiated Th1 clones and was shown to be produced by Th2 lymphocytes (FIORENTINO et al. 1989). Follow-up studies demonstrated that IL-10 can be produced by a large variety of cells including B cells and macrophages, and that the latter group of cells were a major target for IL-10 (FIORENTINO et al. 1991; MOORE et al. 1993). The main mechanism by which IL-10 diminishes cytokine synthesis by NK cells and lymphocytes is through the inhibition of monokine (especially IL-12) synthesis by macrophages (FIORENTINO et al. 1991; D'ANDREA et al. 1993; HSIEH et al. 1993). Thus, previous exposure to IL-10 results in the modulation of cytokine synthesis by macrophages stimulated with microbial products.

Several studies suggested that induction of IL-10 synthesis by macrophages enables different pathogens to evade the immune response and establish chronic infection in the vertebrate host (FINKELMAN et al. 1991; SHER et al. 1991; SILVA et al. 1992). We have assessed the role of IL-10 synthesis in regulating IL-12 and Th1-type cytokines in vivo as well as establishing the parasite:host equilibrium during acute infection with *T. gondii*. Our approach

Inflammatory infiltrates in the liver of
mice acutely infected with *Toxoplasma gondii*

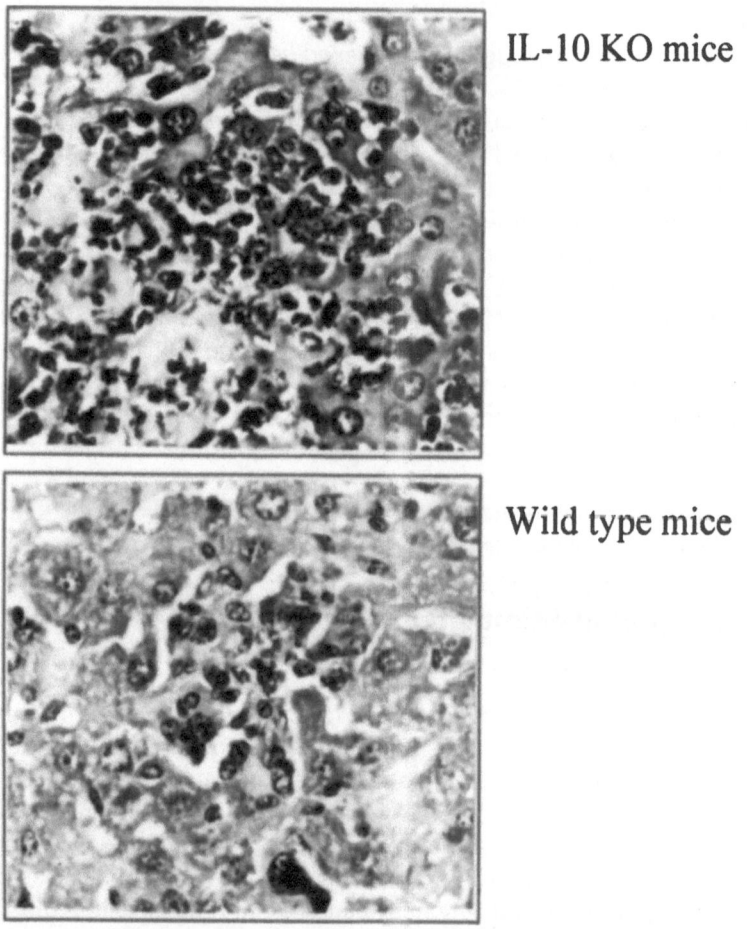

IL-10 KO mice

Wild type mice

Fig. 2. Liver histopathology observed during acute infection of IL-10 knockout mice (IL-10KO). IL-10KO (*top panel*) and wild-type mice (*bottom panel*) were infected with the ME49 strain of *T. gondii* and livers harvested at 6 days post-infection. The infected IL-10KO mice showed clear evidence of enhanced tissue pathology as demonstrated by the increased frequency and intensity of cellular infiltration and, at later time points, increased necrosis in the liver (GAZZINELLI et al. 1996)

was to infect IL-10 knockout (IL-10KO) mice (GAZZINELLI et al. 1996) with an avirulent strain of the parasite and analyze cytokine responses as well as host resistance to this pathogen. All infected IL-10KO mice died by 14 days post-infection in contrast to the infected wild-type animals, which all survived. The IL-10KO mice infected with *T. gondii* showed a dramatic increase in serum levels of IL-12 and IFN-γ as compared with uninfected IL-10KO or infected wild-type mice. Surprisingly, parasite-specific mRNAs in tissues of IL-10KO mice were similar or decreased in the IL-10KO mice compared with those in wild-type animals, and no evidence of increased parasite expansion was observed in histological sections from the IL-10KO animals (GAZZINELLI et al. 1996). Nevertheless, the infected IL-10KO mice showed clear evidence of enhanced tissue pathology, as demonstrated by the increased frequency and intensity of cellular infiltration (Fig. 2) and necrosis in the liver and to a lesser extent in the lungs. Together, these results suggest that the increased mortality of IL-10KO mice is not due to uncontrolled parasite growth but to an abnormal immune response to infection. The histopathology studies demonstrated that infected IL-10KO animals present similar pathology to that observed in wild-type animals under continuous treatment with high doses of IL-12 (GATELY et al. 1994), suggesting that uncontrolled high levels of IL-12 synthesis occuring during microbial infection can be harmful to the host and under some conditions can be lethal. These results argue that IL-10 has an important physiological, regulatory role in controlling cytokine synthesis during the initiation of CMI.

IL-10 has also been shown to be an important regulator of chemokine synthesis by neutrophils (KASAMA et al. 1994). Therefore, it is also possible that this down-regulatory cytokine may either control the intensity and/or shape the profile of chemokine synthesis by different type of cells. Such effects would probably influence inflammation quantitatively and/or qualitatively. It is noteworthy that the IL-10KO mice displayed more intense inflammatory reactions in the liver (Fig. 2) and lungs (data not shown) than the wild-type mice infected with *T. gondii*.

7 Nitric Oxide is an End Product of the IFN-γ-Dependent Pathway Which Controls Lymphocyte Expansion During Acute Infection with *T. gondii*

A second important mechanism by which macrophages regulate the immune response during experimental acute toxoplasmosis is through the generation of nitric oxide. During acute infection with *T. gondii*, T cells are suppressed in their ability to produce IL-2 and proliferate upon stimulation with either parasite antigen or mitogen (CANDOLFI et al. 1994; KHAN et al. 1995). This suppression is mediated by the adherent cells and is partially overcome by neutralization of endogenous IFN-γ, TNF-α, or inhibitors of the nitric oxide synthesis from L-arginine. Since nitric oxide synthesis in macrophages is induced by TNF-α in conjunction with IFN-γ, a lymphocyte product, we believe that this is an important paracrine loop between macrophages and T cells. This paracrine loop is thought to regulate the immune response after strong antigen stimulation. Although not completely understood, our preliminary data suggest that the regulatory effects of nitric oxide on T lymphocytes are at least in part due to the induction of apoptosis, as has been observed in experimental Chagas' disease (Lopes et al. 1995). Furthermore, the initial studies suggest that nitric oxide effects are targeted primarily at CD4$^+$ T lymphocytes. An obvious question is whether this end product from arginine degradation would also regulate expansion of CD8$^+$ T cells and/or NK cells. Based on these observations, several researchers have suggested that an immunodeficient status may occur during acute toxoplasmosis (Candolfi et al. 1994; Khan et al. 1995). However, our current belief is that this immunoregulatory mechanism is crucial for the control of the immune response induced by *T. gondii* to prevent tissue damage caused by an exacerbated CMI. Consistent with this interpretation are the findings that infection with *T. gondii* potentiates both nonspecific and specific CMI, and in fact induces resistance to nonrelated pathogens as well as tumor cells.

8 Conclusions

We believe that the initiation of the immune response during acute toxoplasmosis can be divided into several steps. We propose that the initial event is the induction of cytokine synthesis by macrophages. The monokines released by macrophages exposed to *T. gondii* products will then induce IFN-γ (IL-12) synthesis by NK cells and chemokines (IL-1β and TNF-α) by different types of cells (e.g., macrophages, endothelial cells, and fibroblasts). As our preliminary data suggest, the activation of T cell-independent IFN-γ synthesis may influence the profile of chemokine production during acute toxoplasmosis, since IFN-γ has both positive and negative effects on the synthesis of different chemokines. The events described above will also determine the recruitment and differentiation of T lymphocytes into a Th1 phenotype. Finally, the regulation of this response appears to occur simultaneously with its establishment. Thus, *T. gondii* products trigger macrophages to release IL-10, which regulates cytokine synthesis by macrophages themselves, NK cells, and T cells. In addition, nitric oxide produced by macrophages exposed to IFN-γ and TNF-α has an important regulatory function in controlling T lymphocyte expansion.

References

Baggiolini M, Dahinden CA (1994) CC chemokines in allergic inflammation. Immunol Today 15:127–133

Candolfi E, Hunter CA, Remington JS (1994) Mitogen and antigen specific proliferation of T cells in murine toxoplasmosis is inhibited by reactive nitrogen intermediates. Infect Immun 62:1995–2001

Canning EU (1990) Protozoan infections. Trans R Soc Trop Med Hyg 84 [Suppl]:19-24

D'Andrea A, Aste-Amezaga M, Valiante N, Ma X, Kubin M, Trinchieri G (1993) Interleukin-10 (IL-10) inhibits human lymphocyte interferon-γ production by suppressing natural killer cell stimulatory factor/IL-12 synthesis in accessory cells. J Exp Med 178:1041–1048

Farber JM (1990) A macrophage mRNA selectively induced by γ-interferon encodes a member of the platelet factor 4 family of cytokines. Proc Natl Acad Sci USA 87:5238–5242

Finkelman FD, Pearce EJ, Urban JF, Sher A (1991) Regulation and biological function of helminth induced cytokine responses. In: Ash C, Gallagher C (eds) Immunoparasitology Today. Elsevier, Cambridge, pp A62–A66

Fiorentino DF, Bond MW, Mosmann TR (1989) Two types of mouse T helper cell. IV. Th2 clones secrete a factor that inhibits cytokine production of Th1 clones. J Exp Med 170:2081–2095

Fiorentino DF, Zlotnick T, Mosmann T, Howard M, O'Garra A (1991) IL-10 inhibits cytokine production by activated macrophages. J Immunol 147:3815–3822

Frenkel JK (1988) Pathophysiology of toxoplasmosis. Parasitol Today 4:273–278

Gajewski TF, Fitch FW (1988) Anti-proliferative effect of IFN-γ in immunoregulation. I. IFN-γ inhibits the proliferation of Th2 but not Th1 murine helper T lymphocyte clones. J Immunol 140:4245–4231

Gately MK, Warrier RR, Honasoge S, Carvajal DM, Farherty DA, Connaughton SE, Anderson TD, Sarmiento U, Hubbard BR, Murphy M (1994) Administration of recombinant IL-12 to normal mice enhances cytolytic lymphocyte activity and induces production of IFN-γ in vivo. Int Immunol 6:157–167

Gazzinelli RT, Xu Y, Hieny S, Cheever A, Sher A (1992) Simultaneous depletion of CD4+ and CD8+ T lymphocytes is required to reactivate chronic infection with Toxoplasma gondii. J Immunol 149:175–180

Gazzinelli RT, Denkers EY, Sher A (1993a) Host resistance to Toxoplasma gondii: a model for studying the selective induction of cell-mediated immunity by intracellular parasites. Infect Agents Dis 2:139–149

Gazzinelli RT, Eltoum IA, Wynn T, Sher A (1993b) Acute cerebral toxoplasmosis is induced by in vivo neutralization of TNF-α and correlates with down-regulated expression of inducible nitric oxide synthase and other markers of macrophage activation. J Immunol 151:3672–3681

Gazzinelli RT, Hieny S, Wynn T, Wolf S, Sher A (1993c) IL-12 is required for the T-cell independent induction of IFN-γ by an intracellular parasite and induces resistance in T-deficient hosts. Proc Natl Acad Sci USA 90:6115–6119

Gazzinelli RT, Wysocka M, Hayashi S, Denkers EY, Hieny S, Caspar P, Trinchieri G, Sher A (1994) Parasite-induced IL-12 stimulates early IFN-γ synthesis and resistance during acute infection with Toxoplasma gondii. J Immunol 153:2533–2543

Gazzinelli RT, Wysocka M, Hieny S, Kersten T, Carrera L, Cheever A, Kühn R, Müller W, Trinchieri G, Sher A (1996) In absence of endogenous IL-10 mice acutely infected with Toxoplasma gondii succumb to a lethal immune response associated with increased synthesis of IL-12, IFN-γ and TNF-α. J Immunol (in press)

Grunvald E, Chiaramonte M, Hieny S, Wysocka M, Trinchieri G, Vogel SN, Gazzinelli RT, Sher A (1996) Biochemical characterization and protein kinase C dependency of monokine inducing activities in Toxoplasma gondii. Infect Immun (in press)

Hsieh CS, Macatonia SE, Tripp CS, Wolf S, O'Garra A, Murphy KM (1993) Development of Th1 CD4+ T cells through IL-12 produced by Listeria-induced macrophages. Science 260:547–550

Hunter CA, Remington JS (1994) Regulation of IL-12 and TNF-α induced NK cell production of IFN-γ by IL-1β and TGF-β. Eur Cytokine Netwk 5:145 (abstract)

Hunter CA, Abrams JS, Beaman MH, Remington JS (1993) Cytokine mRNA in the CNS of SCID mice infected with Toxoplasma gondii: importance of T-cell independent regulation of resistance to T. gondii. Infect Immun 61:4038–4045

138   R.T. Gazzinelli et al.

Hunter CA, Subauste CS, Van Cleave VH, Remington JS (1994) Production of gamma interferon by natural killer cells from Toxoplasma gondii-infected SCID mice:regulation by interleukin 10, interleukin 12 and tumor necrosis factor-alpha. Infect Immun 62:2818–2824

Hunter CA, Bermudez L, Beerking H, Waegell W, Remington JS (1995a) Transforming growth factor-β inhibits interleukin-12-induced production of interferon by natural killer cells: a role for transforming growth factor-β in the regulation of T cell-mediated resistance to Toxoplasma gondii. Eur J Immunol 25:994–1000

Hunter CA, Candolfi E, Subauste C, Van Cleave V, Remington JS (1995b) Studies on the role of interleukin-12 in acute murine toxoplasmosis. Immunol 84:16–20

Kasama T, Strietes RM, Lukacs NW, Burdick MD, Kunkel SL (1994) Regulation of neutrophil derived chemokine expression by IL-10. J Immunol 152:3559–3569

Kelvin DJ, Michiel DF, Johnston JA, Lloyd AR, Sprenger H, Oppenheim JJ, Wang JM (1993) Chemokines and serpentines: the molecular biology of chemokine receptors. J Leuk Biol 54:604–612

Khan IA, Matsuura T, Kasper LH (1994) Interleukin-12 enhances murine survival against acute toxoplasmosis. Infect Immun 62:1639–1642

Khan IA, Matsuura T, Kasper LH (1995) IL-10 mediates immunosuppression following primary infection with Toxoplasma gondii in mice. Parasite Immunol 17:185–195

Kreiss JK, Castro KK (1990) Special considerations for managing suspected human immunodeficiency virus infection and AIDS in patients from developing countries. J Inf Dis 162:955–960

Liao F, Rabin, R, Koniaris L, Vanguri P, Yanneli J, Farber JM (1995) The human MIG chemokine: biochemical and functional characterization. J Exp Med 182:1301–1314

Lopes MF, Veiga VF, Santos AR, Fonseca MEF, DosReis GA (1995) Activation-induced CD4+ T cell death by apoptosis in experimental Chagas' disease. J Immunol 154:744–752

Luster ADJ, Unkeless JC, Ravetch JV (1985) γ-Interferon transcriptionally regulates an early-response gene containing homology to platelet proteins. Nature 315:672–676

Meltzer MS, Skilman DR, Gomatos PJ, Kalter DC, Gendelman HE (1990) Role of mononuclear phagocytes in the pathogenesis of human immunodeficiency virus infection. Ann Rev Immunol 8:169–194

Miller MD, Krangel MS (1992) Biology and biochemistry of the chemokines:a family of chemotactic and inflammatory cytokines. Crit Rev Immunol 12:17–46

Moore KW, O'Garra A, Malefyt RW, Vieira P, Mosmann TR (1993) Interleukin-10. Ann Rev Immunol 11:165–190

Ohmori Y, Hamilton T (1994) IFN-γ selectively inhibits lipopolysaccharide-inducible JE/monocyte chemoattractant protein-1 and KC/GRO/melanoma growth-stimulating activity gene expression in mouse peritoneal macrophages. J Immunol 153:2204–2212

Roach TI, Barton CH, Chatterje D, Blackwell JM (1993) Macrophage activation:Lipoarabinomannan from avirulent and virulent strains of Mycobacterium tuberculosis differentially induces early genes cFos, KC, JE and tumor necrosis factor-α. J Immunol 150:1886–1896

Schnyder-Candrian S, Strieter RM, Kunkel SL, Walz A (1995) Interferon-α and Interferon-γ downregulate the production of interleukin-8 and ENA-78 in human monocytes. J Leuk Biol 57:929–935

Scott P (1991) IFN-γ modulates the early development of Th1 and Th2 responses in a murine model of cutaneous leishmaniasis. J Immunol 147:3149–3157

Seder RA, Gazzinelli RT, Sher A, William P (1993) IL-12 acts directly on CD4+ T cells to enhance priming for IFN-γ production and diminishes IL-4 inhibition of such priming. Proc Natl Acad Sci USA 90:10188–10192

Sher A, Fiorentino DF, Caspar P, Pearce E, Mosmann T (1991) Production of IL-10 by CD4+ lymphocytes correlates with down-regulation of Th1 cytokine synthesis in helminth infection. J Immunol 144:2713–2716

Sher A, Oswald IP, Hieny S, Gazzinelli RT (1993) Toxoplasma gondii induces a T-independent IFN-γ response in natural killer cells that requires both adherent accessory cells and tumor necrosis factor-α. J Immunol 150:3982–3989

Silva JS, Morrissey PJ, Grabstein KH, Mohler KM, Anderson D, Reed SG (1992) Interleukin 10 and Interferon-γ regulation of experimental Trypanosoma cruzi infection. J Exp Med 175:169–174

Taub DD, Lloyd AR, Conlon K, Wang JM, Ortaldo JR, Harada A, Matsushima K, Kelvin DJ, Oppenheim JJ (1993) Recombinant human interferon-inducible protein 10 is a chemoattractant for human monocytes and T lymphocytes and promotes T cell adhesion to endothelial cells. J Exp Med 177:1809–1814

Van Damme J (1995) Interleukin-8 and related chemotactic cytokines. In: Thomson A (ed) The cytokine handbook. Academic, New York, pp 185–208

Vanguri P, Farber JM (1990) Identification of CRG-2. An interferon-inducible mRNA predicted to encode a murine monokine. J Biol Chem 265:15049–15057

Vogel, SN (1992) The lps gene. Insights into the genetic and molecular basis of LPS responsiveness and macrophage differentiation. In: Beutler B (ed) Tumor necrosis factors: the molecules and their emerging role in medicine. Raven, New York, pp 485–513

Role of Nitric Oxide-Induced Immune Suppression in Toxoplasmosis During Pregnancy and in Infection by a Virulent Strain of *Toxoplasma gondii*

E. Candolfi, O. Villard, M. Thouvenin, and T.T. Kien

1 Introduction

The mechanisms by which immunocompetent mice resist *Toxoplasma gondii* infection have been partially elucidated. There is general agreement on the schematic cascade of immunological events during the acute stage of the infection: production of interleukin-12 (IL-12) by macrophages stimulates natural killer (NK) cells to produce interferon (IFN)-γ (DENKERS et al. 1993; GAZZINELLI et al. 1993b; SEDER et al. 1993; HUNTER et al. 1994), which in turn activates macrophages to kill *T. gondii* by a combination of oxygen-dependent (WILSON et al. 1980), reactive nitrogen intermediate (RNI)-dependent (ADAMS et al. 1990; GAZZINELLI et al. 1992; LANGERMANS et al. 1992) and tryptophan starvation mechanisms (PFEFFERKORN 1984). IL-12 and IFN-γ are of paramount importance in the protective response to *T. gondii* (SUZUKI et al. 1988; SUBAUSTE and REMINGTON 1991; GAZZINELLI et al. 1993b, KHAN et al. 1994). IL-10 acts as an immunoregulatory cytokine during *Toxoplasma* infection. This cytokine was initially characterized by its production by the Th2 subset of CD4$^+$ T cells and by its ability to suppress IFN-γ production by Th1-type cells (FIORENTINO et al. 1989, 1991). IL-10 inhibits nitric oxide (NO) production in vitro (GAZZINELLI et al. 1992), controls IFN-γ production in severe combined immunodeficiency (SCID) mice infected

Institut de Parasitologie et de Pathologie Tropicale de la Faculté de Médecine de Strasbourg, 3 rue Koeberlé, 67000 Strasbourg, France

with *T. gondii* (SHER et al. 1993; HUNTER et al. 1994) and limits T cell proliferation (CANDOLFI et al. 1995; KHAN et al 1995). These data indicate that, schematically, a Th1-type response is protective and that a Th2-type response is detrimental.

The role of RNI and more generally of NO, and that of the cytokines controlling its production, is not as clear-cut as previously thought. The question was recently reviewed by LIEW (1995). We demonstrated that NO, IFN-γ and (indirectly) IL-12 were involved in the profound suppression of T lymphocyte proliferation observed during the first week of *Toxoplasma* infection (CANDOLFI et al. 1994, 1995; HUNTER et al. 1995b). These data point to NO and IFN-γ having an ambivalent role during the acute stage of *Toxoplasma* infection, as both appear to have protective and permissive actions simultaneously.

Pathogenicity is influenced by the host immune status and by the inherent virulence of the infecting strain of *T. gondii*. Bearing in mind that the significance and clinical consequences of NO-suppressed lymphocyte proliferation have not been established, we postulated that the immunosuppression induced by NO and IFN-γ could lead to increased host susceptibility during pregnancy and/or during infection by virulent strains of *T. gondii*.

We tested this possibility by analyzing lymphocyte proliferation and the production of NO and cytokines in a model of pregnant mice infected with *Toxoplasma* and in mouse models infected with avirulent or intermediately virulent strains.

2 Pregnancy Increases Mouse Susceptibility to *Toxoplasma* Infection and Elicits High Levels of NO, IFN-γ and IL-10

The way in which the "fetal allograft" escapes the maternal immune response is unclear, but early studies have shown that cellular and humoral immunity is depressed during pregnancy (HOLLAND et al. 1984; LEDERMAN 1984). During pregnancy, there is increased susceptibility to intracellular pathogens such as *Listeria*, *T. gondii* (LUFT and REMINGTON 1982), and *Plasmodium falciparum* (BRABIN 1983). However, maternal cellular immune status during pregnancy is controversial (WEINBERG 1984). A recent study showed that the predominant response to parasites during pregnancy was Th2-like (LIN et al. 1993; WEGMANN et al. 1993). Such a Th2-type response might increase susceptibility to intracellular pathogens.

We orally infected pregnant BALB/c mice 11.5 days after conception with cysts from the human avirulent strain PRU, in contrast to ROBERTS and ALEXANDER (1992) who used a strain isolated from a rabbit. On day 18 after conception, the mice were killed, and spleen cell proliferation and the production of NO and cytokines were determined in the antigen- and Con A-stimulated supernatants of cultured spleen cells. Fifty-three percent of the fetuses and 91%

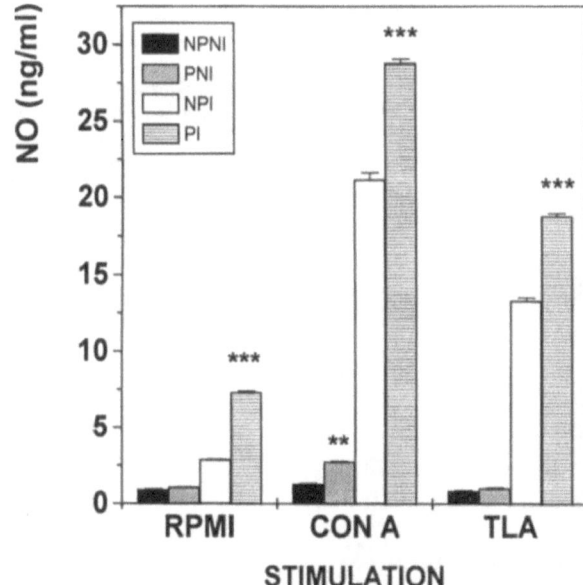

Fig. 1. Nitric oxide (*NO*) production by spleen cells (48 h cultures) at day 7 post infection and from nonpregnant noninfected BALB/c mice (*NPNI*), pregnant noninfected mice (day 18 post conception; *PNI*), nonpregnant infected mice (*NPI*), pregnant infected mice (day 18 post conception; *PI*). Cells were stimulated by Con A (2 µg/ml) and *Toxoplasma* lysate antigen (*TLA*; 1 µg/ml). Levels of significance (Student's *t* test): * $p<0.05$, ** $p<0.01$, *** $p<0.001$

of the placentas were infected. Pregnant mice had a larger lung parasite load on day 7 post infection (PI) and a larger brain load on day 30 PI than nonpregnant animals (333 ± 129 cysts/g per brain versus 87 ± 55 cysts/g per brain; $p<0.05$). Moreover, NO levels were higher in spleen cell supernatants of pregnant mice (Fig. 1). However, these higher NO levels in pregnant mice seem not to control the parasite proliferation in that model. BOHNE and coworkers (1994) have demonstrated that the switch from the tachyzoite to the bradyzoite stage is driven by NO, meaning that high levels of NO in pregnant mice might accelerate the tachyzoite–bradyzoite interconversion.

These high levels of NO were accompanied by a strong Th1-like immune response in the pregnant mice: IFN-γ levels were increased and a shift towards a Th1-type response was illustrated by the IL-4/IFN-γ ratio (Fig. 2). Surprisingly, pregnant mice seemed to be more susceptible to the infection than nonpregnant mice in the Th1-type environment. NO and IFN-γ inhibit mitogenic and specific lymphocyte proliferation during the acute stage of *Toxoplasma* infection (CANDOLFI et al. 1994, 1995), although in our model, pregnancy did not modify spleen cell proliferation, which was observed to be already deeply depressed during the acute stage of infection.

In consequence, other factors capable of explaining increased susceptibility to *T. gondii* during pregnancy were explored, such as a decreased IL-2 production, an increased IL-4 and/or IL-10 production, an increased TNF-α or IL-6 production, or a reduced humoral response. In contast to our results, SHIRIHATA et al. (1993) recently found that IFN-γ and IL-2 production by spleen cells in vitro was reduced in pregnant mice infected with *T. gondii*. Moreover, IL-2 was undetectable at the time of infection, and treatment with IL-2 or IFN-γ reduced the susceptibility of pregnant mice to *Toxoplasma*. Several re-

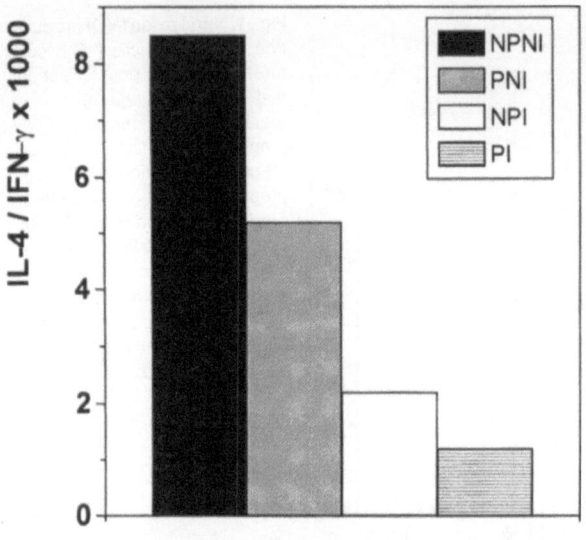

Fig. 2. IL-4/IFN-γ ratio obtained from Con A-stimulated spleen cells at day 7 post infection from nonpregnant noninfected BALB/c mice (*NPNI*), pregnant noninfected mice (day 18 post conception; *PNI*), nonpregnant infected mice (*NPI*), pregnant infected mice (day 18 post conception; *PI*). Cytokine levels were assessed in supernatants of 48 h spleen cell cultures by ELISA

search teams have reported decreased production of IL-2 in acute toxoplasmosis (CHARDÈS et al. 1993; HAQUE et al. 1994; CANDOLFI et al. 1995). In a malaria mouse model recently reviewed by TAYLOR-ROBINSON (1995), the author observed that the decrease in IL-2 synthesis was due to NO, a finding that might explain the immunosuppression observed during the acute phase of malaria infection. However, our results differ from those previously published, especially by SHIRIHATA and coworkers (1993), as IL-2 production was not influenced by pregnancy in our mouse model. The model of infection used by SHIRIHATA et al. (1993; intraperitoneal infection with tachyzoites of a virulent strain) differs greatly from the natural route of infection we chose, and this may partly explain the differences observed between the two models.

High levels of IL-10 were found in infected pregnant mice. However, we found no evidence of previously described mechanisms of immunosuppression driven by IL-10, such as decreased NO (GAZZINELLI et al. 1992), IL-2 (HAQUE et al. 1994) and lymphocyte proliferation (CANDOLFI et al. 1995; KHAN et al. 1995). We cannot rule out the possibility that IL-10 acts as a suppressive cytokine by another mechanism (OSWALD et al. 1992) but, if it does, the previously described mechanisms do not appear to be involved.

It has been reported that TNF-α and IL-6 have a detrimental role during toxoplasmosis (BLACK et al. 1989; GRAU et al. 1992; BEAMAN et al. 1992, 1994). However, high levels of TNF-α and IL-6 are produced by spleen cells from both infected pregnant and nonpregnant mice only when stimulated with crude *Toxoplasma* lysate antigen (TLA). This absence of differences between the increase of proinflammatory cytokine production in pregnant and nonpregnant mice rules out the possibility that these cytokines play a role in the increased susceptibility observed during pregnancy.

During pregnancy, the humoral response to vaccination appears to be intact (BAKER et al. 1988), but the response to placental antigens appears to involve immunoglobulin G1 (IgG1; Th2-type) responses more than IgG2a (Th1-type) responses (BELL and BILLINGTON 1980). In our study, pregnancy did not appear to influence antibody production. Moreover, there was no evidence of a particular Ig class switch. This implies that the increased sensitivity to toxoplasmosis during pregnancy is not due to a deficiency in the humoral immune response.

Finally, this study failed to explain the increased susceptibility to *Toxoplasma* infection in pregnant mice. Other possibilities are NO inhibition of macrophage major histocompatibility class (MHC) II expression (SICHER et al. 1994) and inhibition of the cytotoxic capacities of immune cells such as lymphokine-activated killer cells, NK, or CD8$^+$ T lymphocytes. During pregnancy, other cytokines are important as growth factors, such as granulocyte/macrophage colony-stimulating factor (GM-CSF), epidermal growth factor (EGF), IL-3 and colony-stimulating factor-1 (CSF-1; CHAOUAT et al. 1993). Some cytokines can also act as growth factors for parasites (BARCINSKI and COSTA-MOREIRA 1994), although this remains to be proven in the case of *T. gondii* (BEAMAN et al. 1992). A potent immunosuppressive cytokine, transforming growth factor-β2 (TGF-β2), is secreted by decidual cells and may act as a local immunosuppressor (CLARK et al. 1990). This has already been shown in vitro, where IFN-γ production by NK cells is inhibited by TGF-β (HUNTER et al. 1995a). Levels of hormones such as β-estradiol are increased during pregnancy and can influence IFN-γ production in vitro (GRASSO and MUSCETTOLA 1990). Interestingly, the highest levels of β-estradiol are found in the placental maternal–fetal interface (MENENDEZ 1995), indicating local regulation of the immune response.

3 Influence of Strain Virulence on T Cell Proliferation and Th1/Th2 Equilibrium

3.1 Virulence and Lethality Are Associated with Normal T Cell Proliferative Capacities and with an Early Th2-Type Immune Response Followed by a Marked Th1-Type Response

The ability of *T. gondii* to kill mice has been used to characterize strains. However, only three main groups of strains are found throughout the world: virulent, intermediately virulent, and avirulent (DARDÉ et al. 1992). As pathogenicity in mouse models is strikingly different, we used this feature to explore the immune response and to check for a correlation between strain virulence and immune suppression mediated by NO and IFN-γ.

We used the oral route of infection and two types of *T. gondii* strains: (i) the intermediately virulent strain C56, which kills mice 10–12 days after infection, and (ii) one avirulent strain (the French PRU strain).

On day 6 PI, compared with mice infected with the avirulent strain, mice infected with the intermediately virulent strain showed (i) relatively normal spleen cell proliferation when stimulated by concanavalin A (Con A), (ii) increased NO production (Fig. 3), (iii) a moderate increase in IFN-γ production by spleen cells, (iv) increased serum IFN-γ levels, and (v) increased production of IL-10, TNF-α and IL-6 by spleen cells.

Interestingly, on day 9 PI, the Con A-induced proliferative capacity of spleen cells from mice infected with the intermediately virulent strain was similar to that of uninfected mice, while it was totally suppressed in mice infected with the avirulent strain. Mice infected with the intermediately virulent strain showed a reduced capacity to produce NO, high IFN-γ production (134 ng/ml in TLA-stimulated spleen cell supernatants) and high IL-10 production (215 ng/ml in TLA-stimulated spleen cell supernatants). TNF-α and IL-6 levels did not increase significantly.

The IL-10/IFN-γ ratio showed that protection correlated with an early Th1-type response followed by a switch to a Th2-type response and that virulence was correlated to an early Th2 type response followed by an overwhelming Th1 type response (Fig. 4).

It has already been reported that *T. gondii*-susceptible mice produce more IFN-γ than resistant mice (McLeod et al. 1989), and similar observations have been made with mice infected with *Plasmodium berghei* and *Plasmodium vinkei* (Kremsner et al. 1992; Waki et al. 1992). The authors suggested that overproduction of Th1 cytokines, in this case IFN- γ, could be detrimental for the host by leading to an uncontrolled inflammatory process (Clark et al. 1991).

It has recently been suggested that the simultaneous peak production of IFN-γ and IL-10 might explain the increased susceptibility of female relative to male mice (Roberts et al. 1995). Indeed, male mice produce IFN-γ first, and this controls parasite multiplication; IL-10 then downregulates the Th1 response. It has also been clearly demonstrated that mice lacking IL-10 production die from chronic enterocolitis (Kühn et al. 1993), and that IL-10 regulates the overwhelming inflammatory process associated with autoimmune diseases (Kennedy et al. 1992). As a result, IL-10 may have an important role in controlling the production of proinflammatory mediators (IFN-γ, IL-6, TNF-α, NO, reactive oxygen intermediates, and prostaglandins) that can be detrimental when uncontrolled (Liew et al. 1991; Gerard et al. 1993; Liew 1995). Taylor-Robinson (1995), who recently reviewed the immunological process in a mouse model of malaria, suggested a protective role for the Th2 type response. Therefore, IL-4 or IL-10 production are not wholly detrimental during infection and may control the IFN-γ-driven inflammatory process.

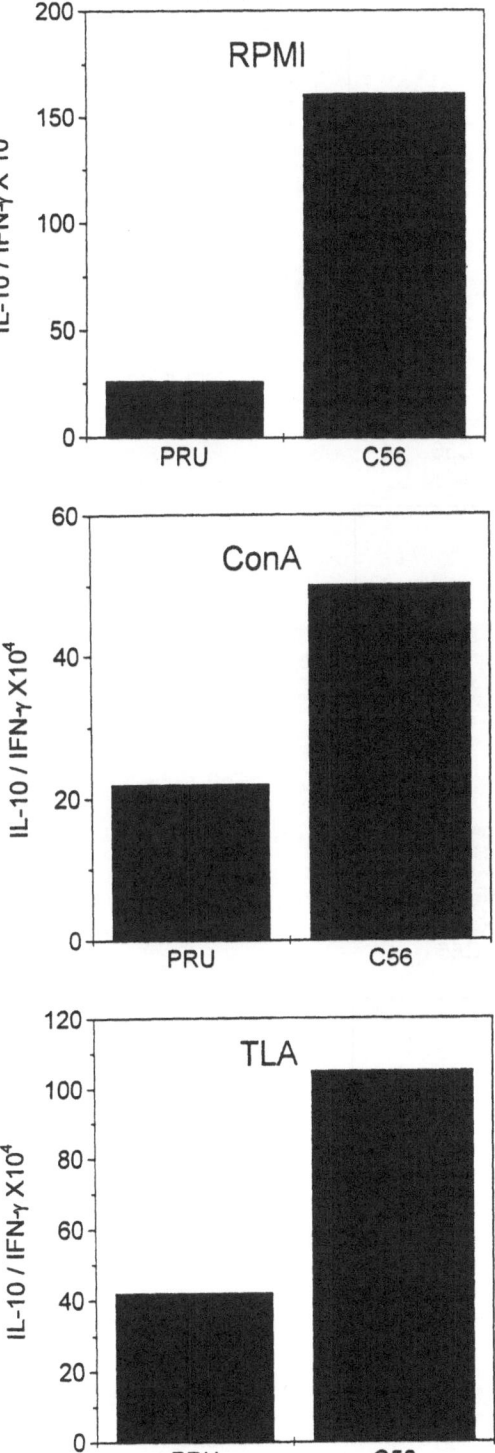

Fig. 3. IL-10/IFN-γ ratio obtained from nonstimulated (*RPMI*) or Con A- or *Toxoplasma* lysate antigen (*TLA*)-stimulated spleen cells from C57/Bl6 mice infected by an intermediately virulent strain (C56 strain, ten cysts orally) and an avirulent strain (PRU strain, ten cysts orally) of *Toxoplasma gondii*. Spleen cells were collected at day 6 post infection, and cytokine levels were assessed by enzyme-linked immunosorbent assay (ELISA)

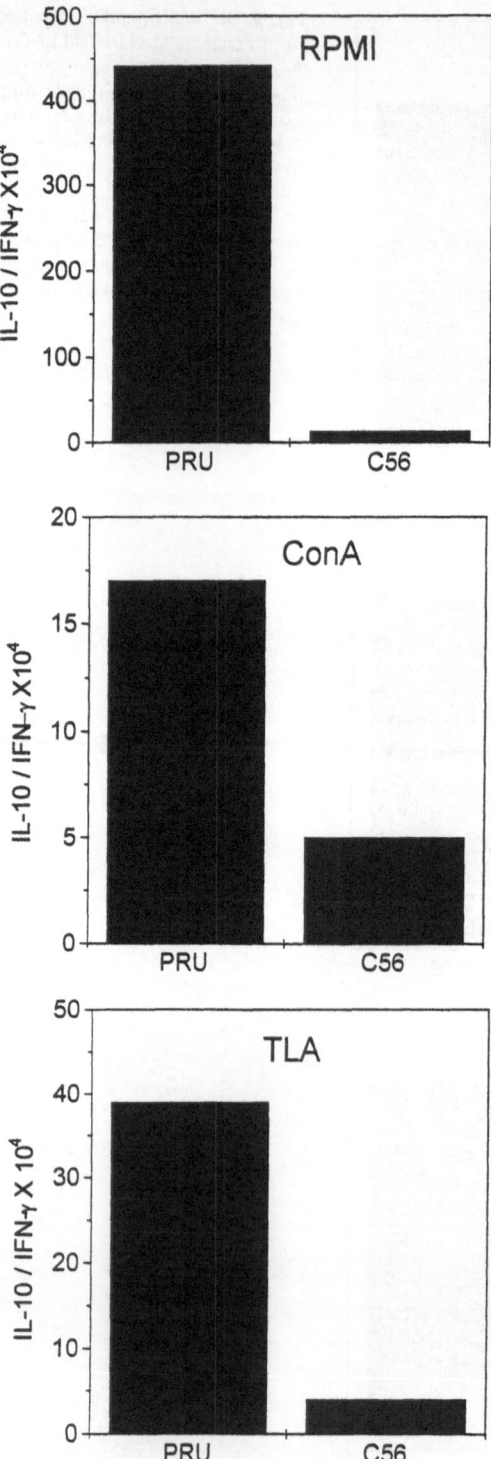

Fig. 4. IL-10/IFN-γ ratio obtained from nonstimulated (*RPMI*) or Con A- or *Toxoplasma* lysate antigen (*TLA*)-stimulated spleen cells from C57/Bl6 mice infected by an intermediately virulent strain (C56 strain, ten cysts orally) and an avirulent strain (PRU strain, ten cysts orally) of *Toxoplasma gondii*. Spleen cells were collected at day 9 post infection, and cytokine levels were assessed by enzyme-linked immunosorbent assay (ELISA). Mice infected with the C56 strain died between day 9 and day 12 post infection. Mice infected with the PRU strain survived

3.2 NO Controls Both Parasite and Lymphocyte Proliferation in Vivo

We showed in the previous section that infection with an intermediately virulent strain leads to early NO production and late but marked IFN-γ and IL-10 production associated with normal lymphocyte proliferative responses. NO and IFN-γ are involved in inhibiting lymphocyte proliferation during the acute stage of infection with an avirulent strain of *Toxoplasma* (CANDOLFI et al. 1994, 1995). These data appear to conflict with reports of protection conferred by NO and IFN-γ in recent years (SUZUKI et al. 1988; ADAMS et al. 1990; GAZZINELLI et al. 1992; LANGERMANS et al. 1992). It remains to be determined whether their antimicrobial and immunosuppressive effects occur in vivo in both types of avirulent and intermediately virulent *T. gondii* strain infection described above. The involvement of IFN-γ is relatively clear, as inhibition of its activity by monoclonal antibodies in vivo leads to the early death of animals infected with *T. gondii*, suggesting that the major role of IFN-γ is to orchestrate the protective response to the parasite (SUZUKI et al. 1988). The role of NO in vivo remains to be determined. We thus administered aminoguanidine (CORBETT et al. 1992; BECKERMAN et al. 1993), an inhibitor of NO production, to the two mouse models of *Toxoplasma* infection described above.

Aminoguanidine treatment led to increased mortality, when mice were infected with the intermediately virulent strain C56, even when concomitantly treated with sulfadiazine, an anti-*Toxoplasma* drug (Fig. 5). No increase in parasite load was observed in the lungs or brains of such mice. In contrast, after aminoguanidine treatment, mortality remained unchanged in mice which were infected with the avirulent ME49 strain of *T. gondii*. However, aminoguanidine treatment resulted in significantly larger numbers of cysts in the brains of ME49-infected mice (Fig. 6). This increased susceptibility to *T.gondii* in C56- and ME49-infected mice treated with aminoguanidine points to a crucial role of NO in controlling *Toxoplasma* infection. The ability of NO to control intracellular microorganisms has previously been observed in animal models of *Listeria* (BECKERMAN et al. 1993) and *Mycobacterium tuberculosis* infection (CHAN

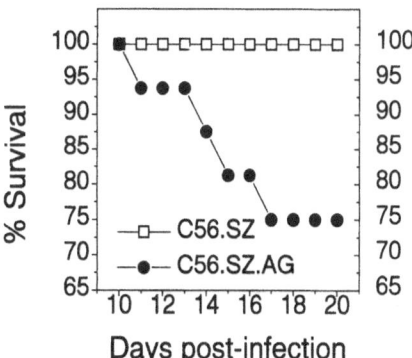

Days post-infection

Fig. 5. Survival rate of CBA/J mice infected with an intermediately virulent strain of *Toxoplasma gondii* (C56 strain, ten cysts orally) and treated by sulfadiazine (200 mg/ml of sterile drinking water from day 3 to day 10, then 400 mg/ml from days 11–21 post infection) and an inhibitor of nitric oxide production, aminoguanidine (*AG*; 100 mg/kg daily)

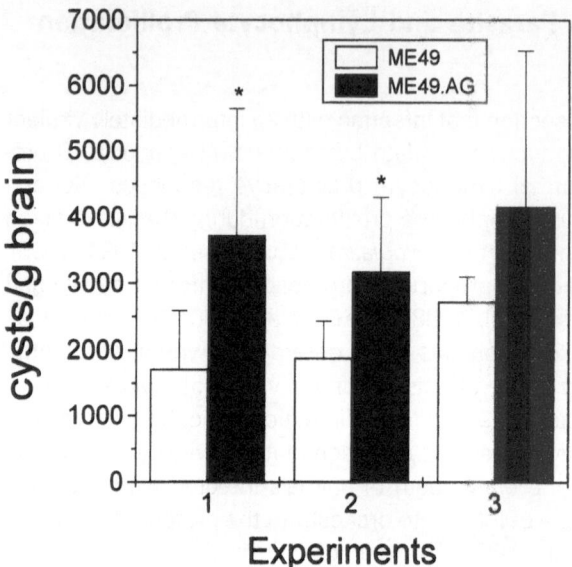

Fig. 6. Cysts brain load of CBA/J mice infected with an avirulent strain of *Toxoplasma gondii* (ME49 strain, ten cysts orally) at day 21 post infection and treated by an inhibitor of nitric oxide production, aminoguanidine (*AG*; 100 mg/kg daily from days 1–21 post infection). Levels of significance (Student *t* test) in three different experiments (*n*=5): * $p<0.05$, ** $p< 0.01$, *** $p< 0.001$

et al. 1995). However, aminoguanidine treatment did not increase parasite load in the lungs or brains of mice, which had been infected with intermediately virulent strain C56. This may suggest that the observed increase in susceptibility is not due to an increased parasite load, but rather to immunopathological phenomena.

Similarly, treatment of mice with aminoguanidine led to a recovery of splenocyte proliferative responses on day 7 PI. In mice infected with the intermediately virulent strain C56, this recovery was marked (90% of recovery in Con A-stimulated spleen cells), whereas it was only partial in mice infected with the avirulent strain ME49 (28% of recovery in Con A-stimulated spleen cells). The observation that aminoguanidine was more potent in mice infected with the intermediately virulent strain could explain the better recovery of proliferative responses in these animals, confirming the negative correlation between lymphocyte proliferation and parasite virulence or susceptibility to infection.

Taken together, these in vivo and in vitro data provide additional evidence that NO is important in controlling lymphocyte and *Toxoplasma* proliferation during the acute stage of infection, and that the recovery of lymphocyte proliferation does not correlate with increased resistance to the infection.

4 Conclusion

The suggestion that NO-induced immunosuppression of the lymphocyte proliferative response leads to increased susceptibility to *T. gondii* infection is
not supported by our studies. On the contrary, profoundly decreased lymphocyte proliferation does not appear to be wholly detrimental to the host but
rather correlates with host survival. Indeed, maintained lymphocyte proliferation
correlates with susceptibility to the infection. In conclusion, through these
two mechanisms of control of parasite and lymphocyte proliferation, NO is
of paramount importance for the survival of the host to *Toxoplasma* infection.

Pregnancy did not enhance the reduced proliferative capacities observed
during the acute stage of *Toxoplasma* infection, even if infected pregnant mice
demonstrate higher levels of NO production in a Th1-type milieu. The reasons
for the increased susceptibility of pregnant hosts to *T. gondii* infection thus
remain to be determined. Levels of NO, IFN-γ and IL-10 are increased in pregnant infected mice, indicating, as suggested by ROBERTS and ALEXANDER (1992)
that concomitant production of IFN-γ and IL-10 may lead to a mild increase
in susceptibility. Other causes of this increased susceptibility during pregnancy
might lie in the placenta and/or fetus.

The analysis of the immune response to two different strains of *Toxoplasma*
throws light on the timing of the protective response. Indeed, protection appears to be dependent upon an early Th1-type response that is immediately
counterbalanced by a Th2-type response associated with a depressed lymphocyte proliferation to mitogens and *Toxoplasma* antigen. NO and IFN-γ may
thus be present early in the infection to ensure protection, but this response
needs to be replaced by a Th2-type reaction later in the infection if immunopathological phenomena are to be avoided. The precise timing of Th1- and
Th2-type cytokine production seems to have a marked influence on *Toxoplasma*
pathogenicity. Further studies should be set up for exploring the influence of
strain-specific antigens on the immune response.

Acknowledgments. We wish to thank Pr. Michel Kremer for helpful comments and valuable dis-
cussion in the preparation of this manuscript, as well as Elizabeth Antoni and Gisèle Kurth for
excellent technical assistance during this work.

References

Adams L, Hibbs J, Taintor R, Krahenbuhl J (1990) Microbiostatic effect of murine-activated mac-
rophages for Toxoplasma gondii. J Immunol 144:2725–2729
Baker C, Rench M, Edwards M, Carpentier R, Hays B, Kasper D (1988) Immunization of pregnant
women with a polysaccharide vaccine of group B streptococcus. N Engl J Med 3:1180–1185
Barcinski M, Costa-Moreira M (1994) Cellular response of protozoan parasites to host-derived cy-
tokines. Parasitol Today 10:352–355
Beaman M, Wong S-Y, Remington J (1992) Cytokines, Toxoplasma and intracellular parasitism.
Immunol Rev 127:97–117

Beaman M, Hunter C, Remington J (1994) Enhancement of intracellular replication of Toxoplasma gondii by IL-6. Interactions with IFN-γ and TNF-α. J Immunol 153:4583–4587

Beckerman K, Rogers H, Corbett J, Schreiber R, Unanue E (1993) Release of nitric oxide during the T-cell-independent pathway of macrophage activation. Its role in resistance to Listeria monocytogenes. J Immunol 150:888–895

Bell CB, Billington WD (1980) Major anti-paternal alloantibody induced by murine pregnancy is non complement-fixing IgG1. Nature 288:387–388

Black C, Israelski D, Suzuki Y, Remington J (1989) Effect of recombinant tumour necrosis factor on acute infection in mice with Toxoplasma gondii or Trypanosoma cruzi. Immunology 68:570–574

Bohne W, Heesemann J, Gross U (1994) Reduced replication of Toxoplasma gondii is necessary for induction of bradyzoite-specific antigens: a possible role for nitric oxide in triggering stage conversion. Infect Immun 62:1761–1767

Brabin B (1983) An analysis of malaria in pregnancy in Africa. Bulletin WHO 61:1005–1016

Candolfi E, Hunter C, Remington J (1994) Mitogen and antigen specific proliferation of T cells in murine toxoplasmosis is inhibited by reactive nitrogen intermediates. Infect Immun 62:1995–2001

Candolfi E, Hunter C, Remington J (1995) Roles of gamma interferon and other cytokines in suppression of the spleen cell proliferative response to concanavalin A and Toxoplasma antigen during acute toxoplasmosis. Infect Immun 63:751–756

Chan J, Tanaka K, Carroll D, Flynn J, Bloom B (1995) Effects of nitric oxide synthase inhibitors on murine infection with Mycobacterium tuberculosis. Infect Immun 63:736–740

Chaouat G, Menu E, Delage G, Djaian V, Assal Messiani A, Ropert S (1993) Cytokines et reproduction. In: Cavaillon J (ed) Les cytokines. Masson, Paris, pp 371–384

Chardès T, Velge-Roussel F, Mevelec P, Mevelec MN, Buzoni-Gatel D, Bout D (1993) Mucosal and systemic cellular immune responses induced by Toxoplasma gondii antigens in cyst orally infected mice. Immunology 78:421–429

Clark D, Flanders K, Banwatt D, Millar-Book W, Manuel J, Stredonska-Clark J, Rowley B (1990) Murine pregnancy decidua produces a unique immunosuppressive molecule related to transforming growth factor β-2. J Immunol 144:3008–3014

Corbett J, Tilton R, Chang K, Hasan K, Ido Y, Wang J, Sweetland M, Lancaster J, Williamson J, McDaniel M (1992) Aminoguanidine, a novel inhibitor of nitric oxide formation, prevents diabetic vascular dysfunction. Diabetes 41:552–556

Dardé M, Bouteille B, Pestre-Alexandre M (1992) Isoenzyme analysis of 35 Toxoplasma gondii isolates and the biological and epidemiological implications. J Parasitol 78:786–794

Denkers E, Gazzinelli R, Martin D, Sher A (1993) Emergence of NK1.1+ cells as effectors of IFN-γ dependent immunity to Toxoplasma gondii in MHC class I-deficient mice. J Exp Med 178:1465–1472

Fiorentino DF, Bond MW, Mosmann TR (1989) Two types of mouse T helper cell. IV. Th2 clones secrete a factor that inhibits cytokine production by Th1 clones. J Exp Med 170:2081–2095

Fiorentino DF, Zlotnik A, Vieira P, Mosmann TR, Howard M, Moore KW, O'Garra A (1991) IL-10 acts on the antigen-presenting cell to inhibit cytokine production by Th1 cells. J Immunol 146:3444–3451

Gazzinelli R, Oswald IP, James ST, Sher A (1992) IL-10 inhibits parasite killing and nitrogen oxide production by IFN-gamma-activated macrophages. J Immunol 148:1792–1796

Gazzinelli R, Denkers E, Sher A (1993a) Host resistance to Toxoplasma gondii: model for studying the selective induction of cell-mediated immunity by intracellular parasites. Infect Agents Dis 2:139–149

Gazzinelli R, Hieny S, Wynn T, Wolf S, Sher A (1993b) Interleukin 12 is required for the T-lymphocyte-independent induction of interferon-γ by an intracellular parasite and induces resistance in T-cell-deficient hosts. Proc Natl Acad Sci USA 90:6115–6119

Gerard C, Bruyns C, Marchant A, Abramowicz D, Vandenabeele P, Grasso A (1993) Interleukin-10 reduces the release of tumor necrosis factor and prevents lethality in experimental endotoxemia. J Exp Med 177:547–550

Grasso G, Muscettola M (1990) The influence of beta-estradiol and progesterone on interferon gamma production in vitro. Int J Neurosci 51:315–317

Grau T, Tacchini-Cottie F, Piguet P-F (1992) Is TNF beneficial or deleterious in toxoplasmic encephalitis. Parasitol Today 8:322–324

Haque S, Khan I, Haque A, Kasper L (1994) Impairment of the cellular immune response in acute murine toxoplasmosis: regulation of Interleukin-2 production and macrophage-mediated inhibitory effects. Infect Immun 62:2908–2916

Holland D, Bretscher P, Russel A (1984) Immunologic and inflammatory responses during pregnancy. J Clin Lab Immunol 14:177–179

Hunter C, Subauste C, Van Cleave V, Remington J (1994) Production of gamma interferon by natural killer cells from Toxoplasma gondii-infected SCID mice: regulation by interleukin-10, interleukin-12, and tumor necrosis factor alpha. Infect Immun 62:2818–2824

Hunter C, Bermudez L, Beernink H, Waegell W, Remington J (1995a) Transforming growth factor-β inhibits interleukin-12-induced production of interferon-γ by natural killer cells: a role for transforming groth factor-β in the regulation of T cell-independent resistance to Toxoplasma gondii. Eur J Immunol 25:994–1000

Hunter C, Candolfi E, Subauste C, Van Cleave V, Remington J (1995b) Studies on the role of interleukin-12 in acute murine toxoplasmosis. Immunology 84:16–20

Kennedy M, Torrance D, Picha S, Molher K (1992) Analysis of cytokine mRNA expression in the central nervous system of mice with experimental autoimmune encephalomyelitis reveals that IL-10 mRNA expression correlates with recovery. J Immunol 149:2496–2505

Khan I, Matsura T, Kasper L (1994) Interleukin 12 enhances murine survival against acute toxoplasmosis. Infect Immun 62:1639–1642

Khan I, Matsuura T, Kasper L (1995) IL-10 mediates immunosuppression following primary infection with Toxoplasma gondii in mice. Parasite Immunol 17:185–195

Kremsner P, Neifer S, Chaves M, Rudolph R, Bienzle U (1992) Interferon-γ induced lethality in the late phase of Plasmodium vinckei malaria despite effective parasite clearance by chloroquine. Eur J Immunol 22:2873–2878

Kühn R, Löhler J, Rennick D, Rajewski K, Müller W (1993) Interleukin-10-deficient mice develop chronic enterocolitis. Cell 75:263–274

Langermans J, Van der Hulst M, Nibbering P, Hiemstra P, Fransen P, Van Furth R (1992) IFN-γ induced L-arginine dependent toxoplasmastatic activity in murine peritoneal macrophages is mediated by endogenous tumor necrosis factor. J Immunol 148:568–574

Lederman M (1984) Cell-mediated immunity and pregnancy. Chest 86:6S–9S

Liew F (1995) Regulation of lymphocyte functions by nitric oxide. Curr Opin Immunol 7:396–399

Liew L, Severn A, Schmidt J, Salter M, Moncada S (1991) A possible novel pathway of regulation by murine T helper type-2 (Th2) cells of a Th1 cell activity via the modulation of the induction of nitric oxide synthase on macrophages. Eur J Immunol 21:2489–2491

Lin H, Mosmann T, Guilbert L, Tuntipopipat S, Wegmann T (1993) Synthesis of T helper 2-type cytokine at the maternal-fetal interface. J Immunol 151:4562–4573

Luft B, Remington J (1982) Effect of pregnancy on resistance to Listeria monocytogenes and Toxoplasma gondii infections in mice. Infect Immun 38:1164–1171

McLeod R, Skamene E, Brown CR, Eisenhauer PB, Mack DG (1989) Genetic regulation of early survival and cyst number after peroral Toxoplasma gondii infection of AxB/BxA recombinant inbred and B10 congenic mice. J Immunol 143:3031–3034

Menendez C (1995) Malaria during pregnancy: a priority area of malaria research and control. Parasitol Today 11:178–183

Oswald I, Gazzinelli R, Sher A, James S (1992) Il-10 synergizes with IL-4 and TGF-β to inhibit macrophages cytotoxic activity. J Immunol 148:3578–3582

Pfefferkorn E (1984) Interferon gamma blocks the growth of Toxoplasma gondii in human fibroblasts by inducing the host cells to degrade tryptophan. Proc Natl Acad Sci USA 81:908–912

Roberts C, Alexander J (1992) Studies on a murine model of congenital toxoplasmosis: vertical disease transmission only occurs in BALB/c mice infected for the first time during pregnancy. Parasitology 104:19–23

Roberts C, Brewer J, Alexander J (1994) Congenital toxoplasmosis in the BALB/c mouse: prevention of vertical disease transmission and fetal death by vaccination. Vaccine 12:1389–1394

Roberts C, Cruickshank S, Alexander J (1995) Sex-determined resistance to Toxoplasma gondii is associated with temporal differences in cytokine production. Infect Immun 63:2549–2555

Seder R, Gazzinelli R, Sher A, Paul W (1993) Interleukin 12 acts directly on CD4[+] T cells to enhance priming for interferon-γ production and diminishes interleukin 4 inhibition of such priming. Immunology 90:10188–10192

Sher A, Oswald I, Hieny S, Gazzinelli R (1993) Toxoplasma gondii induces a T-independent IFN-γ response in natural killer cells that requires both adherent accessory cells and tumor necrosis factor-α. J Immunol 150:3982–3989

Shirahata T, Muroya N, Ohta C, Goto H, Nakane A (1993) Enhancement by recombinant human interleukin 2 of host resistance to Toxoplasma gondii infection in pregnant mice. Microbiol Immunol 37:583–90

Sicher S, Vazqeuez M, Lu C (1994) Inhibition of macrophage Ia expression by nitric oxide. J Immunol 153:1293–1300

Subauste C, Remington J (1991) Role of gamma interferon in Toxoplasma gondii infection. Eur J Clin Microbiol Infect Dis 10:58–67

Suzuki Y, Orellana M, Schreiber R, Remington J (1988) Interferon-γ: the major mediator of resistance against Toxoplasma gondii. Science 240:516–518

Taylor-Robinson A (1995) Regulation of immunity to malaria: valuable lessons learned from murine models. Parasitol Today 11:334–342

Waki S, Uehara S, Kanbe K, Ono K, Suzuki M, Nariuchi H (1992) The role of T cells in pathogenesis and protective immunity to murine malaria. Immunology 75:646–651

Wegmann T, Hui L, Guilbert L, Mosmann T (1993) Bidirectional cytokine interactions in the maternal–fetal relationship: is successful pregnancy a Th2 phenomenon? Immunol Today 14:353–356

Weinberg E (1984) Pregnancy-associated depression of cell-mediated immunity. Rev Infect Dis 6:814–831

Wilson C, Tsai V, Remington J (1980) Failure to trigger the oxidative metabolic burst by normal macrophages. J Exp Med 151:328–346

Cytokine Production by Human Cells after *Toxoplasma gondii* Infection

H. Pelloux and P. Ambroise-Thomas

1 Introduction

Although the pathophysiology of toxoplasmosis is not fully understood, it is now clear that the cytokine network plays a major role in the cellular mechanisms involved in the control of *Toxoplasma gondii* infection (BEAMAN et al. 1992). This is of main importance in the central nervous system, where most of the clinical lesions are localised in case of congenital toxoplasmosis or in case of reactivations in immunocompromised patients (HUNTER and REMINGTON 1994). Most of the research works carried out up to now have focused on mouse models or have analyzed murine cells in vitro (HUNTER and REMINGTON 1994). Among these different studies, only some of them were devoted to cytokine production after *T. gondii* infection. Even fewer studies have analyzed the role of cytokines in human individuals or in human cell populations in vitro. In this review, we will attempt to summarize the different results obtained

Département de Parasitologie-Mycologie Médicale et Moléculaire, EP CNRS 78, Faculté de Médecine, Université J Fourier Grenoble I, BP 217, 38043 Grenoble cedex, France

for cytokine secretion and/or expression after *T. gondii* infection in human models, both in vivo and in vitro.

2 Interferon-Gamma (IFN-γ)

Interferon-gamma (IFN-γ), which was the first cytokine described 40 years ago, has a wide range of activities, including macrophage activation. Up to now, it has been the most extensively studied cytokine in the field of toxoplasmosis.

2.1 In Vivo Studies

Studies in humans reporting the dosages of IFN-γ deal with the two main clinical problems caused by *T. gondii* in man, namely, congenital toxoplasmosis and toxoplasmic reactivations in immunocompromised patients. RAYMOND et al. (1990) studied the production of IFN-γ in acute acquired and in congenital toxoplasmosis. Testing sera from infected pregnant women and cord blood, they demonstrated that IFN-γ was produced in significant amounts during the acute stage of acquired toxoplasmosis, and in the course of congenital infection. Interestingly, it appeared that IFN-γ was produced only before the appearance of *T. gondii*-specific immunoglobulin G (IgG). In contrast to IFN-γ, IFN-α was never detected in these samples. The authors therefore suggested that the detection of IFN-γ could be an aid in the diagnosis of congenital toxoplasmosis. However, since detection of IFN-γ is not specific for *T. gondii* infection, this parameter should not be used alone for definitive diagnosis of congenital toxoplasmosis, but rather could be associated with other *Toxoplasma*-specific biological tests.

The secretion of IFN-γ was compared in patients suffering from toxoplasmic encephalitis (TE) and acquired immunodeficiency syndrome (AIDS) and in those with acquired toxoplasmic lymphadenopathy (CANESSA et al. 1992). It could be shown that secretion of IFN-γ was impaired in AIDS patients, probably as a result of CD4$^+$ cell reduction. No IFN-γ was detected in the cerebrospinal fluid of AIDS patients with TE. In this study, the levels of IFN-γ in sera of immunocompetent patients remained at high levels even 3– 9 months after acute infection, which contradicts the results of RAYMOND et al. (1990), who showed a rapid decrease in IFN-γ levels after primary infection in pregnant women.

2.2 In Vitro Studies

In vitro studies analyzing IFN-γ secretion are more common than in vivo studies. It was demonstrated a long time ago that the supernatants of immune lymphocytes or of lymphocytes stimulated with toxoplasmic antigens could induce the ability of human monocyte-derived macrophages to resist intracellular infection with *T. gondii* (BEAMAN et al. 1992). Other studies provided evidencethat CD4[+] and CD8[+] T cell clones from *T. gondii* seropositive patients produce IFN-γ after stimulation with *T. gondii* antigens or irradiated tachyzoites of the RH strain (SKLENAR et al. 1986, CANESSA et al. 1988). Moreover, IFN-γ was often produced in higher amounts when T cell clones originated from an asymptomatic subject than from a symptomatic patient (SKLENAR et al. 1986). These results seem logical, when the protective role of IFN-γ is considered in the case of infection with *T. gondii* (HUNTER et al. 1994). KHAN et al. (1988) showed that mononuclear cells from *T. gondii*-positive donors could be stimulated to produce IFN-γ by the purified membrane protein SAG1 (P30). Studies using monoclonal antibodies demonstrated that the IFN-γ secretion by CD4[+] cells and monocytes is dependent on interleukin-2 (IL-2), IL-2 receptor, IL-1 and class II major histocompatibility (MHC) antigens, and independent of tumor necrosis factor alpha (TNF-α) and TNF-β (KELLY et al. 1989). Importantly, the ability of lymphocytes to produce IFN-γ in response to *T. gondii* antigens has been shown to be impaired in lymphocytes derived from AIDS patients (MURRAY et al 1984). These in vitro findings are in accordance with in vivo ones, suggesting that the impairment of IFN-γ production in AIDS patients could be the major cause of toxoplasmic reactivations. More recently, some human CD4[+] and CD8[+] T-cell clones from *T. gondii*-positive individuals have been shown to produce IFN-γ after stimulation with a soluble antigenic fraction of *T. gondii* (SAAVEDRA and HÉRION 1991). These data are in accordance with more recent ones, which demonstrated IFN-γ secretion by *T.gondii*-specific CD4[+] clones from latently infected human patients after mitogen or antigen stimulation (PRIGIONE et al. 1995). CD4[+] lines that are cytotoxic for *T. gondii*-infected human melanocytes have been shown to produce IFN-γ after antigen-specific stimulation. Furthermore, the autocrine IFN-γ production enhanced the CD4[+] toxicity for *T. gondii*-infected melanocytes (YANG et al. 1995). Taken together, all these findings indicate that IFN-γ is of major importance for the human immune system during *T. gondii* infection. This cytokine probably represents a key molecule for the resistance against *T. gondii*, since IFN-γ can induce anti-toxoplasmic activity in various cell populations (MURRAY et al 1984; EALES at al 1987; CHAO et al 1994). It also appears that, at least in the case of *T. gondii* infection, IFN-γ production depends not only on *T. gondii* itself, but also on other cytokines and on the human leukocyte antigen (HLA) system.

3 Tumor Necrosis Factor Alpha

TNF-α was first detected by its ability to induce necrosis of tumors. Then, it appeared to be one of the most important cytokines involved in the acute phase of inflammation. Certain functions such as induction of fever, cachexia, macrophage activation, and B lymphocyte proliferation have been described for TNF-α, but only a few studies have investigated TNF-α secretion in human models after *T. gondii* infection. CANESSA et al. (1992) could not demonstrate any difference in vivo in TNF-α secretion among AIDS patients with TE, immunocompetent patients with acute toxoplasmosis, or controls. However, this observation does not exclude the possibility that TNF-α is produced at the site of infection such as the brain. It is possible that cytokines may not be detected in peripheral blood even if they are secreted in infected tissues.

In vitro, the results concerning TNF-α production are quite contradictory. Some studies underline the fact that *T. gondii* does not seem to trigger TNF-α production in human cells. KELLY et al. (1989) demonstrated that *T. gondii* antigens were unable to induce TNF-α secretion of monocytes or peripheral blood mononuclear cells. A monocytic cell-line, THP1, expressed and secreted only minimal amounts of TNF-α after phagocytosis of RH-strain tachyzoites, unlike what happened during phagocytosis of *Mycobacterium tuberculosis* (FRIEDLAND et al. 1993; FRIEDLAND 1995). Using another model, increased amounts of expressed or secreted TNF-α could not be detected in human monocytes, monocyte-derived macrophages, and astrocytoma-derived cells after infection with virulent or chronic strains of *T. gondii* (PELLOUX et al. 1992, 1994b,c). However, in contrast to human astrocytoma-derived cells, both human monocytes and macrophages specifically exhibited a capability to secrete high amounts of TNF-α after exposure to human sera with high titers of *T. gondii*-specific IgG. This underlines the differences which could exist for cytokine secretion between different human cell populations after exposure to *T. gondii*. Furthermore, it seems possible that monocytes and monocyte-derived macrophages cannot secrete TNF-α after direct exposure to *T. gondii* antigens or tachyzoites, but can secrete TNF-α indirectly in the presence of an additional stimulus such as specific IgG.

In contrast, other studies focused on the fact that *T. gondii* can induce monokine production in humans (BALA et al. 1994). Stimulation with soluble tachyzoite antigens induced TNF-α secretion by peripheral blood mononuclear cells of both human immunodeficiency virus (HIV)-positive and HIV-negative patients with *T. gondii* infection, as well as by monocytes of healthy individuals, who were seronegative for HIV and *T. gondii*.

The conflicting results of in vitro studies using human cells indicate that further studies on TNF-α secretion during toxoplasmosis are necessary. More precisely, it is of major importance to understand why TNF-α secretion can be induced by *T. gondii* in some models but not in others. TNF-α could, like IFN-γ, be a key molecule in the pathophysiology of toxoplasmosis, particularly in immunocompromised patients. It has been shown that endogenouslyse-creted TNF-α can act on both *T. gondii* multiplication and HIV replication (BALA et al. 1994, CHAO et al 1994). It is noteworthy that the results concerning TNF-α are contradictory not only in vitro in human cells, but also in mouse models (HUNTER and REMINGTON 1994).

4 Other Cytokines

4.1 Interleukin-1

IL-1 is another monokine for which the results in the literature seem to differ. It is involved in the regulation of inflammation, and particularly in the acute phase response. IL-1β could not be detected in sera of *T. gondii*-infected patients with or without AIDS (CANESSA et al. 1992). Interestingly, *T. gondii* antigens stimulated IL-1 secretion in vitro by peripheral blood mononuclear cells and monocytes (KELLY et al. 1989; BALA et al. 1994), while viable tachyzoites of different strains of *T. gondii* did not induce any IL-1α production or over-expression in human monocytes, macrophages, or astrocytoma-derived cells in another model (PELLOUX et al. 1994a,b,c). Taken together, these results could indicate that *T. gondii* antigens may induce IL-1 secretion, whereas entire parasites may not.

4.2 Interleukin-2

Along with IFN-γ, IL-2 is the second Th1 subset cytokine whose secretion has been studied during *T. gondii* infection. IL-2 secretion was not noted in patients with acute toxoplasmosis or cerebral reactivation (CANESSA et al. 1992). In con-trast, IL-2 was demonstrated to be secreted by CD4[+] clones in vitro, after stimulation with *T. gondii* antigens. This secretion was shown to be associated with IFN-γ production (CANESSA et al. 1988; PRIGIONE et al. 1995). The IFN-γ production induced by *T. gondii* is clearly dependent on IL-2 and IL-2 receptors, contrary to what happens with cytomegalovirus (KELLY et al. 1989). However, human T-cell clones were able to secrete IFN-γ after specific antigen stimu-lation. This stimulation was independent of IL-2, even if the presence of IL-2 enhanced IFN-γ production in most of the tested clones, or triggered it in others (SAAVEDRA and HÉRION 1991). Interestingly, only five T cell clones of 18 secreted IL-2 after antigen stimulation. Of these five clones, four were shown to have a *T. gondii* strain-dependent response in a proliferative assay. The response was weaker when the virulent Wiktor strain was used than with the RH strain, suggesting that the two phenomena are related. In contrast, CD4[+]

cytotoxic T lymphocytes did not produce IL-2, nor IL4, after stimulation with
T. gondii-infected melanoma cells (YANG et al. 1995).

4.3 Interleukin-4 and Interleukin-5

These two Th2-cytokines have recently been shown to be secreted in vitro
by *T. gondii*-specific CD4$^+$ T cell clones after mitogen and antigen stimulations
(PRIGIONE et al. 1995). Since this Th2 secretion profile was associated with a
Th1 profile of secretion (IFN-γ and IL-2), these *T. gondii*-specific CD4$^+$ clones
were classified as Th0 (PRIGIONE et al. 1995).

4.4 Interleukin-6

IL-6 is a cytokine that is involved in B cell and cytotoxic T cell activation.
Studies on its production during *T. gondii* infection have shown that the results
seem to depend on the model that has been used. No difference was found
in vivo between IL-6 levels in sera of both AIDS and non-AIDS patients with
toxoplasmosis (CANESSA et al. 1992). This monokine has been shown to be
secreted by human peripheral blood mononuclear cells and monocytes after
exposure to tachyzoite antigens (BALA et al. 1994). However, other studies did
not show evidence of any increase in IL-6 secretion or expression after phagocytosis of *T. gondii* tachyzoites by monocytic cells (FRIEDLAND et al. 1993) or
infection of human monocytes, macrophages, or astrocytoma-derived cells
(PELLOUX et al. 1994a,b,c). As for TNF-α, an increased secretion of IL-6 by
human monocytes was demonstrated after stimulation with specific anti-*T.
gondii* sera, but this was not observed in monocyte-derived macrophages
(PELLOUX et al. 1994a). An interesting fact should be noted: by using mainly
animal models, TNF-α and IL-6 have been described to have opposite influences
on *T. gondii* growth (BEAMAN et al. 1992, 1994; HUNTER et al 1994). However,
studies of human cells demonstrated that secretion of TNF-α and IL-6 seems
to be induced by the same stimuli (BALA et al. 1994; PELLOUX et al. 1994a).
These observations could indicate that not the presence or absence of cytokine
secretion is important for the control of *T. gondii* infection, but rather the
balance of cytokines with opposite influences on *T. gondii*.

4.5 Interleukin-8

In comparison with controls, monocytic THP1 cells, which have phagocytozed
T. gondii tachyzoites in vitro, minimally expressed and secreted IL-8. This finding
is in contrast to what has been observed after phagocytosis of *M. tuberculosis*
(FRIEDLAND et al. 1993).

4.6 Interleukin 10

The Th2 cytokine IL-10 could not be found in serum nor in cerebral spinal fluid of five AIDS patients with cerebral toxoplasmosis, although increased levels of IL-10 could be detected in the cerebral spinal fluid from AIDS patients with *Cryptococcus neoformans* meningitis (GALLO et al. 1994). In vitro, IL-10 was secreted simultaneously with IFN-γ and IL-2 by *T. gondii*-specific CD4[+] T cell clones after mitogen and antigen stimulation (PRIGIONE et al. 1995). These results raise the same questions as the concomitant secretions of TNF-α and IL-6 do: under the same stimulations, human cells seem to be able to secrete cytokines, which have been described in other models to either potentiate the toxoplasmacidal activity of macrophages (IFN-γ and TNF-α) or to inhibit or downregulate the anti-*Toxoplasma* immune response (IL-6 and IL-10) (BEAMAN et al. 1994; GAZZINELLI et al. 1992).

4.7 Interleukin-12

IL-12 was shown to be secreted in vitro by peripheral blood mononuclear cells and monocytes from both HIV-positive and HIV-negative individuals with *T. gondii* infection after stimulation with soluble tachyzoite antigens (BALA et al. 1994).

5 Granulocyte/Macrophage Colony-Stimulating Factor

Studies on granulocyte/macrophage colony-stimulating factor (GM-CSF) production have been carried out in vivo and in vitro. In vivo, no secretion of GM-CSF was noted after *T. gondii* infection or cerebral reactivation (CANESSA et al. 1992). In vitro, GM-CSF was secreted by peripheral blood mononuclear cells and monocytes after stimulation with soluble tachyzoite antigens (BALA et al. 1994).

6 Conclusions

In conclusion, it can be noted that IFN-γ is the only cytokine in humans whose secretion after *T. gondii* infection has been clearly and undoubtedly demonstrated. For the other cytokines, the results obtained differ widely from one study to another. These discrepancies may be explained by the differences which exist between the cell populations under study, the stimulation protocols, the detection techniques, or perhaps the *T. gondii* strains used. This underlines the limits and difficulties of interpreting the results obtained with each model. On the one hand, in vivo measurement of cytokines in humans may give global information but insufficient details to understand why a cytokine is secreted and by which cell population, and which mechanisms are involved. Furthermore, since cytokines mainly act in an autocrine or a paracrine manner, they may not be detected in peripheral blood even if they are produced at the site of infection such as the brain. On the other hand, the results obtained in studies of cytokine production by human cells in vitro may be too restrictive, since the interactions with the entire immune system are left aside. These disadvantages may be avoided in mouse models, in which in vitro experiments can be performed following in vivo infection in order to mimic the situation in humans. However, the use of human in vivo or in vitro models can provide valuable informations about the behavior of any one of the precise human cell-populations after infection with *T. gondii*, the significant relationship which could exist between the presence or absence of a cytokine, and the presence of *T. gondii* cysts or tachyzoites, or the variations of the intercellular messages, which might depend on *T. gondii* infection. These informations could lead to a better understanding of the pathophysiology of toxoplasmosis, and therefore, to prevent and treat more efficiently congenital infection and reactivations in immunocompromised patients.

References

Bala S, Englund G, Kovacs J, Wahl L, Martin M, Sher A, Gazzinelli RT (1994) Toxoplasma gondii soluble products induce cytokine secretion by macrophages and potentiate in vitro replication of a monotropic strain of HIV. J Eukaryot Microbiol 41:7S

Beaman MH, Wong SY, Remington JS (1992) Cytokines, Toxoplasma and intracellular parasitism. Immunol Rev 127:97–117

Beaman MH, Hunter CA, Remington JS (1994) Enhancement of intracellular replication of Toxoplasma gondii by IL-6. J Immunol 153:4583–4587

Canessa A, Pistoia V, Roncella S, Merli A, Melioli G, Terragna A, Ferrarini M (1988) An in vitro model for Toxoplasma infection in man. Interaction between CD4$^+$ monoclonal T cells and macrophages results in killing of trophozoites. J Immunol 140:3580–3588

Canessa A, Del Bono V, Miletich F, Pistoia V (1992) Serum cytokines in toxoplasmosis:increased levels of interferon-γ in immunocompetent patients with lymphadenopathy but not in AIDS patients with encephalitis. J Infect Dis 165:1168–1170

Chao CC, Gekker G, Hu S, Peterson PK (1994) Human microglial defense against Toxoplasma gondii. The role of cytokines. J Immunol 152:1246–1252

Eales LJ, Moshtael O, Pinching AJ (1987) Microbicidal activity of monocyte derived macrophages in AIDS and related disorders. Clin Exp Immunol 67:227–235

Friedland JS (1995) Chemokines and human infection. Clin Sci 88:393–400

Friedland JS, Shattock RJ, Johnson JD, Remick DG, Holliman RE, Griffin GE (1993) Differential cytokine gene expression and secretion after phagocytosis by human monocytic cell-line of Toxoplasma gondii compared with Mycobacterium tuberculosis. Clin Exp Immunol 91:282–286

Gallo P, Sivieri S, Rinaldi L, Yan XB, Lolli F, De Rossi A, Tavolato B (1994) Intrathecal synthesis of interleukin-10 (IL-10) in viral and inflammatory diseases of the central nervous system. J Neurol Sc 126:49–53

Gazzinelli R, Oswald IP, James SL, Sher A (1992) IL-10 inhibits parasite killing and nitrogen oxide production by IFN-γ activated macrophages. J Immunol 148:1792–1796

Hunter CA, Remington JS (1994) Immunopathogenesis of toxoplasmic encephalitis. J Infect Dis 170:1057–67

Hunter CA, Subauste CS, Remington JS (1994) The rôle of cytokines in toxoplasmosis. Biotherapy 7:237–247

Kelly CD, Russo CM, Rubin BY, Murray HW (1989) Antigen-stimulated human interferon-gamma generation:role of accessory cells and their expressed or secreted products. Clin Exp Immunol 77:397–402

Khan IA, Eckel ME, Pfefferkorn ER, Kasper LH (1988) Products of γ interferon by cultured human lymphocytes stimulated with a purified membrane protein (P30) from Toxoplasma gondii. J Infect Dis 157:979–984

Murray HW, Rubin BY, Masur H, Roberts RB (1984) Impaired production of lymphokines and immune (gamma) interferon in the acquired immunodeficiency syndrome. N Engl J Med 310:883–9

Pelloux H, Chumpitazi BFF, Santoro F, Polack B, Vuillez JP, Ambroise-Thomas P (1992) Sera of patients with high titers of immunoglobulin G against Toxoplasma gondii induce secretion of tumor necrosis factor alpha by human monocytes. Infect Immun 60:2672–2676

Pelloux H, Pernod G, Ricard J, Renversez JC, Ambroise-Thomas P (1994a) Interleukin-6 is secreted by human monocytes after stimulation with anti-Toxoplasma gondii sera. J Infect Dis 169:1181–1182

Pelloux H, Ricard J, Bracchi V, Markowicz Y, Verna JM, Ambroise-Thomas P (1994b) Tumor necrosis factor alpha, interleukin 1 alpha, and interleukin 6 mRNA expressed by human astrocytoma cells after infection by three different strains of Toxoplasma gondii. Parasitol Res 80:271–276

Pelloux H, Ricard J, Nissou MF, Renversez JC, Vuillez JP, Meunier A, Ambroise-Thomas P (1994c) Infection with Toxoplasma gondii does not alter TNF-α and IL-6 secretion by a human astrocytoma cell line. Mediat Inflam 3:291–295

Prigione I, Facchetti P, Ghiotto F, Tasso P, Pistoia V (1995) Toxoplasma gondii-specific CD4[+] T cell clones from healthy, latently infected humans display a Th0 profile of cytokine secretion. Eur J Immunol 25:1298–1305

Raymond J, Poissonnier MH, Thulliez PH, Forestier F, Daffos F, Lebon N (1990) Presence of gamma interferon in human acute and congenital toxoplasmosis. J Clin Microbiol 28:1434–1437

Saavedra R, Hérion P (1991) Human T-cell clones against Toxoplasma gondii:production of interferon-γ, interleukin-2, and strain cross reactivity. Parasitol Res 77:379–385

Sklenar I, Jones TC, Alkan S, Erb P (1986) Association of symptomatic human infection with Toxoplasma gondii with imbalance of monocytes and antigen-specific T-cell subset. J Infect Dis 153:315–324

Yang TH, Aosai F, Norose K, Ueda M, Yano A (1995) Enhanced cytotoxicity of IFN-γ-producing CD4[+] cytotoxic T lymphocytes specific for T. gondii-infected human melanoma cells. J Immunol 154:290–298

Characterization of Human T Cell Clones Specific for *Toxoplasma gondii*

V. Pistoia[1], P. Facchetti[1], F. Ghiotto[1],
M.F. Cesbron-Delauw[2], and I. Prigione[1]

1 Introduction

Toxoplasma gondii is a ubiquitous intracellular pathogen that infects all nucleated cells (as reviewed by WONG and REMINGTON 1993). In the immunocompetent individual, a balanced interaction between the immune system and *T. gondii* allows the pathogen to survive in the intracellular milieu without damaging the host. In the setting of defective cell-mediated immunity, uncontrolled multiplication of *T. gondii* tachyzoites may take place, leading to toxoplasmic encephalitis (LUFT and REMINGTON 1992). Detailed knowledge of the immune response to *T. gondii* is needed to unravel the subtle pathogenetic mechanisms of toxoplasmic infection. In this chapter, emphasis will be placed on antigen-specific cell-mediated responses to viable *T. gondii* tachyzoites, as assessed in chronically infected healthy humans. These studies provide a model for a better comprehension of altered *T. gondii*-specific cellular immunity in severely immunodeficient patients.

[1]Laboratory of Oncology, Scientific Institute G. Gaslini, Largo G. Gaslini 5, 16148 Genoa, Italy
[2]Laboratoire de la Toxoplasmose, Institute Pasteur Lille, 1, rue du Prof. Calmette, 59019 Lille Cédex, France

2 In Vitro Analysis
of *T. gondii*-Specific Cellular Immunity

A good way to investigate T cell-mediated immunity to *T. gondii* is to challenge peripheral blood mononuclear cells (PBMC) from infected individuals with viable *T. gondii* tachyzoites or their lysates, and to determine cellular proliferation by [^{3}H]thymidine incorporation (CANESSA et al. 1988). *T. gondii* infection is documented by the detection of specific serum antibodies; in our experience, a close correlation exists between the presence of such antibodies and the ability of T lymphocytes to mount a specific response to viable *T. gondii* tachyzoites (CANESSA et al. 1988). In contrast, when tachyzoite lysates are used, T cell proliferation can occasionally take place in the absence of specific serum antibodies.

Bulk cultures of antigen-specific T cells are comprised of a mixture of clones recognizing discrete peptides originating from the intracellular processing and presentation of different antigens. Furthermore, the same cultures may contain a variable number of activated, nonspecifically recruited T lymphocytes. For these reasons, clonal analysis is the only approach which allows to investigate at the single cell level fine specificity, human leukocyte antigen (HLA) restriction, patterns of cytokine production, and cytolytic activity of antigen-reactive T lymphocytes.

We have established the following experimental system to generate *T. gondii*-specific T cell clones (CANESSA et al. 1988). PBMC from healthy, latently infected individuals were cultured with *T. gondii* tachyzoites at a 20:1 ratio. All of the experiments reported here were carried out using live tachyzoites. In detail, tachyzoites recovered from mouse peritoneum were repeatedly passed through Vero cells and then used for cell stimulation. Although tachyzoites were initially irradiated before being mixed with PBMC, this procedure was abandoned thereafter. Overgrowth of tachyzoites with complete disappearance of PBMC was observed in seronegative, but not in seropositive individuals.

After 7 days in culture, lymphoid blasts were purified on a Percoll density gradient and cloned by limiting dilution in the presence of autologous irradiated PBMC as antigen-presenting cells (APC), *T. gondii* tachyzoites and recombinant interleukin-2 (IL-2). Proliferating clones were expanded in medium containing IL-2 and tested for antigen specificity and HLA restriction before being used for further experiments (CANESSA et al. 1988).

In this model system, viable *T. gondii* tachyzoites infect circulating PBMC and stimulate the emergence of specific T cell blasts, in analogy to the events occurring in vivo upon *T. gondii* infection.

2.1 Characterization
of *T. gondii*-Specific Human T Cell Clones

All of the T cell clones raised by the aforementioned protocol had a CD4[+] immunophenotype, and their proliferative responses to *T. gondii* tachyzoites were restricted for HLA class II antigens (CANESSA et al. 1988). Since T cell-dependent macrophage activation is instrumental for the clearing of *T. gondii* infection (NATHAN et al. 1983), we investigated whether supernatants from antigen-stimulated CD4[+] clones inhibited the intracellular replication of the parasite in monocyte-derived macrophages, as assessed by [³H]uracyl incorporation. All of the clone supernatants proved capable of suppressing the multiplication of *T. gondii* tachyzoites, and an IFN-γ-specific neutralizing antiserum abolished this toxoplasmacidal activity (CANESSA et al. 1988). These findings confirmed, at the clonal level, previous pivotal observations pointing to IFN-γ as the key mediator of the resistance to *T. gondii* infection (NATHAN et al. 1983; SUZUKI et al. 1988).

Both murine (MOSMANN et al. 1986) and human (DEL PRETE et al. 1991; HAANEN et al. 1991; SALGAME et al. 1991) CD4[+] T cell clones can be subdivided into three major groups, i.e., T helper (Th) 1, Th2 or Th0, depending on the spectrum of cytokines they produce. Thus, human Th1 cell clones which can be raised in vitro by stimulation with mycobacterial antigens (DEL PRETE et al. 1991; SALGAME et al. 1991) release large amounts of IFN-γ and IL-2, but little or no IL-4 or IL-5. In contrast, Th2 cell clones that are derived by in vitro stimulation with allergens or helminthic antigens (DEL PRETE at al. 1991) display an opposite profile of cytokine secretion, i.e., they produce abundant IL-4 and IL-5 with little or no IFN-γ or IL-2. Such polarized Th cell responses represent the ends of a spectrum in which the majority of antigen-specific CD4[+] T cell clones behave as Th0 cells, i.e., they secrete the four cytokines (i.e. IL-2, IFN-γ, IL-4 and IL-5) in variable but roughly comparable amounts.

The Th1 or Th2 phenotype of antigen-specific T cell clones is acquired and can be manipulated in vitro by various stimuli, especially cytokines. For example, the addition of IFN-γ (MAGGI et al. 1992) or IL-12 (MANETTI et al. 1993) to allergen-specific bulk cultures favors the outgrowth of Th1 rather than Th2 cell clones; conversely, IL-4 stimulates protein-purified derivative (PPD)-specific CD4[+] cells to give rise to Th2 rather Th1 clones (MAGGI et al. 1992). There is some experimental evidence supporting the concept that Th0 cells represent precursors of Th1 or Th2 lymphocytes (STREET et al. 1990).

It is assumed that Th1 lymphocytes mediate immunity to various intracellular microorganisms. This result can be achieved through numerous mechanisms such as: (i) release of IFN-γ, which activates macrophages to inhibit pathogen replication (NATHAN et al. 1983; SUZUKI et al. 1988; CANESSA et al. 1988); (ii) killing of infected cells including APC (CURIEL et al. 1993; PRIGIONE et al. 1995); and (iii) direct killing of extracellular pathogens (KHAN et al. 1990).

Starting from these premises, we investigated the cytokine profiles of 20 *T. gondii*-specific T cell clones raised from in vitro infected PBMC (see above)

(PRIGIONE et al. 1995). Cloned T lymphocytes were stimulated with phytohe-magglutinin (PHA) and phorbol myristate acetate (PMA) or with *T. gondii* ta-chyzoites in the presence of autologous APC. Supernatants were then tested for the presence of IL-2, IFN-γ, IL-4 and IL-5 by bioassays or enzyme-linked immunosorbent assays (ELISA). All of the clones displayed a Th0 pattern of cytokine production independent of the nature of the stimuli, with an IL-4/IFN-γ ratio ranging from 0.5 to 10 (PRIGIONE et al. 1995). Figure 1 shows the results obtained with eight representative T cell clones following nonspecific stimu-lation with PHA and PMA. Notably, in four clones that had been incubated with autologous APC and viable tachyzoites, a trend towards an increased production of IFN-γ was observed as compared with the results obtained following PHA-PMA stimulation (PRIGIONE et al. 1995). These findings were ex-tended by the analysis of cytokine mRNAs by reverse transcription-polymerase chain reaction (RT-PCR). The Th0 phenotype of *T. gondii*-specific cell clones was unchanged by supplementing with rIL-4 or rIFN-γ, the bulk cultures from which clones were subsequently derived (PRIGIONE et al. 1995). In further ex-periments, it was found that all the *T. gondii*-specific T cell clones tested produced IL-10 in culture supernatants upon stimulation (PRIGIONE et al. 1995).

Finally, most of our clones lysed the P 815 murine mastocytoma cell line in an assay which involves the triggering of the CD3-TcR complex by a CD3 monoclonal antibody. More importantly, a minority of clones specifically killed *T. gondii*-infected autologous lymphoblastoid cell lines (LCLs) that had been raised by in vitro infection with Epstein-Barr virus (EBV) (PRIGIONE et al. 1995).

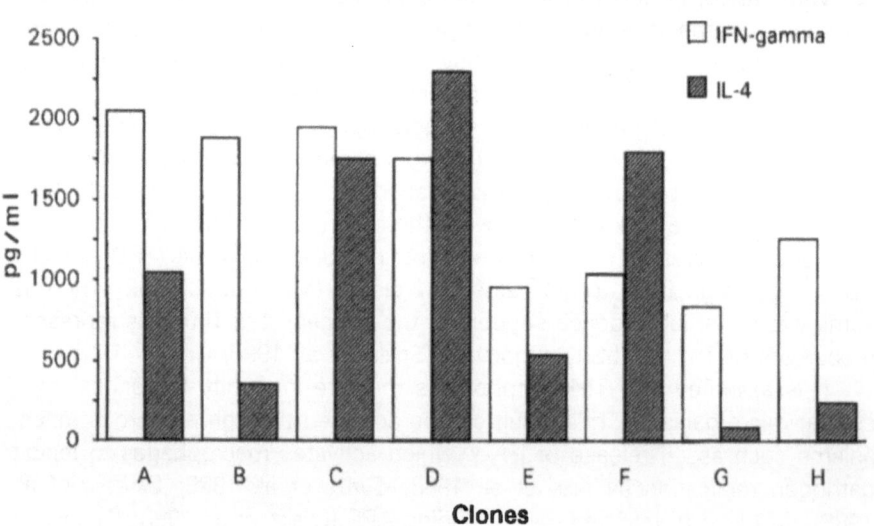

Fig. 1. Production of IFN-γ and IL-4 by *T. gondii*-specific CD4[+] T cell clones. Clones were stimulated with phytohemagglutinin and phorbol myristate acetate as reported (PRIGIONE et al. 1995), and culture supernatants were subsequently tested for the presence of the two cytokines by enzyme-linked immunosorbent assay (ELISA). Results are in pg/ml

MAGGI et al. (1994) have recently reported on the generation and characterization of a large panel of *T. gondii*-specific CD4$^+$ T cell clones from healthy individuals and patients infected with human immunodeficiency virus (HIV). Based upon the patterns of cytokines secreted, 60% of CD4$^+$ cell clones from donors (seronegative for HIV) were classified as Th0, whereas the remainder were defined as Th1. In contrast, in HIV-infected patients, all of the *T. gondii*-specific clones were Th0. Altogether, these results are consistent with our data. A major difference between the two studies is that MAGGI et al. (1994) employed a *T. gondii* extract as antigen for T cell clone generation, whereas we used live *T. gondii* tachyzoites.

Such methodological differences may lead to the hypothesis that *T. gondii* tachyzoites drive the generation of CD4$^+$ cell clones with a predominant Th0 pattern of cytokine secretion. On the other hand, PBMC from healthy, chronically infected individuals produced IL-2 and IFN-γ, but not IL-4 or IL-5, following stimulation with *T. gondii* tachyzoite antigens in bulk cultures (GAZZINELLI et al. 1995).

2.2 A Model of the Immune Response to *T. gondii* During Chronic Human Infection

Based upon our results and those from other laboratories, the sequence of events taking place during the course of chronic *T. gondii* infection in healthy individuals can be envisaged. As will be discussed below, certain antigens may be continuously released into the circulation by *T. gondii* bradyzoites (as well as tachyzoites, see below). Furthermore, from time to time, bradyzoites can give rise to transient parasitemia due to the rupture of tissue cysts (WONG and REMINGTON 1993). It is conceivable that both mechanisms contribute to boost *T. gondii*-specific, T cell-mediated memory responses. In the chronically infected, healthy individual, it can be hypothesized that CD4$^+$ T cell clones with a Th0 profile of cytokine secretion play a fundamental role in the anamnestic immune responses to the parasite.

Such cells potentiate the toxoplasmacidal activity of macrophages through the release of IFN-γ (NATHAN et al. 1983; SUZUKI et al. 1988; CANESSA et al. 1988). Under the influence of CD4$^+$ cell-derived IL-2, the clonal expansion of antigen-specific CD4$^+$ and CD8$^+$ T lymphocytes and the proliferation of lymphokine activated killer (LAK) cells takes place. The latter lymphocytes have been shown to lyse *T. gondii*-infected cells in a major histocompatibility complex (MHC)-unrestricted manner (SUBAUSTE et al. 1992). Both CD8$^+$ and, to a lesser extent CD4$^+$, T cells can hamper parasite multiplication by killing *T. gondii* infected APC or extracellular parasites. On the other hand, most of the clones release IL-10 upon in vitro stimulation (PRIGIONE et al. 1995). Since IL-10 has been shown to inhibit the IFN-γ-mediated activation of murine macrophages against *T. gondii* (GAZZINELLI et al. 1992), IL-10 production might serve the function of limiting the magnitude of the immune response against the parasite.

IL-4 release by *T. gondii*-specific T cell clones may be aimed at stimulating antibody responses against the pathogen; in this respect, it is of note that *T. gondii*-specific immunoglobulin E (IgE) have been recently detected in acutely infected patients (PINON et al. 1995). Since eosinophilia is an unusual finding in acute or chronic *T. gondii* infection, the role of IL-5 in the host response to the parasite warrants further investigation.

The situation is likely to be different in acute toxoplasmosis; there is evidence that circulating CD8[+] α/β T cells increase in acutely infected patients (SKLENAR et al. 1986) and that also γ/δ T cells contribute to the response against the pathogen (DE PAOLI et al. 1992; SUBAUSTE et al. 1995). Furthermore, in the earliest phase of acute murine infection, *T. gondii* tachyzoites trigger natural killer (NK) cells to release IFN-γ through a pathway which involves the production of TNF-α and IL-12 by macrophages (SHER et al. 1993; GAZZINELLI et al. 1993). The NK cell-derived IFN-γ, in turn, activates macrophages to inhibit the intracellular replication of *T. gondii* tachyzoites when specific T cell-mediated responses to *T. gondii* have not yet been developed. As far as chronic asymptomatic toxoplasmosis is concerned, no information is available on the contribution of NK cells to the protection from the parasite during reinfection episodes. Likewise, in human studies with healthy seropositive donors, the generation of *T. gondii*-specific CD8[+] T cell clones has usually been unsuccessful (CANESSA et al. 1988; CURIEL at al. 1993; PRIGIONE et al. 1995); this finding makes it difficult to evaluate the actual role of such cells in the memory responses to the parasite.

3 Initial Characterization of the Fine Specificity of CD4[+] Human T Cell Clones Directed to *T. gondii*

In a recent series of experiments, we have undertaken the characterization of the fine specificity of *T. gondii*-reactive T cell clones generated as described above. According to a working hypothesis formulated by CAPRON and DESSAINT, the so-called excretory secretory antigens (ESAs) of *T. gondii* would be involved in the maintenance of the immunological memory against the parasite, since they are expressed both at the tachyzoite and the bradyzoite stages (CAPRON and DESSAINT 1988). More recently, ESAs were found to be comprised of dense granule proteins (GRA), which have been partly purified and cloned (CESBRON-DELAUW 1994). The main ESAs so far identified are the following: GRA-1 (23 Kd), GRA-2 (28 Kd), GRA-3 (30 Kd), GRA-4 (40 Kd), GRA-5 (21 Kd) and GRA-6 (32 Kd) (review in CESBRON-DELAUW 1994, and this volume).

Based upon these considerations, we evaluated the reactivity of a panel of *T. gondii*-specific T cell clones against ESAs present in cell-free medium from *T gondii* tachyzoites. As shown in Fig. 2, which demonstrates the results obtained with eight representative clones of the 16 tested, all of the clones

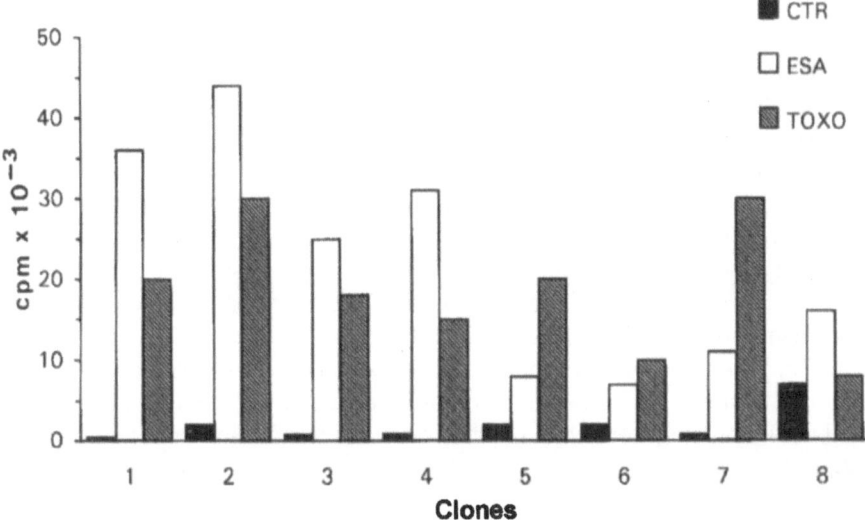

Fig. 2. Proliferative responses of *T. gondii*-specific T cell clones to viable tachyzoites (*TOXO*) and to excretory secretory antigens (*ESA*; *CTR*, controls). Cells were cultured for 48 h with antigens or medium alone and pulsed with [³H]thymidine for an additional 12 h. Results are in cpm/min

proliferated in response to ESA stimulation, indicating that GRA antigens could conceivably be involved in the anamnestic response to *T. gondii*. These results are consistent with previous observations showing that ESAs are immunogenic during human and experimental infection and protect susceptible nu/nu rats from challenge with *T. gondii* (DARCY et al. 1988). Notably, in preliminary experiments, we have found that some ESA-reactive T cell clones specifically recognize GRA1 or GRA2 antigens of *T. gondii* (data not shown).

4 Conclusions

Our studies have shown that *T. gondii*-specific CD4+ T cell clones could be generated from the peripheral blood of chronically infected, healthy individuals. These clones have a Th0 profile of cytokine secretion and could activate macrophages to inhibit the intracellular replication of *T. gondii* tachyzoites by an IFN-γ dependent mechanism; some *T. gondii*-specific T cell clones were able to lyse tachyzoite-infected APC. T cell clones raised by in vitro infection with *T. gondii* tachyzoites recognize ESAs, supporting the hypothesis that these molecules are good stimulators of the long-term immunological memory against the parasite (CAPRON and DESSAINT 1988). It should be stressed that, due to the similarities between our experimental model and the events occurring during in vivo infection, these preliminary observations may open new perspectives for the preparation of a *T. gondii* vaccine.

However, a number of questions remains to be addressed. It is not clear what the functional properties and antigenic specificities of T cell clones raised from patients with acute toxoplasmosis are, and whether T cell clones raised by ESA stimulation display the same features as those generated by in vitro infection with *T. gondii* tachyzoites. In addition, the T cell receptor variable (V) region repertoire of *T. gondii*-specific T cell clones should be analyzed. Finally, it would be important to know what kind of dysregulation occurs in the immune response to *T. gondii* in immunodeficient patients who develop toxoplasmic encephalitis (CANESSA et al. 1992; MAGGI et al. 1994; GAZZINELLI et al. 1995). These issues are now under investigation in our laboratories. The answers to these and other fundamental questions will help to clarify how the host immune system copes with *T. gondii* in normal and immunodeficient subjects.

Acknowledgements. We thank Prof. André Capron for encouragement and suggestions and Dr. Andrea Canessa for collaboration and suggestions.

References

Canessa A, Pistoia V, Merli A, Melioli G, Terragna A, Ferrrarini M (1988) An in vitro model of Toxoplasma infection in man. Interaction between CD4$^+$ monoclonal T cells and macrophages results in killing of trophozoites. J Immunol 140:3580–3588

Canessa A, Delbono V, Miletich F, Pistoia V (1992) Serum cytokines in toxoplasmosis. Increased levels of interferon-gamma in immunocompetent patients with lymphadenopathy but not in AIDS patients with encephalitis. J Infect Dis 165:1168–1170

Capron A, Dessaint JP (1988) Vaccination against parasitic diseases:some alternative concepts for the definition of protective antigens. Ann Inst Pasteur/Immunol 139:109–115

Cesbron-Delauw MF (1994) Dense-granule organelles of Toxoplasma gondii: their role in the host-parasite relationship. Parasitol Today 10:293–296

Curiel TJ, Krug EC, Purner MB, Poignard P, Berens RL (1993) Cloned human CD4$^+$ cytotoxic T lymphocytes specific for Toxoplasma gondii lyse tachyzoite-infected target cells. J Immunol 151:2024–2031

Darcy F, Deslée D, Santoro F, Charif H, Auriault C, Decoster A, Duquesne V, Capron A (1988) Induction of a protective antibody-dependent response against toxoplasmosis by in vitro excreted/secreted antigens from tachyzoites of Toxoplasma gondii. Parasite Immunol 10:53–60

Del Prete GF, De Carli M, Mastromauro C, Biagiotti R, Macchia D, Falagiani P, Ricci M, Romagnani S (1991) Purified protein derivative of Mycobacterium tuberculosis and excretory-secretory antigen(s) of Toxocara canis expand in vitro human T cells with stable and opposing (type I T helper and type II T helper) profile of cytokine production. J Clin Invest 88:346–350

De Paoli P, Basaglia G, Gennari D, Crovatto M, Modolo ML, Santini G (1992) Phenotypic profile and functional characteristics of human gamma and delta T cells during acute toxoplasmosis. J Clin Microbiol 30:729–731

Gazzinelli RT, Oswald IP, James SL, Sher A (1992) IL-10 inhibits parasite killing and nitrogen oxide production by IFN-γ-activated macrophages. J Immunol 148:1792–1796

Gazzinelli RT, Hieny S, Wynn T, Wolf S, Sher A (1993) Interleukin-12 is required for the T-lymphocyte independent induction of interferon-γ by an intracellular parasite and induces resistance in T-cell deficient hosts. Proc Natl Acad Sci 90:6115–6118

Gazzinelli RT, Bala S, Stevens R, Baseler M, Wahl L, Kovacs J, Sher A (1995) HIV infection suppresses type 1 lymphokine and IL-12 responses to Toxoplasma gondii but fails to inhibit the synthesis of other parasite-induced monokines. J Immunol 155:1565–1574

Haanen JBAG, de Waal Malefijt R, Res PCM, Kraakman EM, Ottenhoff THM, de Vries RRP, Spits H (1991) Selection of a human T helper type 1-like T cell subset by mycobacteria. J Exp Med 174:583–592

Khan IA, Smith KA, Kasper LH (1990) Induction of antigen-specific human cytotoxic T cells by Toxoplasma gondii. J Clin Invest 85:1879–1886

Luft B, Remington JS (1992) Toxoplasmic encephalitis in AIDS. Clin Infect Dis 15:211–219

Maggi E, Parronchi P, Manetti R, Simonelli C, Piccinni M-P, Santoni Rugiu F, De Carli M, Ricci M, Romagnani S (1992) Reciprocal regulatory effects of IFN-γ and IL-4 on the in vitro development of human Th1 and Th2 clones. J Immunol 148:2142–2147

Maggi E, Mazzetti M, Ravina A, Annunziato F, De Carli M, Piccinni MP, Manetti R, Carbonari M, Pesce AM, Del Prete G, Romagnani S (1994) Ability of HIV to promote a TH1 to TH0 shift and to replicate preferentially in TH2 and TH0 cells. Science 265:244–248

Manetti R, Parronchi P, Giudizi MG, Piccinni M-P, Maggi E, Trinchieri G, Romagnani S (1993) Natural killer cell stimulatory factor (interleukin-12 [IL-12]) induces T helper type 1 (Th1)-specific immune responses and inhibits the development of IL-4 producing Th cells. J Exp Med 177:1–14

Mosmann TT, Cherwinski H, Bond MW, Giedlin MA, Coffman RL (1986) Two types of murine helper T-cell clones. I. Definition according to profiles of lymphokine activities and secreted proteins. J Immunol 136:2348–2357

Nathan CF, Murray HW, Wiebe MF, Rubin BY (1983) Identification of interferon-gamma as the lymphokine that activates human macrophage oxidative metabolism and antimicrobial activity. J Exp Med 158:670–688

Pinon JM, Foudrinier F, Mougeot G, Marx C, Aubert D, Toupance O, Niel G, Danis M, Camerlynck P, Remy G, Frottier J, Jolly D, Bessieres MH, Richard-Lenoble D, Bonhomme A (1995) Evaluation of risk and diagnostic value of quantitative assay for anti-Toxoplasma gondii immunoglobulin A (IgA), IgE, and IgM and analytical study of specific IgG in immunodeficient patients. J Clin Microbiol 33:878–884

Prigione I, Facchetti P, Ghiotto F, Tasso P, Pistoia V (1995) Toxoplasma gondii-specific CD4+ T cell clones from healthy, latently infected humans display a Th0 profile of cytokine secretion. Eur J Immunol 25:1298–1305

Salgame P, Abrams JS, Clayberger C, Goldstein H, Convi J, Modlin RL, Bloom BR (1991) Differing lymphokine profiles of functional subsets of human CD4 and CD8 T cell clones. Science 254:279–282

Sher A, Oswald IP, Hieny S, Gazzinelli RT (1993) Toxoplasma gondii induces a T-independent IFN-γ response in NK cells which requires both adherent accessory cells and TNF-α. J Immunol 150:3982–3989

Sklenar I, Jones TC, Alkan S, Erb P (1986) Association of symptomatic human infection with Toxoplasma gondii with imbalance of monocytes and antigen-specific T-cell subsets. J Infect Dis 153:315–322

Street NE, Schumacher JH, Fong TAT, Bass H, Fiorentino D, Leverah JA, Mosmann TR (1990) Heterogeneity of mouse helper T cells: Evidence from bulk cultures and limiting dilution cloning for precursors of Th1 and Th2 cells. J Immunol 144:1629–1639

Subauste CS, Dawson L, Remington JS (1992) Human lymphokine-activated killer cells are cytotoxic against cells infected with Toxoplasma gondii. J Exp Med 176:1511–1519

Subauste C, Chung JY, Do D, Koniaris AH, Hunter CA, Montoya JG, Porcelli S, Remington JS (1995) Preferential activation and expansion of human peripheral blood γδ T cells in response to Toxoplasma gondii in vitro and their cytokine production and cytotoxic activity against T. gondii-infected cells. J Clin Invest 96:610–619

Suzuki Y, Orellana MA, Schreiber RD, Remington JS (1988) Interferon-gamma:the major mediator of resistance against Toxoplasma gondii. Science 240:516–518

Wong SY, Remington JS (1993) Biology of Toxoplasma gondii. AIDS 7:299–316

Toxoplasma Proteins Recognized by Protective T Lymphocytes

H.G. Fischer, G. Reichmann, and U. Hadding

1 Introduction:
T Cell Response Against *T. gondii*

As an intracellular parasite, *Toxoplasma gondii* evades direct attack by the humoral immune response. Host defense reactions which efficiently strike the pathogen within the parasitophorous vacuole are cell-mediated and result either in lysis of the host cell or in its activation to inhibit intracellular parasite growth. Several populations of immune cells are crucially involved in both effector pathways: natural killer cells, $CD8^+$ T lymphocytes and $CD4^+$ T cells of the helper 1 (Th1) subtype. Each of these cell types is able to produce interferon-γ (IFN-γ), the prerequisite mediator of protective immunity to *T. gondii*. IFN-γ triggers antiparasitic activity in macrophages and other cells. The relevance of immune-mediated host cell lysis for the control of acute or chronic infection still needs to be clarified.

Evidence for the participation of both the $CD4^+$ and the $CD8^+$ T cell compartment in *T. gondii*-induced immunity has been obtained from depletion experiments in the murine system. Additional evidence comes from adoptive transfer experiments with isolated parasite-specific T cells. Furthermore, the detection of T cell cytokine expression directly proved that T cell activation

Institut für Medizinische Mikrobiologie und Virologie, Heinrich-Heine-Universität, Universitätsstraße 1, 40225 Düsseldorf, Germany

occurs during the acute phase and also in the course of progressive toxo-plasmic encephalitis. Whereas activation of Th2 cells apparently characterizes mucosal immune reaction and accompanies brain pathology, it is generally accepted that the induction of a Th1-polarized cellular immune response con-fers resistance against *T. gondii*.

2 *Toxoplasma* Proteins Known to Elicit T Cell-Dependent Immunity

Antigen-directed T cell activation is major histocompatibility complex (MHC)-restricted and based on the cognate cell surface interaction between an antigen-presenting cell and a T lymphocyte whose receptor matches the presented antigen. As a rule, CD4$^+$ T cells recognize MHC class II, CD8$^+$ T cells MHC class I molecules which bear the processed antigenic peptide within the binding groove.

Several *T. gondii* proteins of both membraneous and cytoplasmic origin have been demonstrated to induce antigenic activation of T cells (Table 1). Interestingly, four of these five antigens which exhibit T cell stimulatory proper-ties are secretory products of the parasite. Besides, a toxoplasmic superantigen has been detected in the mouse by the expansion of Vβ5 T cell receptor-ex-pressing CD8$^+$ cells (DENKERS et al. 1994).

2.1 Surface Antigen 1 (SAG1/P30)

SAG1, probably the most immunogenic *Toxoplasma* antigen, induces strong T and B cell responses in infected animals, and also IFN-γ production in cultured human lymphocytes (KHAN et al. 1988). Successful vaccination of mice was achieved by using the purified antigen combined with the adjuvant QuilA or integrated into liposomes via the subcutaneous and intraperitoneal immuni-zation route, respectively (KHAN et al. 1991; BÜLOW and BOOTHROYD 1991). A SAG1-specific CD8$^+$ T cell clone from such animals exhibited cytolytic activity against target cells infected with *T. gondii* (KASPER et al. 1992) and was protective when transferred into nonimmune recipient mice (KHAN et al. 1994). In infected rats, most SAG1-reactive T cells recognized peptide 238–256, which is distinct from the B cell epitopes mapped on this molecule (GODARD et al. 1994).

Table 1. *Toxoplasma* proteins recognized by T cells and mediating protective effects in vitro or in vivo

Antigen	Preparation used	Antigen-specific T cells challenged	T cell-dependent protective effects	References
SAG1	Affinity-purified SAG1/P30	Murine CD8+ splenocytes or T cell clone	Cytolytic activity for infected macrophages	KASPER et al. 1992
		Murine CD8+ splenocytes or T cell clone	Reduced mortality and brain cyst number following vaccination or adoptive transfer	KHAN et al. 1991 KHAN et al. 1994
	SAG1/P30 peptides and multiple peptide construct	Rat lymph node and spleen cells	Prolonged survival and induction of specific antibodies in nude rats following adoptive transfer	DARCY et al. 1992
			Induction of specific antibodies following vaccination	GODARD et al. 1994
ROP2	Recombinant ROP2/P54	Human Th1 clone	Release of IFN-γ and IL-2	SAAVEDRA and HÉRION 1991 SAAVEDRA et al. 1991
GRA1	GRA1/P24 peptide	Rat lymph node cells	Reduced mortality of athymic rats following adoptive transfer	DUQUESNE et al. 1991
GRA2	Affinity-purified GRA2/P28	Murine CD4+ splenocytes	Prolonged survival and induction of antibodies following vaccination	BRINKMANN et al. 1993
GRA4	Purified GRA4/P40, GRA4/P40 peptide	Murine mesenteric and splenic T cells	(Not determined)	CHARDÈS et al. 1993

2.2 Rhoptry Protein 2 (ROP2/P54)

A panel of human *Toxoplasma*-specific T cell clones which produce interleukin-2 (IL-2) and IFN-γ has been isolated from a chronically infected donor. One of these CD4+ clones recognized recombinant P54 in context with human leukocyte antigen (HLA)-DPw4 (SAAVEDRA and HÉRION 1991; SAAVEDRA et al. 1991). The antigen was identical with ROP2 based on its subcellular localization and immunoblot analyses (HÉRION et al. 1993). Besides the respective T cell epitope on the ROP2 peptide 197–215, two other epitopes have been defined on peptides 393–410 and 501–524, against which human T cell reactivity is directed (SAAVEDRA et al. 1995).

2.3 Dense Granule Proteins

Several *T. gondii* dense granule antigens (GRA) display T cell immunogenicity in rodents. By using a panel of GRA1 peptides to stimulate T lymphocytes from infected rats, three T cell epitopes could be distinguished. The adoptive transfer of T cells specific for peptide 170–193 into athymic rats resulted in a long lasting immunity – meaning protection against twofold reinfection (Du-QUESNE et al. 1991).

Vaccination of mice with purified GRA2 induced complete protection against lethal *Toxoplasma* infection (SHARMA et al. 1984). In parallel, CD4$^+$ T cells were preferentially activated and responded to restimulation with this antigen in vitro (BRINKMANN et al. 1993). The T cell immunogenicity of GRA2 was further confirmed by the detection of primed antigen-specific T cells in *Toxoplasma*-infected animals.

T cell antigenicity of GRA4, another dense granule protein, was shown in a similar way: Purified GRA4 as well as peptide 229–242, which carries a predicted T cell epitope, proved to be stimulatory for mesenteric and splenic T lymphocytes from infected mice (CHARDÈS et al. 1993).

3 Experimental Strategies to Identify Toxoplasmic T Cell Antigens

With regard to future vaccination studies, immunodominant T cell antigens are suitable candidates to be used as a vaccine subunit. Up to now, the identification of such *T. gondii* antigens was only achieved with the aid of defined monoclonal antibodies but alternative strategies have been employed recently (see Sects. 3.2–3.4). In particular, the development of a method for screening recombinant antigens from cDNA expression libraries with T cell clones (see Sect. 3.2) will allow a broad investigation of the parasite-specific T cell repertoire and, as a result, will facilitate the search for the target proteins.

3.1 Use of Antibody-Prescreened Antigen Preparations in T Cell Assays

Poly- or monoclonal antibodies represent an irreplaceable tool for the isolation and purification of the toxoplasmic T cell antigens identified so far. Antigenicity towards T cells has been overall determined by measuring the antigen-specific proliferative response or lytic activity of isolated T cells in co-culture with appropriate antigen-presenting cells. Because the T and B cell repertoires do not necessarily overlap and T and B cell epitopes were often mapped to different sites of a given antigen, some T cell antigens are not detected by antibodies.The failure of infected animals into which protective T cells of defined epitope specificity had been transferred to induce a GRA1-directed antibody response may indicate such difference between the T and B cell repertoire concerned (DUQUESNE et al. 1991). Likewise, a preselection of recombinant antigenic fragments from DNA expression libraries with antibodies will limit the pool of antigens recognized in subsequent screening with T cells.

3.2 Direct Screening of Recombinant Proteins with T Cell Clones

Recently, a *Leishmania* antigen has been identified by using T cells as a primary probe to screen an epitope-tagged cDNA expression library (MOUGNEAU et al. 1995). In this system, large pools of different recombinant antigens could be tested at once following purification. Furthermore, in subsequent T cell stimulation assays, antigen uptake by the presenting cells was facilitated by adding a monoclonal antibody directed against the fusion partner protein. Although methodically extensive, this approach circumvents the antibody-caused restriction for probing T cell responses.

3.3 T Cell Blot Analysis of Electrophoretically Separated Proteins

In order to distinguish separated antigenic fractions from complex protein mixtures, e.g., crude microbial lysates, T cell blot techniques have been used. For example, T cell stimulatory activity has been detected on nitrocellulose particles bearing the P54 antigen (SAAVEDRA et al. 1991). Furthermore, testing of antigenic fractions electroeluted into soluble phase allowed us to determine the molecular weights of the antigens recognized by four distinct *Toxoplasma*-specific Th1 clones (Table 2). Antigen preparations obtained by both procedures must not impair the subsequent T cell bioassay. While T cell blotting has been applied so far as a step of antigen purification or characterization, the method by itself does not allow the identification of separated antigens. Two-dimensional gel electrophoresis of antigenic material prior to its electroelution results in maximum resolution (GULLE et al. 1990), which in combination with microsequencing technique may replace the monoclonal antibodies needed so far to define isolated antigens.

Table 2. Clonotypic specificities of *Toxoplasma*-reactive murine CD4[+] Th1 clones

Clone[a]	T cell receptor		Characteristics of the *T. gondii* antigen recognized		
	Vαβ[b]	MHC class II restriction element	Strains[c]	Stages	MW[d] (kDa)
3TxS	Vα2	I-A$_\alpha$[b]A$_\beta$[k]	v,a	tachyzoites, bradyzoites	26
3Tx11	Vβ14	I-A[b]	v	tachyzoites	58
3Tx14	Vα11/Vβ6	I-E[k]	v,a	tachyzoites, bradyzoites	55
3Tx15	not determined	I-E[k]	v,a	tachyzoites, bradyzoites	40

MHC, major histocompatibility complex; MW, molecular weight
[a] All clones were isolated from (B10×C3H)F$_1$ mice, and belong to the Th1 subtype due to their strong production of IFN-γ, TNF-α and IL-2 activity, while lacking IL-4 or IL-10 synthesis. These T cells mediate toxoplasmastasis in co-culture with syngeneic macrophages and viable tachyzoites (cf. Sect. 4).
[b] As detected with monoclonal antibodies directed against distinct V-elements of αβ T cell receptor.
[c] Mouse-virulent (v) and/or mouse-avirulent (a) *T. gondii* strains, according to the classification in PARMLEY et al. (1994), provided the respective T cell antigen.
[d] As determined in T cell blot analyses of crude tachyzoite lysate which had been fractionated after reducing sodium dodecyl sulfate-polyacrylamide gel electrophoresis.

3.4 Sequence Analysis of T Cell Stimulatory Peptides Isolated from MHC Molecules

An inverse approach to find toxoplasmic T cell antigens has been described recently. AOSAI and coworkers (1994) analyzed antigenic peptides from the MHC class I restriction element for a *T. gondii*-specific human CD8[+] cytotoxic T cell line. Peptides with a length of up to nine amino acids were isolated from HLA-A2 molecules of the antigen-presenting infected target cell, checked for antigenicity and sequence analyzed. Whereas the determined sequences partially matched the predicted binding motif of the corresponding MHC molecule, no homology to three selected toxoplasmic antigens, GRA1, GRA2 and SAG1, was found. Nevertheless, this procedure may lead to the identification of other, possibly unknown antigens for *T. gondii*-specific T lymphocytes.

4 Measurement of T Cell-Dependent Protective Effects In Vitro and in Animals

An antigen-directed cooperation between *T. gondii*-specific human or murine T lymphocytes and infected partner cells was studied in vitro by co-cultures, where CD8[+] or CD4[+] T cell clones mediated lysis of syngeneic infected target macrophages (CURIEL et al. 1993; KASPER et al. 1992; PRIGIONE et al. 1995). Furthermore, the MHC class II-restricted presentation of *Toxoplasma* antigen to murine CD4[+] Th1 clones (cf. Table 2) was found to result in an inhibition of intracellular parasite growth which correlates to the degree of T cell activation (H.G. FISCHER et al., submitted). Such an in vitro system allows the detailed analysis of bidirectional cell–cell interaction and rapid testing of protective capacity of isolated *T. gondii*-specific Th cells as well.

T cell-mediated protection in vivo was determined by a reduction of mortality and/or brain cyst number following adoptive transfer of isolated T cells into infected animals. In the experiments reported, this classical test led to immune resistance. In comparison, the success of vaccination using T cell-recognized antigens obviously varies with the immunization route and the adjuvant employed (KASPER et al. 1985; BÜLOW and BOOTHROYD 1991; KHAN et al. 1991).

Acknowledgements. The authors are grateful to U. Bonifas, K. Buchholz and B. Nitzgen for expert technical assistance and J.F. Dubremetz, U. Gross and H.M. Seitz for providing us with *T. gondii* strains, respectively. The research described here was supported by the Deutsche Forschungsgemeinschaft project B11 SFB 194 and by BMBF grant 01 KI 9455.

References

Aosai F, Yang TH, Ueda M, Yano A (1994) Isolation of naturally processed peptides from a Toxoplasma gondii-infected human B lymphoma cell line that are recognized by cytotoxic T lymphocytes. J Parasitol 80:260–266

Brinkmann V, Remington JS, Sharma SD (1993) Vaccination of mice with the protective F3G3 antigen of Toxoplasma gondii activates CD4[+] but not CD8[+] T cells and induces Toxoplasma specific IgG antibody. Mol Immunol 30:353–358

Bülow R, Boothroyd JC (1991) Protection of mice from fatal Toxoplasma gondii infection by immunization with P30 antigen in liposomes. J Immunol 147:3496–3500

Chardès T, Velge-Roussel F, Mevelec P, Mevelec MN, Buzoni-Gatel D, Bout D (1993) Mucosal and systemic cellular immune responses induced by Toxoplasma gondii antigens in cyst orally infected mice. Immunol 78:421–429

Curiel TJ, Krug EC, Purner MB, Poignard P, Berens RL (1993) Cloned human CD4[+] cytotoxic T lymphocytes specific for Toxoplasma gondii lyse tachyzoite-infected target cells. J Immunol 151:2024–2031

Darcy F, Maes P, Gras-Masse H, Auriault C, Bossus M, Deslee D, Godard I, Cesbron MF, Tartar A, Capron A (1992) Protection of mice and nude rats against toxoplasmosis by a multiple antigenic peptide construction derived from Toxoplasma gondii P30 antigen. J Immunol 149:3636–3641

Denkers EY, Caspar P, Sher A (1994) Toxoplasma gondii possesses a superantigen activity that selectively expands murine T cell receptor Vβ5-bearing CD8[+] lymphocytes. J Exp Med 180:985–994

Duquesne V, Auriault C, Gras-Masse H, Boutillon C, Darcy F, Cesbron-Delauw MF, Tartar A, Capron A (1991) Identification of T cell epitopes within a 23-kDa antigen (P24) of Toxoplasma gondii. Clin Exp Immunol 84:527–534

Godard I, Estaquier J, Zenner L, Bossus M, Auriault C, Darcy F, Gras-Masse H, Capron A (1994) Antigenicity and immunogenicity of P30-derived peptides in experimental models of toxoplasmosis. Mol Immunol 31:1353–1363

Gulle H, Schoel B, Kaufmann SHE (1990) Direct blotting with viable cells of protein mixtures separated by two-dimensional gel electrophoresis. J Immunol Meth 133:253–261

Hérion P, Hernández-Pando R, Dubremetz JF, Saavedra R (1993) Subcellular localization of the 54-kDa antigen of Toxoplasma gondii. J Parasitol 79:216–222

Kasper LH, Currie KM, Bradley MS (1985) An unexpected response to vaccination with a purified major membrane tachyzoite antigen (P30) of Toxoplasma gondii. J Immunol 134:3426–3431

Kasper LH, Khan IA, Ely KH, Bülow R, Boothroyd JC (1992) Antigen-specific (P30) mouse CD8+ T cells are cytotoxic against Toxoplasma gondii-infected peritoneal macrophages. J Immunol 148:1493–1498

Khan IA, Eckel ME, Pfefferkorn ER, Kasper LH (1988) Production of γ interferon by cultured human lymphocytes stimulated with a purified membrane protein (P30) from Toxoplasma gondii. J Infect Dis 157:979–984

Khan IA, Ely KH, Kasper LH (1991) A purified parasite antigen (P30) mediates CD8+ T cell immunity against fatal Toxoplasma gondii infection in mice. J Immunol 147:3501–3506

Khan IA, Ely KH, Kasper LH (1994) Antigen-specific CD8+ T cell clone protects against acute Toxoplasma gondii infection in mice. J Immunol 152:1856–1860

Mougneau E, Altare F, Wakil AE, Zheng S, Coppola T, Wang ZE, Waldmann R, Locksley RM, Glaichenhaus N (1995) Expression cloning of a protective Leishmania antigen. Science 268:563–566

Parmley SF, Gross U, Sucharczuk A, Windeck T, Sgarlato GD, Remington JS (1994) Two alleles of the gene encoding surface antigen P22 in 25 strains of Toxoplasma gondii. J Parasitol 80:293–301

Prigione I, Facchetti P, Ghiotto F, Tasso P, Pistoia V (1995) Toxoplasma gondii-specific CD4+ T cell clones from healthy, latently infected humans display a Th0 profile of cytokine secretion. Eur J Immunol 25:1298–1305

Saavedra R, Hérion P (1991) Human T-cell clones against Toxoplasma gondii: production of interferon-γ, interleukin-2, and strain cross-reactivity. Parasitol Res 77:379–385

Saavedra R, de Meuter F, Decourt JL, Hérion P (1991) Human T cell clone identifies a potentially protective 54-kDa protein antigen of Toxoplasma gondii cloned and expressed in Escherichia coli. J Immunol 147:1975–1982

Saavedra R, Becerril MA, Dubeaux C, Lippens R, Hérion R, Bollen A (1995) Identification of T-cell epitopes in the ROP2 protein antigen of Toxoplasma gondii. 9th International Congress of Immunology, 23–29 July 1995, San Francisco, p 347 (abstract)

Sharma SD, Araujo FG, Remington JS (1984) Toxoplasma antigen isolated by affinity chromatography with monoclonal antibody protects mice against lethal infection with Toxoplama gondii. J Immunol 13:2818–2820

Immunological Control of *Toxoplasma gondii* and Appropriate Vaccine Design

J. Alexander[1], H. Jebbari[1], H. Bluethmann[2], A. Satoskar[1], and C.W. Roberts[1,3]

1 Introduction

Toxoplasma gondii, by all conceivable ecological and epidemiological measurements and analysis, has to be regarded as an extremely successful parasite. It has been found in high incidence in most warm blooded vertebrates and has a generally global distribution. Approximately 30% of the world's human population is infected, although the prevalence varies markedly from country to country (Jackson and Hutchison 1989).

Toxoplasmosis is of major clinical and veterinary importance, being a cause of congenital disease and abortion in humans and their domestic animals. Post-natally acquired infection, although normally mild to asymptomatic, can be severe to life-threatening in immunosuppressed individuals (Luft and Remington 1992; Ambroise-Thomas and Pelloux 1993). Generally, following recovery from an early mild to subclinical phase of infection, associated with the rapidly dividing tachyzoite stage, individuals are thought to harbour cysts containing bradyzoites, particularly within skeletal and heart muscle and the central nervous system (CNS), for life. This results in a state of premunition and immunity to reinfection which is presumed to be lifelong. This assumption of life-long

---

[1]Department of Immunology, University of Strathclyde, The Todd Center, 31 Taylor Street, Glasgow G4 0NR, UK
[2]F. Hoffmann La Roche, 4002 Basel, Switzerland
[3]Present Address: Division of Infectious Diseases, Michael Reese Hospital and Medical Center, 2429 South Ellis Avenue, Chicago, Jl 60616–3390, USA

immunity to reinfection is reinforced by the remarkable clonality of *T. gondii* (DARDÉ et al. 1992; SIBLEY and BOOTHROYD 1992; SIBLEY et al. 1994; L.D. SIBLEY and D.K. HOWE, this volume). Thus, despite the existance of a meiotic phase of the life cycle in cats, only a few clonal lineages have been identified with little evidence of the recombination that would be expected if cats were to feed on hosts which had succumbed to more than one *T. gondii* infection.

The challenge, therefore, for the immunologist is firstly to characterise the protective immune response generated initially by a primary acute infection and thereafter maintained by a chronic infection, secondly to identify those antigens from the various life-cycle stages to which the immune response is generated, and finally to use this knowledge to develop appropriate vaccine strategies. While vaccines to prevent foetal damage and abortion would be of the utmost medical and veterinary importance, those capable of limiting the number of cysts in meat products or reducing oocyst shedding by cats would also have significant beneficial public and veterinary health implications. The following review summarises the rapid progress that has been made in our understanding of the immunological control of *T. gondii*. Using this information highlights the problems, as well as indicating some solutions, in designing suitable vaccines.

2 The Immune Response

2.1 Protective Immune Response

Following oral ingestion of oocysts or cysts, *T. gondii* infection is characterised by an early acute phase of infection followed by a long-lasting, probably lifelong chronic phase, each associated with a different life-cycle stage (DUBEY 1994). Thus, while tachyzoites comprise early infection, multiplying largely within mononuclear phagocytes in almost every tissue of the host, the later chronic stage consists of encysted bradyzoites predominantly found in muscle and the CNS. Studies, particularly in mice, indicate that immunity during the different phases of infection is under different genetic (MCLEOD et al. 1989; BLACKWELL et al. 1993) and changing immunoregulatory controls (GAZZINELLI et al. 1993a). Thus there is little, if any, correlation between mortality during early infection and the eventual cyst burdens during chronic infection in the wide variety of mouse strains examined to date (MCLEOD et al. 1989; BLACKWELL et al. 1993). Survival in the acute phase of infection has been shown to be under the influence of at least five different genes, one of which is located in the major histocompatibility complex (MHC). The innate immune response involving the interaction of natural killer (NK) cells and macrophages plays a vital role during this stage of infection (GAZZINELLI et al. 1993a; HUNTER and REMINGTON 1994). The general consensus of opinion would suggest that *T. gondii* induces tumour necrosis factor alpha (TNF-α) and interleukin 12 (IL-12) production by macrophages which together induce NK cells to release interferon gamma (IFN-γ) (GAZZINELLI et al. 1993b). This cytokine, probably in synergy with macrophage-produced TNF-α (SIBLEY et al. 1994), activates macrophages to kill parasites by a combination of oxygen-dependent and -independent mechanisms (PFEFFERKORN and GUYRE 1984; HUGHES 1988; LANGERMANS et al. 1992). Much of the information to date has been derived from in vivo and in vitro studies using recombinant cytokines and neutralising cytokine antibodies. However, the current availability of mice genetically depleted of cytokines or their receptors now allows clear confirmation of the contribution each cytokine makes

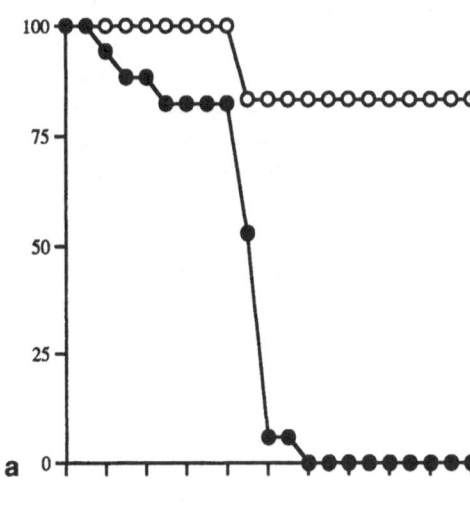

Fig. 1. The percentage survival of IFN-γR or TNF-αR1 disrupted mice and their wild-type counterparts following infection orally with 20 *T.gondii* cysts (RRA strain). *Top:* Age and sex-matched IFN-γR$^{-/-}$ (*closed circles*) and IFN-γR$^{+/+}$ (*open circles*) 129SV mice. *Bottom:* Age and sex-matched TNF-αR1$^{-/-}$ (*closed circles*) and TNF-αR1$^{+/+}$ (*open circles*) 129 SVJ mice

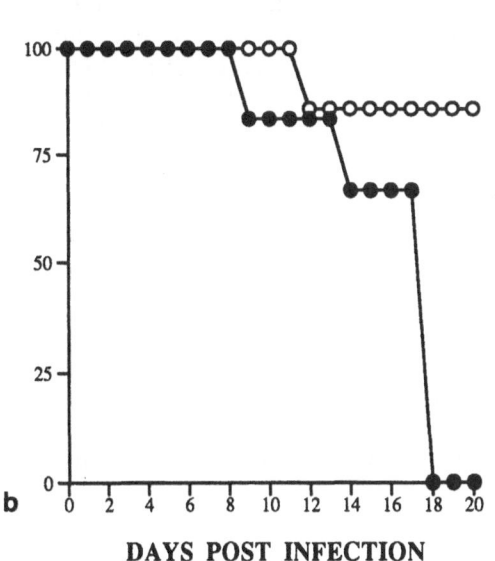

% SURVIVAL

DAYS POST INFECTION

to the development of immunity. The protective effects of IFN-γ and TNF-α during the early stages of infection can be clearly seen in those mice lacking the IFN-γR (Fig. 1a) and TNF-αR1 (Fig. 1b). All gene-disrupted mice, but not the control mice, die very rapidly after infection. Significantly, virtually all IFN-γR "knockout" mice die before there are any deaths in TNF-αR1 "knockout" mice, indicating that a possible sequential importance in the production of these cytokines exists.

The results appear to confirm the observations of ROTHE et al. (1993) that signalling through both TNF-αR1 and IFN-γR is essential for the non-specific effector phase against intracellular pathogens. Nevertheless, a detrimental role for TNF-α during *T. gondii* infection has been suggested (GRAU et al. 1992). This poses the intriguing possibility that detrimental effects are a result of signalling through TNF-αR2. This is an area of future investigation in our laboratory.

The pro-inflammatory cytokines IL-12 (HSIEH et al. 1993; SCOTT 1993) and IFN-γ (GAJEWSKI and FITCH 1991) have also been clearly demonstrated as playing crucial co-stimulatory roles in the developing adaptive immune response. This is due to their triggering of the expansion of the Th1 subset of CD4$^+$ lymphocytes which in turn produces IL-2 to drive the expansion of CD8$^+$ lymphocytes (DENKERS et al. 1993b). IL-12 has also been shown to promote type 1 (IFN-γ producing) cytolytic CD8$^+$ T cell development (CROFT et al. 1994), further emphasising the potential role of the innate immune system in directing the evolution of the adaptive immune response. Adoptive transfer and in vivo depletion studies in various mouse strains confirm the paramount importance of CD8$^+$ T cells in mediating immunity, not only during acute infection but also in preventing high cyst burdens and toxoplasmic encephalitis (BROWN and MCLEOD 1990; GAZZINELLI et al. 1991; KHAN et al. 1991; PARKER et al. 1991; SUBAUSTE et al. 1991). In fact, consistent with the demonstration of the importance of CD8$^+$ T cells in controlling the late cyst stage phase of infection, genetic control has been clearly mapped to the MHC genes at the L locus encoding the class I molecules (BLACKWELL et al. 1993; BROWN et al. 1995). It has been suggested that CD8$^+$ T cells may mediate their effect through IFN-γ production (GAZZINELLI et al. 1993a). This cytokine not only induces macrophage microbicidal activity (SUZUKI et al. 1988; LANGERMANS et al. 1992), but may promote the expression of bradyzoite antigens (BOHNE et al. 1993) as well as preventing cyst rupture (MCCABE et al. 1984). A role for CD8$^+$ T cells in class I restricted cytolytic responses has also been demonstrated in humans (YANO et al. 1989) and mice (DENKERS et al. 1993a,b). Furthermore, CD8$^+$ T cells appear to have direct tachyzoite cytolytic activity (KHAN et al. 1988). However, a role for the CD4$^+$ lymphocyte population cannot be discounted as MHC-controlled cytolytic activity has also been demonstrated by human CD4$^+$ T cell clones (CURIEL et al. 1993), and the Th1 subset has been implicated as playing a synergistic role in CD8$^+$ T cell-mediated protective immunity in mice (GAZZINELLI et al. 1991; PARKER et al. 1991).

2.2 Exacerbative Immune Response

It is currently accepted and logical, given the information summarised above, that in general the Th2 subset of $CD4^+$ T cells and their products such as IL-4 (HUNTER et al. 1992; VILLIARD et al. 1995), IL-10 (GAZZINELLI et al. 1992), and IL-6 (BEAMAN et al. 1994; SUZUKI et al. 1994) which downregulate Th1 cell responses and antagonize the effect of their products, may play disease-promoting roles during *T. gondii* infection. Thus IL-4 and IL-10, which are known to antagonize the production and effects of IFN-γ such as the induction of nitric oxide (GAZZINELLI et al. 1992) and TNF-α production (ESSNER et al. 1989; GAUTUM et al. 1992), have been implicated as contributory factors leading to death in the acute phases of infection. IL-4 also favours the expansion of Type 2 or "Th2-like" $CD8^+$ T cells, which are not cytolytic (CROFT et al. 1994). Consequently, in our laboratories we have demonstrated the early presence of mRNA transcripts for IL-4 (HUNTER et al. 1992) in the brains of mice with progressive toxoplasmic encephalitis while IL-10 has been shown to be present in higher concentrations in the lymph nodes and CNS of mice susceptible to toxoplasmic encephalitis than their resistant counterparts (HUNTER et al. 1994).

2.3 Immune Response – Unresolved Problems

Several recent studies (McLEOD et al. 1989; BURKE et al. 1994; DENKERS et al. 1994; ROBERTS et al. 1994) indicate that the dichotomy between a developing protective and a disease-promoting response may not be quite as polarized as suggested by the work summarised above. Overproduction or inappropriate production of certain pro-inflammatory mediators such as IFN-γ (McLEOD et al. 1989; DENKERS et al. 1994) and TNF-α (BLACK et al. 1990) may increase mortality and/or morbidity. Thus McLEOD et al. (1989) observed that splenocytes isolated from mouse strains genetically susceptible to acute mortality produced more IFN-γ when stimulated with *T. gondii* antigen than splenocytes from resistant strains. Similarly, DENKERS and co-workers (1994) found that those mice most responsive to *T. gondii* superantigen-driven IFN-γ production also had the highest mortality levels. Overproduction of IFN-γ and TNF-α have also been associated with mortality during murine malaria infection (TAVERNE 1993). These cytokines individually and collectively stimulate macrophages to produce reactive oxygen and nitrogen intermediates (SIBLEY et al. 1991; LANGERMANS et al. 1992) which, while important in killing parasites can also be toxic to the host if produced in large quantities (CLARK et al. 1991). Consequently, IL-4 and IL-10 production may not be completely detrimental to the host as these cytokines may act directly on the macrophage and play an important role in regulating potentially dangerous inflammatory responses. Thus, IL-10-deficient mice fail to survive the early phase of a *T. gondii* infection (GAZZINELLI et al. 1996) while significantly increased mortality during acute infection occurs in IL-4-deficient mice compared with their wild-type counterparts (Table 1) (ROBERTS et al. 1995).

Table 1. Comparison of percentage survival over 15 days of male or female IL-4$^{-/-}$ and IL-4$^{+/+}$ mice infected orally with 20 *T. gondii* cysts (RRA strain)

Experiment	sex	% Survival IL-4$^{+/+}$	IL-4$^{-/-}$
1	female	83.3	0
2	female	50	23.1
3	male	78.9	73.9
4	male	66.6	66.6
5	male	100	83.3
6	male	87.5	80.0

Survival significantly greater in the wild-type IL-4$^{+/+}$ mice ($p<0.025$) using the Mantel Haenzal test.

In addition, during the later stages of chronic infection and only following a massive reduction in brain cyst numbers in *T. gondii* infected B10 mice, mRNA transcripts for IL-10 are found within the brain (BURKE et al. 1994). It is proposed that IL-10 downregulates inflammatory mediators and thus limits pathology in a similar manner to that described during the resolution of experimental autoimmune encephalitis (KENNEDY et al. 1992).

The recent emphasis on cell-mediated immunological control of *T. gondii* has tended to ignore the possibility that antibodies may also influence the course of infection. There are several examples of immunity being passively transferred by antibodies or adoptively transferred by B cells (PAVIA et al. 1992). Immunoglobulin A (IgA) in particular may play a role in protecting mucosal surfaces from parasite invasion (CHARDÈS et al. 1993, McLEOD et al. 1993). Furthermore, it has been clearly demonstrated that antibodies can influence interactions between tachyzoites and macrophages; by binding to the parasite they have been shown under certain circumstances to inhibit uptake by the host cell (GRIMWOOD and SMITH 1992), but if ingestion takes place they have been shown to allow lysosomes to fuse to the parasitophorous vacuole (JONES et al. 1975) thus killing the intracellular parasites.

3 Stage-Specific Antigens and Vaccination

It is widely recognised that immunity to a disease such as malaria, in which more than one life-cycle stage is present within the host, is stage-specific (PHILLIPS 1994). Not only are different immunological mechanisms directed against different stages but these responses are directed against stage-specific immunodominant antigens. Is this also true for *T. gondii* infections? Certainly the immune responses protecting mice against early infection and mortality are not identical to those controlling cyst burdens (McLEOD et al. 1989). Furthermore, recent evidence from murine and human studies would indicate that the tachyzoite and bradyzoite cyst antigens recognised by serological techniques are largely stage-specific with little recognition of shared antigens (KASPER 1989; ZHANG and SMITH 1995; ZHANG et al. 1995, J. E. SMITH et al., this volume). In fact, overall there is generally very little recognition of cyst antigens and much of the immunological activity is directed against tachyzoite antigens (ZHANG and SMITH 1995; ZHANG et al. 1995).

Not surprisingly there is already strong evidence that stage-specific antigen recognition and stage-specific immunity can influence the success of vaccination. For example, oral immunization and intraduodenal immunization with tissue cysts or bradyzoites of the non-oocyst forming mutant T-263 protected cats totally against oocyst shedding following challenge (FREYRE et al. 1993). Although vaccination by either route with T-263 tachyzoites induced seroconversion, only partial protection against oocyst shedding was provided. The route of vaccine administration may also be crucial in mice. Thus, McLEOD et al. (1988) found that intraintestinal immunization of mice with the temperature-sensitive mutant TS-4, but not subcutaneous immunization, was effective in reducing the incidence of congenital transmission. A number of studies further demonstrated the limitations of using one life-cycle stage as a basis of a vaccine. For example, a live, attenuated tachyzoite vacccine used in sheep (BUXTON et al. 1991) and a soluble tachyzoite vaccine in mice (ROBERTS et al. 1994), while preventing abortion following challenge during pregnancy, did not prevent vertical disease transmission and cyst formation. In addition, a further study immunizing mice with tachyzoite antigens mostly P30 (SAG1) and P22 (SAG2), in immunostimulatory complexes (ISCOMs) (LUNDEN et al. 1993) protected against challenge with this life-cycle stage and also against oocysts but was not protective against challenge with tissue cysts. Consequently, we compared the protective ability of vaccines comprising equal quantities of either killed tachyzoites or cysts or both to protect mice from developing large cyst burdens following challenge. Only the cyst and the cyst/tachyzoite antigen mix significantly reduced brain cyst burdens (Fig. 2). Preliminary results from western blot analysis of plasma sampled from the mice in these experiments indicate that vaccination with different life-cycle stages or mixes of different life-cycle stages alters the hierarchy of antigen recognition during active infection (H. JEBBARI et al., unpublished). This will undoubtedly modify the outcome of disease.

Apart from the results reported here, all vaccine studies to date against toxoplasmosis in the intermediate host have utilised attenuated *T. gondii* tachyzoites or antigens and subunits derived from this life-cycle stage. While foetal death and abortion have been attributed to rapidly dividing tachyzoites, the CNS lesions and chorioretinitis observed as a result of congenital infection are caused directly by the cysts containing bradyzoites or by reactivation of the disease from this so-called dormant stage (reviewed in JACKSON and HUTCHISON 1989). In a similar manner, fatal toxoplasmic encephalitis often occurs in transplant and AIDS patients also as a result of reactivation from this dormant

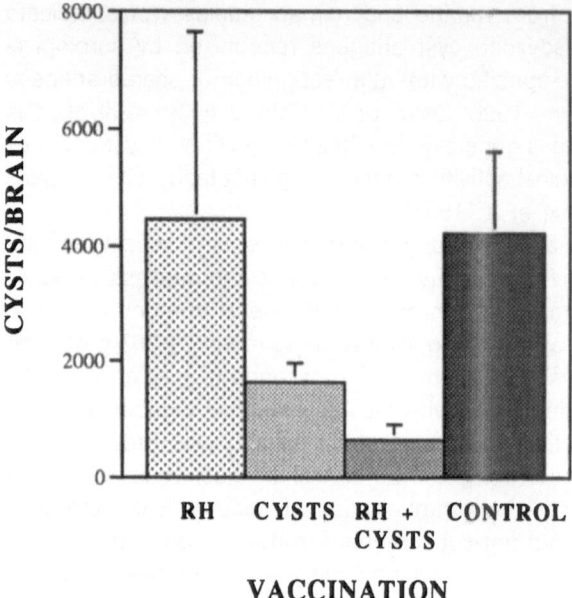

Fig. 2. The brain cyst burdens (± standard error) in female BALB/K mice on day 28 post-infection with 20 *T. gondii* cysts orally. Mice were vaccinated subcutaneously 4 weeks and 2 weeks before infection with either killed 5×10^5 RH tachyzoites or 1000 killed cysts or a mixture of 2.5×10^5 tachyzoites and 500 cysts. Antigen was emulsified with Freund's Complete Adjuvant followed by Freund's Incomplete Adjuvant for the first and second vaccinations, respectively. Vaccination with cyst antigen and combined cyst/tachyzoite antigen resulted in significantly reduced cyst burdens following infection ($p < 0.05$ and $P < 0.01$ respectively by Mann-Whitney U test)

stage (LUFT and REMINGTON 1992; AMBROISE-THOMAS and PELLOUX 1993). A vaccine capable of inducing immunity to the cyst stage is, therefore, worthy of serious investigation.

4 Appropriate Vaccine Design

A live, attenuated tachyzoite vaccine which successfully prevents abortion (BUXTON 1993) is commercially available for sheep and a similar vaccine has proved reasonably successful experimentally in limiting infection in pigs (DUBEY 1994; DUBEY et al. 1994). Vaccination with live organisms in humans is unlikely. In addition, these vaccines, while perhaps reducing cyst burdens following challenge or natural infection, do not completely prevent this life-cycle stage from being established. In humans, vaccines comprising sub-unit or recombinant antigens or synthetic peptides are more likely to be approved for use. Two tachyzoite antigens, the major immunodominant membrane antigen P30 (SAG1; BÜLOW and BOOTHROYD 1991; KHAN et al. 1991; GODARD et al. 1994) and the secretory antigen P24 (DUQUESNE et al. 1991), but as yet no cyst antigens, have been identified as likely vaccine candidates. Vaccinating rats with P24 or synthetic peptides derived from the primary sequence, reduced mortality in these animals infected with the virulent RH strain (DUQUESNE et al. 1991). Similarly, vaccination with liposomal entrapped P30 (SAG1) or synthetic multi-antigenic peptides derived from it have been shown to increase the survival of mice and rats following challenge infection (BÜLOW and BOOTHROYD 1991; DARCY et al. 1992). However, vaccination with the P30 antigen as well as other antigen preparations can exacerbate infection and increase mortality if used in conjunction with Freund's Complete Adjuvant (FCA; KASPER et al. 1985; ALEXANDER et al. 1993). Nevertheless, there is evidence from the present study that a vaccine regime using FCA can be used successfully in a cyst antigen based preparation. Vaccination with unadjuvanted soluble tachyzoite antigen (STAg) may also be detrimental, as it resulted in increased foetal death in a murine model of congenital infection (ROBERTS et al. 1994). However, vaccination with STAg entrapped in lipid vesicles reduced *T. gondii*-induced abortion in the same study. Thus, antigens may not necessarily be inherently protective or exacerbative, but rather it is how they are presented to the immune system that influences the development of a protective response. Unfortunately, the only adjuvants licensed for use in humans are the aluminium gels which polarize the immune response towards Th2 cell activation (GRUN and MAURER 1989) and do not stimulate cytotoxic T lymphocytes (CTL) responses (LINDBLAD 1995). This adjuvant would, therefore, be totally inappropriate for *Toxoplasma* vaccines. However, FCA does not necessarily promote protective immunity against *T. gondii* (KASPER et al. 1985; ALEXANDER et al. 1993) and yet this adjuvant stimulates Th1 cell activation (GRUN and MAURER 1989), which intuitively should induce protection against *Toxoplasma* infection. Why then have lipoidal vesicles and ISCOMs been continuously more successful as adjuvants in experimental vaccines than FCA? Lipid vesicles (LOPES and CHAIN 1992; REDDY et al. 1992; ZHOU et al. 1992) and ISCOMs (HEEG et al. 1991) have been shown to stimulate $CD8^+$ T cell responses, perhaps by targeting entrapped antigen to macrophages which have been implicated as accessory cells in the generation of $CD8^+$ cytolytic T cells to exogenous antigen (DEBRICK et al. 1991; ZHOU et al. 1992). In contrast, there is little evidence that emulsion systems such as FCA are capable of inducing a similar response (ROBERTS et al. 1994).

The activation of $CD8^+$ T cells as well as Th1 cells may, therefore, be pivotal to the successful control of *T. gondii* infection following vaccination. It has often been suggested that recombinant cytokines such as IL-12 could provide the adjuvant activity in new generation vaccines, and, indeed, this cytokine has proved tremendously successful in vaccination studies on *Leishmania major* (SCOTT 1993). However, protection against this organism is entirely Th1 lymphocyte mediated. Whether IL-12 on its own, given its ability to direct both Th1 and type 1 $CD8^+$ T cell development (CROFT et al. 1994), could successfully adjuvant a vaccine against *T. gondii*, is an intriguing question. Perhaps a boosting vaccine containing IL-2 would be needed to further drive CD8+ T cell expansion? In addition, the antigen may have to be targeted to the appropriate antigen presenting cell, the most likely target being the macrophage.

5 Conclusions

The prospects for the development of successful vaccines against *T. gondii* are extremely encouraging. Live, attenuated vaccines have proved successful in limiting infection in the cat and domestic livestock and several "potentially" protective antigens have been identified. However, future vaccination development will not only have to take account of all the life-cycle stages that need to be targeted but will have to consider which immune responses need to be generated and in which tissue sites. These problems can probably be overcome by delivering by the appropriate route or routes of inoculation, a suitably adjuvanted vaccine containing a cocktail of life-cycle stage-specific antigens.

Acknowledgments. The work was supported by the Medical Research Council and Tenovus Scotland. Craig W. Roberts is a Glaxo-Jack Research Lecturer and a Fulbright Scholar. Heather Jebbari is in receipt of a BBSRC studentship. Abhay Satoskar is in receipt of an ORS Award. The authors wish to thank Dr. Michel Aguet, Genentech Inc., USA, for his kind permission to use IFN-γR knockout mice, and are most grateful to Dr. James Brewer for reading the final manuscript.

References

Alexander J, Roberts CW, Brewer JM (1993) Progress towards the development of a vaccine against congenital toxoplasmosis: identification of protective antigens and the selection of appropriate adjuvants. In: Smith JE (ed) Toxoplasmosis. Springer, Berlin Heidelberg New York, (NATO ASI Series H. Cell Biology vol 78), pp 217–229

Ambroise-Thomas P, Pelloux H (1993) Toxoplasmosis – congenital and in immunocompromised patients: a parallel. Parasitol Today 9:61–62

Beaman MH, Hunter CA, Remington JS (1994) Enhancement of intracellular replication of Toxoplasma gondii by IL-6: Interactions with IFN-gamma and TNF-alpha. J Immunol 153:4583–4587

Black CM, Bermudez LEM, Young LS, Remington JS (1990) Co-infection of macrophages modulates interferon-γ and tumor necrosis factor-induced activation against intracellular pathogens. J Exp Med 172:977–980

Blackwell JM, Roberts CW, Alexander J (1993) Influence of genes within the MHC on mortality and brain cyst development in mice infected with Toxoplasma gondii: kinetics of immune regulation in BALB H-2 congenic mice. Parasitol Immunol 15:317–324

Bohne W, Heesemann J, Gross U (1993) Induction of bradyzoite-specific Toxoplasma gondii antigens in gamma interferon-treated mouse macrophages. Infect Immun 61:1141–1145

Brown CR, McLeod R (1990) Class I MHC genes and CD8+ T cells determine cyst number in Toxoplasma gondii infection. J Immunol 145:3438–3441

Brown CR, Hunter CA, Estes RG, Beckmann J, Forman J, David C, Remington JS, McLeod R (1995) Definitive identification of a gene that confers resistance against Toxoplasma cyst burden and encephalitis. Immunology 85:419–428

Bülow R, Boothroyd JC (1991) Protection of mice from fatal Toxoplasma gondii infection by immunization with P30 antigen in liposomes. J Immunol 147:3496–3500

Burke JM, Roberts CW, Hunter CA, Murray M, Alexander J (1994) Temporal differences in the expression of mRNA for IL-10 and IFN-γ in the brains and spleens of C57BL10 mice infected with Toxoplasma gondii. Parasitol Immunol 16:305–314

Buxton D (1993) Toxoplasmosis: the first commercial vaccine. Parasitol Today 9:335–337

Buxton D, Thomson K, Maley S, Wright S, Bos HJ (1991) Vaccination of sheep with a live incomplete strain (S48) of Toxoplasma gondii and their immunity to challenge when pregnant. Vet Rec 129:89–93

Chardès T, Velge-Roussel F, Mevelec P, Mevelec M-N, Buzoni-Gatei D, Bout D (1993) Mucosal and systemic cellular immune responses induced by Toxoplasma gondii antigens in cyst orally infected mice. Immunology 78:421–429

Clark IA, Rockett KA, Crowden WB (1991) Proposed link between cytokines, nitric oxide and human cerebral malaria. Parasitol Today 7:205–207

Croft M, Carter L, Swain SL, Dutton RW (1994) Generation of polarized antigen-specific CD8 effector populations: reciprocal action of interleukin IL-4 and IL-12 in promoting type 2 versus type 1 cytokine profiles. J Exp Med 180:1715–1728

Curiel TJ, Krug EC, Purner MB, Poignard P, Berens RL (1993) Cloned human CD4$^+$ cytotoxic T lymphocytes specific for Toxoplasma gondii lyse tachyzoite-infected target cells. J Immunol 151:2024–2031

Darcy F, Maes P, Gras-Masse H, Auriault C, Bossus M, Deslee D, Godard I, Cesbron M-F, Tartar A, Capron A (1992) Protection of mice and nude rats against toxoplasmosis by a multiple antigenic peptide construction derived from Toxoplasma gondii P30 antigen. J Immunol 149:3636–3641

Dardé ML, Riaki H, Bouteille B, Pestre-Alexandre M (1992) Isoenzyme analysis of 35 Toxoplasma gondii isolates and the biological and epidemiological implications. J Parasitol 78:786–794

Debrick JE, Campbell PA, Staerz FD (1991) Macrophages as accessory cells for class I MHC restricted immune responses. J Immunol 142:2846–2851

Denkers EY, Gazzinelli RT, Hieny S, Caspar P, Sher A (1993a) Bone marrow macrophages process exogenous Toxoplasma gondii polypeptides for recognition by parasite-specific cytolytic T lymphocytes. J Immunol 150:517–526

Denkers EY, Sher A, Gazzinelli RT (1993b) T cell interactions with Toxoplasma gondii: implications for processing of antigen for class I-restricted recognition. Res Immunol 144:51–57

Denkers EY, Caspar P, Sher A (1994) Toxoplasma gondii possesses a superantigen activity that selectively expands murine T cell receptor Vβ5-bearing CD8$^+$ lymphocytes. J Exp Med 180:985–994

Dubey JP (1994) Toxoplasmosis. J Am Vet Med Assoc 205:1593–1598

Dubey JP, Baker DG, Davis SW, Urban JD, Sken SK (1994) Persistence of immunity to toxoplasmosis in pigs vaccinated with a nonpersistent strain of Toxoplasma gondii. Am J Vet Res 55:982–987

Duquesne VC, Auriault C, Gras-Masse H, Boutillon C, Darcy F, Cesbron-Delauw M-F, Tartar A, Capron A (1991) Identification of T cell epitopes within a 23-kD antigen (P24) of Toxoplasma gondii. Clin Exp Immunol 84:527–534

Essner R, Rhoades K, McBride WH, Morton DL, Economou JS (1989) IL-4 downregulates IL-1 and TNF gene expression in human monocytes. J Immunol 142:3857–3862

Freyre A, Choromanski L, Fishback JL, Popiel I (1993) Immunisation of cats with tissue cysts, bradyzoites, tachyzoites of the T-263 strain of Toxoplasma gondii. J Parasitol 79:716–719

Gajewski TF, Fitch FW (1991) Differential activation of murine Th1 and Th2 clones. Res Immunol 142:19–23

Gautum S, Tebo JM, Hamilton TA (1992) IL-4 suppresses cytokine gene expression induced by IFN-γ and/or IL-2 in murine peritoneal macrophages. J Immunol 148:1725–1730

Gazzinelli RT, Hakim FT, Hieny S, Shearer GM, Sher A (1991) Synergistic role of CD4$^+$ and CD8$^+$ T lymphocytes in IFN-γ production and protective immunity induced by an attenuated Toxoplasma gondii vaccine. J Immunol 146:286–292

Gazzinelli RT, Oswalk IP, James SL, Sher A (1992) IL-10 inhibits parasite killing and nitrogen oxide production by IFN-γ activated macrophages. J Immunol 148:1792–1796

Gazzinelli RT, Denkers EY, Sher A (1993a) Host resistance to Toxoplasma gondii: model for studying the selective induction of cell-mediated immunity by intracellular parasites. Infect Agents Dis 2:139–149

Gazzinelli RT, Hieny S, Wynn T, Wolf S, Sher A (1993b) IL-12 is required for the T-cell independent induction of IFN-γ by an intracellular parasite and induces resistance in T-deficient hosts. Proc Natl Acad Sci 90:6115–6119

Gazzinelli RT, Wysocka M, Hieny S, Kersten T, Carrera L, Cheever A, Kühn R, Müller W, Trinchieri G, Sher A (1996) In absence of endogenous IL-10 mice acutely infected with Toxoplasma gondii succumb to a lethal immune response associated with increased synthesis of IL-12, IFN-γ and TNF-α. J Immunol (in press)

Godard I, Estaquier J, Zenner L, Bossus M, Auriault C, Darcy F, GrasMasse H, Capron A (1994) Antigenicity and immunogenicity of P30-derived peptides in experimental models of toxoplasmosis. Mol Immun 31:1353–1363

Grimwood J, Smith JE (1992) Toxoplasma gondii: the role of a 30-KDa surface protein in host cell invasion. Exp Parasitol 74:106–111.

Grau T, Tacchini-Cotti F, Piguet P-F (1992) Is TNF beneficial or deleterious in toxoplasmic encephalitis? Parasitol Today 8:322–374

Grun JL, Maurer PH (1989) Different T helper subsets elicited in mice utilising two different adjuvant vehicles: the role of endogenous IL-1 in proliferative responses. Cell Immunol 121:131–145

Heeg K, Kuon W, Wagner H (1991) Vaccination of class I restricted murine CD8+ cytotoxic lymphocytes towards soluble antigens: ISCOM-OVA complexes enter the class I MHC restricted antigen pathway and allow sensitisation against the immunodominant peptide. Eur J Immunol 21:1525–1527

Hsieh CS, Macatonia SE, Tripp CS, Wolf S, O'Garra A, Murphy KM (1993) Development of Th1 CD4+ T cells through IL-12 produced by Listeria-induced macrophages. Science 260:547–549

Hughes HPA (1988) Oxidative killing of intracellular parasites mediated by macrophages. Parasitol Today 4:340–347

Hunter CA, Remington JS (1994) Immunopathogenesis of toxoplasmic encephalitis. J Infect Dis 170:1057–67.

Hunter CA, Roberts CW, Alexander J (1992) Kinetics of cytokine mRNA production in the brains of mice with progressive toxoplasmic encephalitis. Eur J Immunol 22:2317–2322.

Hunter CA, Litton MJ, Remington JS, Abrams JS (1994) Immunocytochemical detection of cytokines in the lymph nodes and brains of mice resistant or susceptible to toxoplasmic encephalitis. J Infect Dis 170:939–945

Jackson MH, Hutchison WM (1989) The prevalence and source of Toxoplasma infection in the environment. Adv Parasitol 28:55–105

Jones TC, Len L, Hirsch JG (1975) Assessment in vitro of immunity against Toxoplasma gondii. J Exp Med 141:466–482

Kasper LH (1989) Identification of stage specific antigens of Toxoplasma gondii. Infect Immun 57:668–672

Kasper LH, Currie KM, Bradley MS (1985) An unexpected response to vaccination with a purified membrane tachyzoite antigen (P30) of Toxoplasma gondii. J Immunol 134:3426–3431

Kennedy MK, Torrance D, Picha KS, Mohler KM (1992) Analysis of cytokine mRNA expression in the central nervous system of mice with experimental autoimmune encephalomyelitis reveals that IL-10 mRNA expression correlates with recovery. J Immunol 149:2496–2505

Khan IA, Smith KA, Kasper LH (1988) Induction of antigen-specific parasiticidal cytotoxic T cell splenocytes by a major membrane protein (p30) of Toxoplasma gondii. J Immunol 141:3600–3605

Khan IA, Ely KH, Kasper LH (1991) A purified parasite antigen (P30) mediates CD8+ T cell immunity against fatal Toxoplasma gondii infection in mice. J Immunol 47:3501–3506

Langermans JAM, Van der Hulst MEB, Nibbering PH, Hiemstra PS, Fransen L, Van Furth R (1992) IFN-γ induced L-arginine-dependent toxoplasmastatic activity in murine peritoneal macrophages is mediated by endogenous tumor necrosis factor-α. J Immunol 148:568–574

Lindblad EB (1995) Aluminium adjuvants. In: Steward-Tull DES (ed) Theory and practical application of adjuvants. Wiley, Chichester

Lopes LM, Chain, BM (1992) Liposome mediated delivery stimulates a class I restricted cytotoxic T cell response to soluble antigen. Eur J Immunol 22:287–290

Luft BJ, Remington JS (1992) Toxoplasmic encephalitis in AIDS. Clin Infect Dis 15:211–222

Lunden A, Lovgren K, Uggla A, Araujo FG (1993) Immune responses and resistance to Toxoplasma gondii in mice immunized with antigens of the parasite incorporated into immunostimulating complexes. Infect Immun 61:2639–2643

McCabe RE, Luft BJ, Remington JS (1984) Effect of murine interferon gamma on murine toxoplasmosis. J Infect Dis 150:961–962

McLeod R, Frenkel JK, Estes RG, Mack DG, Eisenhauer PB, Gibori G (1988) Subcutaneous and intestinal vaccination with tachyzoites of Toxoplasma gondii and acquisition of immunity to peroral and congenital Toxoplasma challenge. J Immunol 140:1632–1637

McLeod R, Skamene E, Brown CR, Eisenhauer PB, Mack DG (1989) Genetic regulation of early survival and cyst number after peroral Toxoplasma gondii infection of AxB/BxA recombinant inbred and B10 congenic mice. J Immunol 143:3031–3034

McLeod R, Mack DG, Brown C, Skamene E (1993) Secretory IgA antibody to SAG1, H-2 class 1 restricted CD8+ T-lymphocytes and the INT-1 locus in protection against Toxoplasma gondii. In: Smith JE (ed) Toxoplasmosis. Springer, Berlin Heidelberg New York, (NATO ASI Series H. Cell Biology vol 78), pp 131–137

Parker SJ, Roberts CW, Alexander J (1991) CD8$^+$ T cells are the major lymphocyte subpopulation involved in the protective immune response to Toxoplasma gondii in mice. Clin Exp Immunol 84:207–212

Pavia CS, Bittker SJ, Curnick KE (1992) Passive immunization protects guinea pigs from lethal Toxoplasma infection. FEMS Microbiol Immunol 89:97–104

Pfefferkorn ER, Guyre PM (1984) Inhibition of growth of Toxoplasma gondii in cultured fibroblasts by human recombinant gamma interferon. Infect Immun 44:211–216

Phillips RS (1994) Malarial vaccines – a problem solved or simply a promising start. CAB Abstracts 18:459–486

Reddy RF, Zhou F, Nair S, Huang L, Rouse BT (1992) In vivo cytotoxic T lymphocyte induction with soluble proteins administered in liposomes. J Immunol 148:1585–1589

Roberts CW, Brewer JM, Alexander J (1994) Congenital toxoplasmosis in the Balb/c mouse: Prevention of vertical disease transmission and fetal death by vaccination. Vaccine 12:1389–1394

Roberts CW, Cruickshank S, Alexander J (1995) Sex determined resistance to Toxoplasma gondii is associated with temporal differences in cytokine production. Infect Immun 63:2549–2555

Rothe J, Lesslauer W, Lotscher H, Lang Y, Koebel P, Kontgen F, Althage A, Zinkernagel R, Steinmetz M and Bluethmann H (1993) Mice lacking the tumour necrosis factor receptor 1 are resistant to TNF-α mediated toxicity but highly susceptible to infection by Listeria monocytogenes. Nature 364:798–802

Scott P (1993) IL-12: initiation cytokine for cell-mediated immunity. Science 260:496–497

Sibley LD, Boothroyd JC (1992) Virulent strains of Toxoplasma gondii comprise a single clonal lineage. Nature 350:82–85

Sibley LD, Adams LB, Fukutomi Y, Krahenbuhl JL (1991) Tumor necrosis factor-α triggers antitoxoplasmal activity of IFN-γ primed macrophages. J Immunol 147:2340–2345

Sibley LD, Dobrowolski J, Morisaki JH, Heuser JE (1994) Strategies for intracellular survival of microbes. Clin Infect Dis 1:245–264

Subauste CS, Koniaris AH, Remington JS (1991) Murine CD8$^+$ cytotoxic T lymphocytes lyse Toxoplasma gondii-infected cells. J Immunol 147:3955–3959

Suzuki Y, Orellana MA, Schreiber RD, Remington JS (1988) Interferon-γ: the major mediator of resistance against Toxoplasma gondii. Science 240:516–518

Suzuki Y, Yang Q, Conley FK, Abrams JS, Remington JS (1994) Antibody against Interleukin-6 reduced inflammation and numbers of cysts in brains of mice with toxoplasmic encephalitis. Infect Immun 2:2773–2778

Taverne J (1993) Unravelling the cytokine network in malaria. Parasitol Today 9:38–39.

Villiard O, Candolfi E, Despringre JL, Derouin F, Marcellin, Viville S, Kien T (1995) Protective effect of low doses of an anti IL-4 monoclonal antibody in a murine model of acute toxoplasmosis. Parasite Immunol 17:233–236

Yano A, Aosai F, Ohta M, Hasekura H, Sugane K, Hayashi S (1989) Antigen presentation by Toxoplasma gondii-infected cells to CD4$^+$ proliferative T cells and CD8$^+$ cytotoxic cells. J Parasitol 75:411–416

Zhang YW, Smith JE (1995) Toxoplasma gondii: reactivity of murine sera against tachyzoite and cyst antigens via FAST-ELISA. Int J Parasitol 25:637–640

Zhang YW, Fraser A, Balfour AH, Wreghitt TG, Gray JJ, Smith JE (1995) Serological reactivity against cyst and tachyzoite antigens of Toxoplasma gondii determined by FAST-ELISA. J Clin Pathol 48:908–911

Zhou F, Rouse BT, Huang, L (1992) Induction of cytotoxic T lymphocytes in vivo with protein antigen entrapped in membrenous vesicles. J Immunol 149:1599–1604

D

Diagnosis and Treatment of Toxoplasmosis

Detection of IgA anti-P30 (SAG1) Antibodies in Acquired and Congenital Toxoplasmosis

A. Decoster

1 Introduction

Toxoplasmosis, a ubiquitous protozoan infection, is caused by an intracellular parasite, *Toxoplasma gondii*. Generally benign for healthy people, it can be serious in the context of immunodeficiency, especially in the case of acquired immunodeficiency syndrome (AIDS) and of bone marrow or heart transplant patients or in children infected in utero.

Early diagnosis of acute *Toxoplasma* infection in pregnant women is of utmost importance for better management of an efficient anti-toxoplasmic therapy. The diagnosis of acquired infection is based on serological tests which detect immunoglobulin M (IgM) and IgG antibodies, and many techniques have been used to attempt to distinguish between recently acute or chronic toxoplasmosis (DESMONTS et al. 1981; SHARMA et al. 1983; SANTORO et al. 1985; CESBRON et al. 1986; HEDMAN et al. 1989; DANNEMAN et al. 1990). The use of very sensitive methods to detect IgM antibodies has complicated the interpretation of serological results, because the specific IgM antibodies, classically considered to be characteristic of acute toxoplasmosis, can be detected by these methods long after onset of disease (LE FICHOUX et al. 1984).

The diagnosis of congenital infection at birth or in utero is based on the same serological tests and may be performed in conjunction with polymerase chain reaction (PCR), cell culture, and intraperitoneal inoculation of mice with umbilical cord blood, amniotic fluid or placenta by intraperitoneal inoculation

Hôpital St Vincent, Laboratoire de Microbiologie, Boulevard de Belfort, 59044 Lille Cedex, France of mice in order to directly detect the parasite (DESMONTS et al. 1985; DEROUIN et al. 1988; BURG et al. 1989). Conventional tests to detect antibodies against *Toxoplasma* are not immediately helpful for serodiagnosis of congenital infection, because the children's IgG *Toxoplasma* antibodies cannot be distinguished from maternal IgG antibodies acquired across the placenta during pregnancy. The maternally-transmitted IgG antibodies persist for months in the child. A specific immune response of the infant, which indicates infection, is determined by the presence of antibodies which do not cross the placenta, such as IgM or IgA antibodies. A number of techniques have been applied to detect IgA antibodies in *Toxoplasma* infection, including the enzyme-linked immunosorbent assay (TURUNEN et al. 1983; VAN LOON et al. 1983; FAVRE et al. 1984), the immunosorbent agglutination assay (PINON et al. 1986; LE FICHOUX et al. 1987; BESSIERES et al. 1992; PATEL et al. 1993), western blotting (PARTANEN et al. 1984; HUSKINSON et al. 1990; GROSS et al. 1992) and enzyme-linked immunofiltration assay (PINON et al. 1985). Anti-*Toxoplasma* IgA antibodies were shown to be useful human markers in congenital and acute toxoplasmosis. The analysis of a wider range of sera collected from fetuses, newborns and pregnant women allowed us to confirm in all cases the value and reliability of the detection of IgA antibodies directed against P30, the major *T. gondii* surface protein (DECOSTER et al. 1988b, 1991, 1992, 1995).

2 P30 or SAG1: Major *T. gondii* Surface Antigen

The role of various excreted/secreted (ESA) and surface (SAG) antigens in the immunogenicity of *T. gondii* or pathogenesis of toxoplasmosis is currently being explored in a number of laboratories (reviewed in JOHNSON 1985; DARCY et al. 1986). Because of their direct accessibility to the immune system, SAG probably represent the candidate targets for the killing of the extracellular parasites, especially the major radio-iodinated surface membrane protein of *T. gondii*. This protein, termed P30 or SAG1 (SIBLEY et al. 1991) has a molecular weight of 30 kDa (KASPER et al. 1983). It is a stage-specific antigen, since it can be detected only in rapidly dividing and invasive tachyzoites and not in the bradyzoites of tissue cysts or sporozoites from oocysts (KASPER 1989), and it appears to have an important role in both immune and pathogenic mechanisms of the parasite (KASPER and KHAN 1993).

It was demonstrated that immunizations with P30 induced nearly total protection of mice infected with moderately virulent *T. gondii* strains (BÜLOW and BOOTHROYD 1991; KHAN et al. 1991). In support of the importance of P30 in immunity, it was reported that the passive transfer of a monoclonal antibody directed against P30 significantly protected nude rats against *T. gondii* RH strain infection (SANTORO et al. 1987), and that total protection was observed after immunization with P30, which was either incorporated into liposomes

(Bülow and Boothroyd 1991) or used in conjunction with QuilA (Khan et al. 1991). Protection of mice and nude rats against toxoplasmosis by a multiple antigenic peptide construction, which was derived from *T. gondii* P30 antigen, was also reported (Darcy et al. 1992).

3 IgA Antibody Response Against *T. gondii*

Infection with *T. gondii* most commonly occurs via the oral route (Frenkel et al. 1969). Ingested organisms are released from cysts or oocysts within the gastrointestinal tract, invade the intestinal epithelium, and divide as tachyzoites. The local tissues underlying this intestinal epithelium are heavily populated with lymphoid cells from the mucosal immune system, which could rapidly come into contact with the parasites during their intestinal penetration. This mucosal immune system is now well-known as a protective barrier against various agents, and an efficient stimulation of the mucosal response could be highly valuable in controlling infection by *T. gondii* via the oral route. IgA antibodies constitute more than 80% of all antibodies in mucosa-associated tissues (Chardès and Bout 1993). Preliminary work was required to screen *T. gondii* antigens as effective inducers of a mucosal immune response. Using a murine model, McLeod and colleagues (1988) showed that P30 is a major antigen recognized by IgA antibodies in the milk of acutely infected mothers, and it has been suggested that mucosal anti-P30 IgA antibodies may play a protective role against toxoplasmosis. Mice infected perorally develop intestinal IgA antibodies to the major 30-kDa antigen of *T. gondii*. Thus, P30 (SAG1) has an important functional role in infection and elicits an intestinal antibody response following peroral infection (Kasper and Khan 1993). Therefore, the role of secretory IgA to P30 appears to be of both clinical and biological importance in host mucosal immunity.

Consequently, the study of the IgA response has been of particular interest in toxoplasmosis. Several authors have reported a serum IgA response following *T. gondii* infection using mouse (Chardès et al. 1990; Godard et al. 1990) and rat models (Godard et al. 1990) as well as in humans (Turunen et al. 1983; Van Loon et al. 1983; Favre et al. 1984; Partanen et al. 1984; Pinon et al. 1985; Le Fichoux et al. 1987; Huskinson et al. 1990; Stepick-Biek et al. 1990; Bessières et al. 1992; Patel et al. 1993). These studies mainly focused on the detection of anti-P30 IgA antibodies (Decoster et al. 1988b, 1991, 1992, 1995; Gross et al. 1992).

4 IgA Antibody Response Against P30

As demonstrated by immunoblotting, IgA antibodies are present to a variety of antigens during the various stages of toxoplasmosis. The predominant antigens recognized have molecular weights of 5, 30, and 35 kDa. Interestingly, only the antigens of 35 and 30 kDa molecular weight (P30 or SAG1) were recognized by the IgA antibodies from congenitally infected infants (PARTANEN et al. 1984; POTASMAN et al. 1987; HUSKINSON et al. 1990; STEPICK-BIEK et al. 1990).

Our first studies have confirmed the value of measuring the IgA antibody response directed against P30 in congenital and acute acquired toxoplasmosis (DECOSTER at al. 1988b, 1991). Using an immunocapture assay, an anti-P30 IgA antibody response could be observed in the fetuses infected with *T. gondii*, suggesting that this test is additionally useful in the prenatal diagnosis of congenital toxoplasmosis (DECOSTER et al. 1992).

In congenital toxoplasmosis, anti-P30 IgA antibodies are found more frequently in infected fetuses and newborns than anti-P30 IgM (DECOSTER et al. 1991, 1992). The presence of specific IgA proves the presence of congenital infection since IgA, like IgM, does not cross the placental barrier. In relation to the progressive development of a functioning fetal immune system during pregnancy, anti-P30 IgM humoral fetal response seems to be weaker than the anti-P30 IgA response, which is more intense and persisting. The serological profile at birth is dependent on the date of infection in utero (DECOSTER et al. 1991, 1992). When infection occurred during the third trimester of gestation, the infant had, either at birth or in the following weeks, specific IgA and IgM antibodies simultaneously. In contrast, when infection occurred in the first term, the infant was at the end of the acute phase of fetal infection at birth, characterized by the absence of IgM antibodies (which probably appeared during fetal life for a short time and disappeared before birth) and the presence of IgA antibodies. This explains the detection of both IgM and IgA more frequently in the case of a late infection, unless infection occurred very late in the last month of pregnancy. In this case, anti-P30 IgA antibodies were detected later than anti-P30 IgM antibodies, as observed in the onset of an acute acquired toxoplasmosis. This indicates that additional serodiagnosis for toxoplasmosis must be performed after 1 month of life in seronegative newborns, if maternal serodiagnosis suggests a possible infection (MARX-CHEMLA et al. 1990). As observed in some cases of congenitally infected children, the late detection of anti-P30 IgA antibodies (particularly if neither anti-P30 IgA nor anti-P30 IgM were detected at birth) can constitute an additive criterion of infection.

Interestingly, anti-P30 IgA can be detected in fetuses as early as the second term of pregnancy, provided that maternal infection was acquired soon after conception. This observation is fascinating from the immunological point of view. In the absence of antigenic stimulation, the production of all isotypes of immunoglobulins is non-existent in the human fetus; in particular, the level of IgA in 1-year old children is only 20% of that observed for adults (HAYWARD and LAWTON 1977). Although the specific IgA response is generally weaker in fetuses than in newborns, these data demonstrate that *Toxoplasma* infection can evoke production of specific IgA by the fetal immune system.

The anti-P30 IgA response can disappear before birth, showing the importance of the prenatal diagnosis. Fetal blood samples must be collected in the second term of pregnancy (when the fetal immune system is functional) by experienced obstetrical teams (DAFFOS et al. 1985). Several biological tests must be performed to verify that fetus samples are not contaminated with maternal blood, which would interfere with diagnosis of congenital infection on the basis of parasitological and serological results (CEDERQVIST et al. 1977; DESMONTS et al. 1985; DAFFOS et al. 1988).

In the case of negative results, the fetal infection cannot be excluded; in particular, it could be possible that the cord blood was collected before the infection of the fetus has happened. The classical survey and treatment by spiramycin of the pregnant women must be continued until delivery.

In the case of positive results, clinical and ultrasound survey must be intensified, and the termination of pregnancy has to be discussed if clinical signs of congenital infection are observed (hydro- or microcephaly). If the continuation of pregnancy has been decided, the administration of spiramycin must be replaced by a combination of pyrimethamine plus sulfadiazine plus folinic acid in the third term of pregnancy in order to effect greater inhibition of *Toxoplasma* proliferation in the fetal tissues (FORESTIER et al. 1985; COUVREUR et al. 1988; HOHFELD et al. 1989).

During the acute phase of toxoplasmosis, we detected IgA antibodies directed against P30 in sera of all patients, sometimes later than specific IgM but always earlier than specific IgG. In the majority of cases, when IgG continued to rise and IgM antibodies persisted ("residual" IgM), IgA antibodies disappeared earlier (between 6–12 months) and were not detected in the chronic phase of toxoplasmosis, as described in other studies focused on the anti-*Toxoplasma* IgA response (TURUNEN et al. 1983; VAN LOON et al. 1983; FAVRE et al. 1984; PARTANEN et al. 1984; PINON et al. 1985; LE FICHOUX et al. 1987; HUSKINSON et al. 1990; STEPICK-BIEK et al. 1990; PATEL et al. 1993). We observed only a few exceptions corresponding to untreated patients, suggesting that the treatment decreasing the antigenic stimulation could influence the antibody kinetics. Thus, simultaneous detection of the three isotypes, IgG, IgM, and IgA, is of major value in determining the phase of infection, especially during pregnancy. If specific IgM alone is detected, the analysis of a second sample 3 weeks later will confirm (with presence of IgA) or disprove (in the absence of IgA) the possibility of a recent infection, and when specific IgM and IgA are simultaneously detected, the patient must be considered to be in the course of acute toxoplasmosis. As shown by the number of studies focused on the subject, the early diagnosis of acute toxoplasmosis in pregnancy is of utmost importance. Indeed, pregnancy and fetal management are very different depending on whether the acute phase occurred after the date of conception (when the risk of fetal infection exists) or before the date of conception (DES-MONTS et al. 1985; DAFFOS et al. 1988). In the latter case, there is no risk of congenital infection, except in exceptional cases or when the woman is immunocompromised (DESMONTS et al. 1990; FORTIER et al., 1991).

As previously noticed (STEPICK-BIEK et al. 1990; DARCY et al. 1991; PATEL et al. 1993), patients with AIDS-associated primary toxoplasmosis consistently were shown to produce specific IgM and IgA. However, there were no significant differences in the frequency of detectable IgA between AIDS patients with chronic toxoplasmosis and those with secondary reactivation resulting in cerebral or ocular disease.

The method used to detect anti-P30 IgA antibodies is a double-sandwich enzyme-linked immunosorbent assay (ELISA) (DECOSTER et al. 1991), which is known to be more sensitive than direct ELISA (DESMONTS et al. 1981; VAN LOON et al. 1983). The enhanced sensitivity is probably due to use of a monoclonal antibody directed against P30, which is known to elicit a very early and intense antibody response (TURUNEN et al. 1983; SANTORO et al. 1985; CESBRON et al. 1986; DECOSTER et al. 1988a,b; STEPICK-BIEK et al. 1990; GROSS et al. 1992). Detailed studies have shown that the early immune response in which IgM and IgA are involved is primarly directed against membrane antigens of the parasite, whereas antibodies recognizing cytoplasmic antigens are formed as the immune response matures (PARTANEN et al. 1984; HUSKINSON et al. 1990). Therefore, we recommend this method for estimating the time of *Toxoplasma* infection, because of the shorter antibody response that is detected. Due to the enhanced sensitivity of this assay compared with methods which detect IgA antibodies directed against various *Toxoplasma* antigens, we also recommend this assay to be used for the investigation of congenital toxoplasmosis (DECOSTER et al. 1995).

5 Conclusion

The clinical value of the detection of specific anti-*Toxoplasma* IgA antibodies for early diagnosis of acute acquired and congenital *Toxoplasma* infection has been evaluated during the past few years and is now well-established. In acquired toxoplasmosis, the majority of patients are positive for anti-P30 IgA and specific IgM antibodies during the acute phase of infection, whereas no anti-P30 IgA antibodies are detected in the chronic phase, when patients could still be positive for specific IgM ("residual" IgM). Thus, the testing of both IgM and IgA antibodies has proven reliable and of great help to diagnose recently acquired toxoplasmosis during pregnancy. In congenital toxoplasmosis, anti-P30 IgA antibodies were found in fetuses and infected newborns more frequently than anti-P30 IgM antibodies. It appears from our data that the combined testing of both IgM and IgA in the fetus and the newborn is essential for a more efficient diagnosis of congenital infection.

References

Bessières MH, Roques C, Berrebi A, Barre V, Cazaux M, Seguela JP (1992) IgA antibody response during acquired and congenital toxoplasmosis. J Clin Pathol 45:605–608

Bülow R, Boothroyd JC (1991) Protection of mice from fatal Toxoplasma gondii infection by immunization with P30 antigen in liposomes. J Immunol 147:3496–3500

Burg JL, Grover CM, Pouletty P, Boothroyd JC (1989) Direct and sensitive detection of a pathogenic protozoan, Toxoplasma gondii, by polymerase chain reaction. J Clin Microbiol 27:1787–1792

Cederqvist LL, Kimball AC, Ewool LC, Litwin SD (1977) Fetal immune response following congenital toxoplasmosis. Obst Gyn 50:200–204

Cesbron JY, Caron A, Santoro F, Wattre P, Ovlaque G, Pierce RJ, Delagneau JP, Capron A (1986) Une nouvelle méthode ELISA pour le diagnostic de la toxoplasmose: dosage des IgM sériques par immunocapture avec un anticorps monoclonal anti-Toxoplasma gondii. Presse Méd 15:737–740

Chardès T, Bout D (1993) Mucosal immune response in toxoplasmosis. Res Immunol 144:57–60

Chardès T, Bourguin I, Mevelec MN, Dubremetz JF, Bout D (1990) Antibody responses to Toxoplasma gondii in sera, intestinal secretions and milk from orally infected mice and characterization of target antigens. Infect Immun 58:1240–1246

Couvreur J, Desmonts G, Thulliez P (1988) Prophylaxis of congenital toxoplasmosis: effects of spiramycin on placental infection. J Antimicrobiol Chemother 22:193–200

Daffos F, Capella-Pavlovsky M, Forestier F (1985) Fetal blood sampling during pregnancy with use of a needle guided by ultrasound: a study of 606 consecutive cases. Am J Obst Gynecol 153:655–660

Daffos F, Forestier F, Capella-Pavlovsky M, Thulliez P, Aufrani C, Valenti D, Cox LW (1988) Prenatal management of 746 pregnancies at risk for congenital toxoplasmosis. N Engl J Med 518:271–275

Danneman BR, Vaughan WC, Thulliez P, Remington JS (1990) Differential agglutination test for diagnosis of recently acquired infection with Toxoplasma gondii. J Clin Microbiol 28:1928–1933

Darcy F, Santoro F, Charif H, Deslee D, Hacot C, Capron A (1986) Toxoplasma gondii excreted-secreted antigens: first characterization and immunogenicity in human and experimental infections. Immunobiol 173:241–242

Darcy F, Foudrinier F, Mougeot G, Decoster A, Caron A, Marx-Chemla C, Capron A, Pinon JM (1991) Diagnostic value of specific IgA antibodies in AIDS patients with Toxoplasma infection: a bicentric evaluation. Immunology letters 30:345–348

Darcy F, Maes P, Gras-Mass H, Auriault C, Bossus M, Deslee D, Godard I, Cesbron-Delauw MF, Tartar A, Capron A (1992) Protection of mice and nude rats against toxoplasmosis by a multiple antigenic peptide construction derived from Toxoplasma gondii P30 antigen. J Immunol 149:3636–3641

Decoster A, Caron A, Darcy F, Capron A (1988a) Recognition of Toxoplasma gondii excreted and secreted antigens by human sera from acquired and congenital toxoplasmosis: identification of markers of acute and chronic infection. Clin Exp Immunol 73:376–382

Decoster A, Darcy F, Caron A, Capron A (1988b) IgA antibodies against P30 as markers of congenital and acute toxoplasmosis. Lancet 332:1104–1107

Decoster A, Slizewicz B, Simon J, Bazin C, Darcy F, Vittu G, Boulanger C, Champeau Y, Demory JL, Duhamel M, Capron A (1991) Platelia-Toxo IgA, a new kit for early diagnosis of congenital toxoplasmosis by detection of anti-P30 immunoglobulin A antibodies. J Clin Microbiol 29:2291–2295

Decoster A, Darcy F, Caron A, Vinatier D, Houze de l'Aulnoit D, Vittu G, Niel G, Heyer F, Lecolier B, Delcroix M, Monnier JC, Duhamel M, Capron A (1992) Anti-P30 IgA antibodies as prenatal markers of congenital Toxoplasma infection. Clin Exp Immunol 87:310–315

Decoster A, Gontier P, Dehecq E, Demory JL, Duhamel M (1995) Detection of anti-Toxoplasma immunoglobulin A antibodies by Platelia-Toxo IgA directed against P30 and by IMx Toxo IgA for diagnosis of acquired and congenital toxoplasmosis. J Clin Microbiol 33:2206–2208

Derouin F, Thulliez P, Candolfi E, Daffos F, Forestier F (1988) Early prenatal diagnosis of congenital toxoplasmosis using amniotic fluid samples and tissue culture. Eur J Clin Microbiol Infect Dis 7:423–425

Desmonts G, Naot Y, Remington JS (1981) An immunoglobulin M – immunosorbent agglutination assay for diagnosis of infectious diseases: diagnosis of acute congenital and acquired Toxoplasma infections. J Clin Microbiol 13:859–864

Desmonts G, Forestier F, Thulliez P, Daffos F, Capella-Pavlovsky M, Chartier M (1985) Prenatal diagnosis of congenital toxoplasmosis. Lancet 325:500–503

Desmonts G, Couvreur J, Thulliez P (1990) Congenital toxoplasmosis. Five cases with mother-to-child transmission of pre-pregnancy infection. Presse Méd 19:1445–1449

Favre G, Bessières MH, Seguela JP (1984) Dosage des IgA sériques spécifiques de la toxoplasmose par une méthode ELISA. Application à 120 cas. Bull Soc Fr Parasitol 3:139–142

Forestier F, Daffos F, Capella-Pavlovsky M (1985) Intérêts du prélèvement de cordon foetal dans le diagnostic prénatal, la foetologie et la thérapeutique in utero. Ann Biol Clin 43:535–542

Fortier B, Aissi E, Ajana F, Dieusart P, Denis P, Martin de Lassale E, Lecomte-Houcke M, Vinatier D (1991) Spontaneous abortion and reinfection by Toxoplasma gondii. Lancet 338:444–446

Frenkel JK, Dubey JP, Miller N (1969) Toxoplasma gondii: fecal forms separated from eggs of the nematode Toxocara cati. Science 164:432–436

Godard I, Darcy F, Deslee D, Dessaint JP, Capron A (1990) Isotypic profiles of antibody responses to Toxoplasma gondii infection in rats and mice: kinetic study and characterization of target antigens of immunoglobulin A antibodies. Infect Immun 58:2446–2451

Gross U, Roos T, Appoldt D, Heesemann J (1992) Improved serological diagnosis of Toxoplasma gondii infection by detection of Immunoglobulin A (IgA) and IgM antibodies against P30 by using the immunoblot technique. J Clin Microbiol 30:1436–1441

Hayward R, Lawton AR (1977) Induction of plasma cell differentiation of human fetal lymphocytes: evidence for functional immaturity of T and B cells. J Immunol 119:1213–1217

Hedman K, Lappalainen M, Seppada I, Makela O (1989) Recent primary Toxoplasma infection indicated by a low avidity of specific IgG. J Infect Dis 159:736–739

Hohfeld P, Daffos F, Thulliez P, Aufrant C, Couvreur J, Macaleese J, Descombey D, Forestier F (1989) Fetal toxoplasmosis: outcome of pregnancy and infant follow-up after in utero treatment. J Ped 115:765–769

Huskinson J, Thulliez P, Remington JS (1990) Toxoplasma antigens recognized by human immunoglobulin A antibodies. J Clin Microbiol 28:2632–2636

Johnson AM (1985) The antigenic structure of Toxoplasma gondii: a review. Pathology 17:9–19

Kasper LH (1989) Identification of stage-specific antigens of Toxoplasma gondii. Infect Immun 57:668–672

Kasper LH, Khan IA (1993) Role of P30 in host immunity and pathogenesis of T. gondii infection. Res Immunol 144:45–48

Kasper LH, Crabb JH, Pfefferkorn ER (1983) Purification of a major membrane protein of Toxoplasma gondii by immunoadsorption with a monoclonal antibody. J Immunol 130:2407–2412

Khan IA, Ely KH, Kasper LH (1991) A purified parasite antigen (P30) mediates CD8+ T cell immunity against fatal Toxoplasma gondii infection in mice. J Immunol 147:3501–3506

Le Fichoux Y, Marty P, Chan H, Doucet J (1984) Détection des IgM antitoxoplasmiques par ISAGA. A propos de 3786 sérologies. Bull Soc Fr Parasitol 3:15–18

Le Fichoux Y, Marty P, Chan H (1987) Les IgA spécifiques dans le diagnostic de la toxoplasmose. Ann Pédiatr 34:375–379

Marx-Chemla C, Puygauthier-Toubas D, Foudrinier F, Dorangeon PH, Leullier J, Quereux C, Leroux B, Pinon JM (1990) La surveillance immunologique d'une femme enceinte séronégative pour la toxoplasmose doit-elle s'arrêter à l'accouchement? Presse Méd 19:367–368

McLeod R, Frenkel JK, Estes RG, Mack DG, Eisenhauer PB, Gibori G (1988) Subcutaneous and intestinal vaccination with tachyzoites of Toxoplasma gondii and acquisition of immunity to peroral and congenital Toxoplasma challenge. J Immunol 140:1632–1637

Partanen P, Turunen HJ, Paasivuo RTA, Leinikki PO (1984) Immunoblot analysis of Toxoplasma gondii antigens by human immunoglobulins G, M and A antibodies at different stages of infection. J Clin Microbiol 20:133–135

Patel B, Young Y, Duffy K, Tanner RP, Johnson J, Holliman RE (1993) Immunoglobulin-A detection and the investigation of clinical toxoplasmosis. J Med Microbiol 38:286–292

Pinon JM, Thoannes SH, Gruson N (1985) An enzyme-linked immunofiltration assay used to compare infant and maternal antibody profiles in toxoplasmosis. J Immunol Methods 77:15–23

Pinon JM, Thoannes H, Pouletty PH, Poirriez J, Damiens J, Pelletier P (1986) Detection of IgA specific for toxoplasmosis in serum and cerebrospinal fluid using a non-enzymatic IgA-capture assay. Diagn Immunol 4:223–227

Potasman J, Araujo FG, Thulliez P, Desmonts G, Remington JR (1987) Toxoplasma gondii antigens recognized by sequential samples of serum obtained from congenitally infected infants. J Clin Microbiol 25:1926–1931

Santoro F, Afchain D, Pierce JR, Cesbron JY, Ovlaque G, Capron A (1985) Serodiagnosis of Toxoplasma infection using a purified parasite protein (P30). Clin Exp Immunol 62:262–269

Santoro F, Auriault C, Leite P, Darcy F, Capron A (1987) Infection du rat athymique par Toxoplasma gondii. CR Acad Sc Paris, pp. 297–300

Sharma SD, Muliena J, Araujo FG, Erlich HA, Remington JS (1983) Western blot analysis of the antigens of Toxoplasma gondii recognized by human IgM and IgG antibodies. J Immunol 131:977–983

Sibley LD, Pfefferkorn ER, Boothroyd JC (1991) Proposal for a uniform genetic nomenclature in Toxoplasma gondii. Parasitol Today 7:327–328

Stepick-Biek P, Thulliez P, Araujo FG, Remington JS (1990) IgA antibodies for diagnosis of acute congenital and acquired toxoplasmosis. J Inf Dis 162:270–273

Turunen H, Vuorio KA, Leinikki PO (1983) Determination of IgG, IgM and IgA antibody responses in human toxoplasmosis by enzyme-linked immunosorbent assay (ELISA). Scand J Infect Dis 15:307–311

Van Loon AM, Van der Logt JTM, Heessen WFA, Van der Veen J (1983) Enzyme-linked immunosorbent assay that uses labeled antigens for the detection of immunoglobulin M and A antibodies in toxoplasmosis: comparison with indirect immunofluorescence and double-sandwich enzyme-linked immunosorbent assay. J Clin Microbiol 17:997–1004

AIDS-Associated Cerebral Toxoplasmosis: An Update on Diagnosis and Treatment

A. Ammassari, R. Murri, A. Cingolani,
A. De Luca, and A. Antinori

1 Epidemiology and Risk Factors

Toxoplasmic encephalitis (TE) is one of the major opportunistic infections of the central nervous system (CNS) and the most frequent cause of focal brain lesions (FBL) in patients with acquired immunodeficiency syndrome (AIDS) (Luft et al. 1993). CNS toxoplasmosis is rapidly progressive and fatal without treatment and has been reported to be the AIDS index diagnosis in 22%–51% of patients infected with human immunodeficiency virus (HIV) (Leport et al. 1988; Cohn et al. 1989; Dannemann et al. 1991; Zangerle et al. 1991; Porter and Sande 1992).

The occurrence of TE in HIV-positive patients correlates with the prevalence of *Toxoplasma* antibodies among the general population, which largely varies according to geographic localizations and ethnic groups (e.g., 10%–32% in the United States, and 46%–73% in the European countries) (Grant et al. 1990; Dannemann et al. 1991; Zangerle et al. 1991; Israelski et al. 1993; Wallace et al. 1993; Oksenhendler et al. 1994). Observational studies showed that the 1-year estimated rates of TE were 21% in patients with latent *Toxoplasma gondii* infection, but 0% in seronegative individuals (Oksenhendler et al. 1994). In a randomized trial, 91% of patients who developed TE were seropositive for *T.gondii* antibodies (Opravil et al. 1995). Therefore, the finding of a negative

Department of Infectious Diseases, Catholic University, L.go A. Gemelli 8, 00168 Rome, Italy

Toxoplasma-specific serology does not completely exclude the possibility of TE (DANNEMANN et al. 1991; ZANGERLE et al. 1991; PORTER and SANDE 1992).

Besides positive *Toxoplasma* serology, the magnitude of immunodeficiency further enhances the risk of TE in patients with HIV infection (BACELLAR et al. 1994; JACOBSON et al. 1994). After 18 months, the probability of TE for patients with a CD4 cell count below 100/μl was 35%–40% (GIRARD et al. 1993; OPRAVIL et al. 1995). The risk of TE in patients with a low CD4 cell count was 2.3 times greater than in those with a high level of CD4 cells. Patients presenting positive *Toxoplasma* serology, as well as a CD4 cell count below 100/μl, had a 34.4% risk of TE within 1 year (95% Confidence Interval, C.I., 25.5–43.3) (OKSENHENDLER et al. 1994).

Based on the results of prospective clinical trials, it is likely that the extensive use of systemic prophylaxis with agents that are simultaneously active against *T.gondii* and *Pneumocystis carinii* will cause a reduction of the incidence of TE in the next few years (GIRARD et al. 1993; OKSENHENDLER et al. 1994; ANTINORI et al. 1995). Depending on prophylaxis, significantly different incidence rates of TE were shown for patients with <100 CD4 cells/μl (1.6 per 100 person–years with prophylaxis versus 3.7 per 100 person–years without prophylaxis) (BACELLAR et al. 1994). A longitudinal study performed on consecutive patients with enhanced FBL showed a progressive reduction of the probability of TE during the years 1991–1995 (chi square test for linear trend $p=0.03$) (AMMASSARI et al. 1995). This trend correlated with an increased use of simultaneous prophylaxis against *P. carinii* and *T. gondii* ($p=0.04$). The positive predictive value of presumptive diagnosis of TE, defined as the presence of positive *Toxoplasma* serology and contrast-enhancing lesions with computed tomography (CT) or magnetic resonance imaging (MRI) (CENTERS FOR DISEASE CONTROL 1993), decreased from 86% in 1991 to 66% in the years 1994 and 1995. This reduction in the incidence of TE as well as the probable increase in primary central nervous system lymphoma (PCNSL) incidence (AMMASSARI et al. 1995) makes accurate diagnostic criteria for patients with FBL necessary.

2 Diagnosis

2.1 Clinical and Neuroradiological Features

Clinical findings of TE, a disease which usually has a subacute or acute onset, are extremely heterogeneous and include: headache (10%–56%), focal neurologic deficits, such as hemiparesis and cranial-nerve palsy (58%–89%), disorientation (15%–37%), fever (10%–78%), and seizures (15%–35%) (NAVIA et al. 1986; HAVERKOS 1987; COHN et al. 1989). Focal hypodense mass lesions with contrast enhancement in up to 91% of cases and localized most frequently in the basal ganglia and in the corticomedullary junction of cerebral hemispheres are found by using CT. The presence of multiple lesions is highly suggestive of TE (CIRILLO and ROSENBLUM 1991; STEINMETZ et al. 1995).

In order to assess the value of clinical and neuroradiological variables predictive of TE or PCNSL, we conducted a validation cohort study including all patients showing at least one *first-ever* FBL with CT or MRI observed from 1991 to 1995. Out of the 101 patients, 78 were considered as eligible for the analysis because a definitive diagnosis based on histopathologic examination or presumptive diagnosis of TE (CENTERS FOR DISEASE CONTROL 1993) was available. Thirty-seven presumptive cases of TE, 23 PCNSL, 13 progressive multifocal leukoencephalopathies (PML), three tuberculomas, and two mycotic abscesses were observed. The risk of TE or PCNSL according to the presence or absence of different clinical, immunological, and neuroradiological variables calculated at univariate analysis is shown in Table 1. Logistic regression analysis confirmed a highly increased probability of having TE in the presence of contrast enhancement (odds ratio, O.R., 26.1; 95% C.I. 2.4–273.5; $p=0.006$) and positive *Toxoplasma* serology (O.R. 46.8; 95% C.I. 7.7–286.3; $p<0.001$); whereas *Toxoplasma*-specific prophylaxis had a strong protective effect on the occurrence of TE (O.R. 0.14; 95% C.I. 0.03–0.6; $p=0.01$). On the contrary, presence of PCNSL was inversely associated with positive *Toxoplasma* serology (O.R. 0.11;

Table 1. Risk of toxoplasmic encephalitis (TE) and primary central nervous system lymphoma (PCNSL), according to different variables

	TE O.R. (95% C.I.)	p	PCNSL O.R. (95% C.I.)	p
Age >35 years	0.46 (0.16–1.26)	NS	1.75 (0.59–5.29)	NS
Sex (female)	0.57 (0.16–1.83)	NS	1.56 (0.44–5.26)	NS
Intravenous drug use	0.92 (0.34–2.47)	NS	0.52 (0.17–1.53)	NS
CD4 cell count ≤50/µl	1.12 (0.37–3.41)	NS	2.29 (0.63–10.6)	NS
Previous AIDS diagnosis	0.39 (0.13–1.10)	NS	3.72 (1.22–12.0)	0.02
Antiretroviral therapy	0.61 (0.22–1.67)	NS	3.18 (0.96–12.6)	0.04
Anti-*T.gondii* prophylaxis	0.22 (0.07–0.62)	0.003	3.93 (1.27–13.3)	0.01
Positive *T.gondii* serology	28.8 (6.02–281.5)	<0.001	0.17 (0.05–0.54)	0.001
Clinical features at presentation				
– >15 Days from onset	1.36 (0.51–3.65)	NS	0.42 (0.13–1.26)	NS
– Focal signs	1.25 (0.42–3.78)	NS	0.71 (0.22–2.34)	NS
– Abnormal level of consciousness	6.12 (0.64–302.8)	NS	0.46 (0.01–4.45)	NS
Radiographic findings				
– Contrast enhancement	20.1 (2.77–897.9)	<0.001	3.54 (0.71–34.9)	NS
– Multiple lesions	1.75 (0.64–4.85)	NS	0.73 (0.24–2.19)	NS
– Basal ganglia site localization	4.92 (1.74–14.8)	<0.001	1.30 (0.44–3.90)	NS

NS, not significant

95% C.I. 0.04–0.3; $p < 0.001$) and *Toxoplasma*-specific prophylaxis (O.R. 2.7; 95% C.I. 1.1–6.8; $p = 0.03$).

2.2 Amplification of *T. gondii* DNA

In recent years, polymerase chain reaction (PCR) has been developed to detect *T.gondii* DNA from several biological fluids (GROVER et al. 1990; BRETAGNE et al. 1995; WEISS et al. 1991). Moreover, the detection of *T.gondii* DNA in cerebrospinal fluid (CSF) by PCR has been shown to be a reliable diagnostic tool for the diagnosis of TE in AIDS patients (LEBECH et al. 1992; SCHOONDERMARK-VAN DE VEN et al. 1993; NOVATI et al. 1994; DUPON et al. 1995; EGGERS et al. 1995). Sensitivity ranges from 11% (EGGERS et al. 1995) to 100% (LEBECH et al. 1992), depending on two important variables: (i) the set of primers used, and (ii) the time elapsing from the starting of *Toxoplasma*-specific therapy and the collection of CSF by lumbar puncture. In a recent study, an overall sensitivity of 46% was obtained in patients with TE diagnosed presumptively or by histology (NOVATI et al. 1994). This value rose to 100% in patients who had been receiving therapy for less than one week at the time of lumbar puncture.

We have developed a nested-PCR to detect *T. gondii* DNA for the early diagnosis of TE in patients with HIV infection and FBL. We prospectively evaluated 46 HIV-infected patients with FBL detected by CT or MRI. According to presumptive criteria (CENTERS FOR DISEASE CONTROL 1993), 17 patients were affected by TE, and 16 of them improved after *Toxoplasma*-specific therapy. Among the 29 control patients, definite diagnoses consisted of PCNSL (15), PML (8), and brain vasculitis (1). Lumbar puncture was performed after written informed consent had been obtained, and DNA was amplified from 10 µl of unextracted heated CSF. We used this rapid sample preparation method because no difference in DNA amplification has been observed with it in comparison to standard techniques (AURELIUS et al. 1990). Primers were derived from the 35-fold repetitive region B1 (BURG et al. 1989; NOVATI et al. 1994). Amplification products of 97 bp were detected after 2% agarose gel electrophoresis and ethidium bromide staining. Nested-PCR, with a detection limit of ten tachyzoites, identified *T. gondii* DNA in CSF from five out of the 17 patients with proven TE and in none of the 29 negative controls, showing an overall sensitivity of 29.4% (95% C.I. 11.4–56.0), a specificity of 100% (95% C.I. 85.4–100), a positive predictive value of 100% (95% C.I. 46.3–100), and a negative predictive value of 70.7% (95% C.I. 54.3–83.4). We demonstrated that, if the CSF sample is collected during the first week of therapy, diagnostic sensitivity rises to 44% (95% C.I. 15.3–77.3), while if lumbar puncture is delayed, the sensitivity falls to 12.5% (95% C.I. 0.7–53.3). These results confirm that *Toxoplasma*-specific therapy rapidly induces the clearance of the parasite from CSF (NOVATI et al. 1994). The only case of *T. gondii* DNA detected in CSF after the first week of therapy was a patient with autopsy-confirmed relapse of TE, who did not improve during treatment. This case and the favorable response to therapy in seven TE patients with negative *T.gondii* PCR results in CSF after the first week of treatment suggest a correlation between detection of *T.gondii* DNA in CSF and clinical response.

Among other diagnostic techniques, histologic demonstration of tachyzoites in tissue sections remains the standard of definitive diagnosis (CENTERS FOR DISEASE CONTROL 1993). However, the sensitivity of the currently available immunohistological techniques which use labeled antibodies in immunofluorescence or the peroxidase–antiperoxidase test is moderately low. *T. gondii* can also be isolated by inoculation of brain specimens into the peritoneal cavities of mice or by tissue cell culture techniques.

2.3 Management of Patients with Suspected TE

The differential diagnosis between (i) TE, which responds well to empiric therapy (COHN et al. 1989), and (ii) PCNSL, which still needs to be diagnosed histologically (BAUMGARTNER et al. 1990), represents the most challenging question in the management of HIV-infected patients with contrast-enhancing FBL. As an indication of good response to therapy, radiological improvement of TE has been observed in 86% of patients by day 7 and in 91% of cases by day 14 (LUFT and REMINGTON 1988; LUFT et al. 1993). No difference in response rates was seen between patients treated empirically and those who had biopsy-proven TE (COHN et al. 1989). Consequently, based on the high prevalence of TE and the predictive value of contrast enhancement and positive *Toxoplasma* serology (CIMINO et al. 1991), the first choice in the management of FBL until now was to treat all patients empirically with *Toxoplasma*-specific drugs and to perform brain biopsy in the absence of a clinical or radiological response (HAVERKOS 1987; COHN et al. 1989).

A recent study found that the stage of HIV disease is a useful criterion in the differential diagnosis and treatment of contrast-enhancing FBL. Patients developing FBL early after HIV infection should be treated with empiric *Toxoplasma*-specific therapy, whereas those showing intracerebral lesions later in the course of AIDS should be directly selected for brain biopsy, due to an increased risk of PCNSL at this later stage (CHAPPELL et al. 1992). Other investigators base diagnosis of TE on the presence of multiple FBL, arguing that multifocal PCNSL is an improbable event (CIRILLO and ROSENBLUM 1991; ANSON et al. 1992; LEVY et al. 1992). Nonetheless, PCNSL has been shown to be multifocal in up to 80%–100% of cases at autopsy (MORGELLO et al. 1990). Reliability of the different clinical predictive factors strongly depends on the relative frequency of TE and PCNSL among the particular FBL population. In the different studies, this prevalence ratio varies greatly, with such differing results as 0.54 (CIMINO et al. 1991), 0.75 (COHN et al. 1989), 1.6 (CIRILLO and ROSENBLUM 1991), 4.5 (LEVY et al. 1992), and 47.0 (STEINMETZ et al. 1995) being obtained.

An algorithm suggested to improve the clinical management of patients with HIV-related contrast-enhancing FBL is shown in Fig. 1. It is based on published results and personal data, summarizing clinical variables found to be significantly associated with TE and PCNSL such as contrast enhancement,

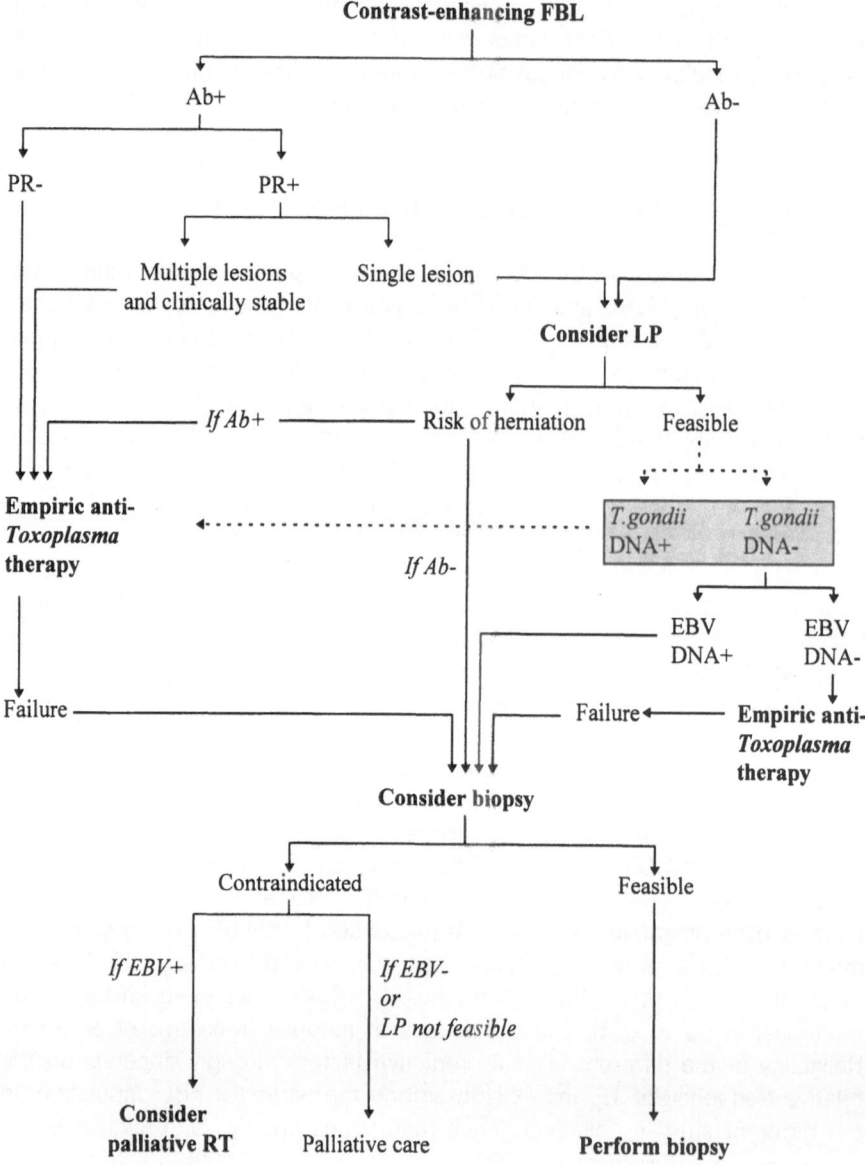

Fig. 1. Algorithm suggested for the management of contrast-enhancing AIDS-related focal brain lesions. (*Ab, Toxoplasma* serology; *PR,* anti-*Toxoplasma* prophylaxis; *LP,* lumbar puncture; *EBV,* Epstein-Barr virus; *RT,* radiotherapy)

Toxoplasma serology, biomolecular tests performed on CSF, and *Toxoplasma*-specific prophylaxis. Personal data have been obtained from the prospective study on FBL in AIDS patients (see above), employing Bayes' theorem to calculate posterior probabilities of TE and PCNSL based on their prior probability (0.47 for TE, 0.30 for PCNSL) and conditional to the outcome of predictive variables.

Our data suggest that the major predictive clinical variables employed for the presumptive diagnosis of TE in the past years, such as the presence of contrast enhancement and positive *Toxoplasma* serology (CIMINO et al. 1991; CENTERS FOR DISEASE CONTROL 1993), are now insufficient to allow an accurate identification of patients who need an empiric *Toxoplasma*-specific treatment. In fact, the probability of TE rises to 0.92 in patients who, besides the two predictive variables, do not take *Toxoplasma*-specific prophylaxis, and falls to only 0.62 in those who receive a prophylactic regimen. At the moment, empiric *Toxoplasma*-specific therapy is mandatory in patients with enhancing FBL, positive *Toxoplasma* serology, and without prophylactic treatment. Moreover, a short empiric treatment trial should be administered also in patients who, despite being on *Toxoplasma*-specific prophylaxis, show multiple CNS lesions on CT/MRI associated with a neurologically stable picture (CIRILLO and ROSENBLUM 1991; ANSON et al. 1992; LEVY et al. 1992).

If a solitary lesion is detected on neuroradiological investigation, the possibility of a safe lumbar puncture should be evaluated. Furthermore, the evaluation of feasibility of lumbar puncture represents also the initial approach to patients with negative *Toxoplasma* serology. Considering that the probability of TE among these seronegative patients is extremely low and that they may be affected by PCNSL and by other treatable disorders, biomolecular tests performed on CSF represent the crucial point in the differential diagnosis of FBL in these patients. Amplification of *T. gondii* DNA from CSF is of limited diagnostic value because of its low sensitivity; but due to the high specificity of this test it is recommended that empiric *Toxoplasma*-specific treatment should be started in the case of a positive result. PCR investigation of Epstein-Barr virus (EBV) DNA in CSF has a high sensitivity (88%–100%) and specificity (98%) for PCNSL (CINQUE et al. 1993; CINGOLANI et al. 1995; DE LUCA et al. 1995). Moreover, amplification of EBV DNA in CSF has a higher sensitivity and needs smaller samples than those required for cytology, which has been proposed by other investigators (FORSYTH et al. 1994). On the contrary, in patients who do not show EBV DNA in CSF, initial management with empiric *Toxoplasma*-specific treatment is reasonable.

Brain biopsy appears to be safe and effective in AIDS patients since it has a morbidity of less than 5% and provides a definitive diagnosis in more than 95% of FBL cases (PELL et al. 1991; CHAPPELL et al. 1992; LEVY et al. 1992; ZIMMER et al. 1992; ANTINORI et al. 1993; IACOANGELI et al. 1994). We found that variables related to HIV infection stage, such as CD4 cell counts or previous AIDS diagnosis (CHAPPELL et al. 1992), are useless for discriminating between PCNSL and TE. In fact, applying to our study the presence of a previous

AIDS-defining event as selection criterion to brain biopsy, nearly 30% of patients who had TE and only 60% of those with PCNSL would have been diagnosed if brain biopsy had been immediately performed. In addition, the mean CD4 cell counts were comparable in patients suffering from these two CNS disorders (Table 1). Brain biopsy should be considered in patients who show a clinical and/or radiological progression after 10–14 days of *Toxoplasma*-specific therapy, in those with positive EBV DNA-amplification from CSF, and in cases of negative *Toxoplasma* serology.

3 Therapy and Prophylaxis

3.1 Acute/Primary Therapy

The combination of pyrimethamine plus sulfadiazine is generally considered as the first-line treatment for acute therapy of TE in AIDS patients, whereas pyrimethamine combined with clindamycin represents a reliable alternative. In two randomized studies performed to compare pyrimethamine (100–200 mg loading dose followed by 50–75 mg/daily) plus sulfadiazine (4 g/daily) to pyrimethamine (100 mg loading dose followed by 50 mg/daily) plus clindamycin (2.4–4.8 g/daily) for primary therapy of TE showed no statistically significant difference in the response rate between the two treatment groups (DANNEMANN et al. 1991; KATLAMA ET AL. 1994). These clinical trials documented a complete or partial response in 77%–79% of patients randomized to receive pyrimethamine-sulfadiazine and 68%–77% in those assigned to pyrimethamine-clindamycin therapy. However, several factors, such as small sample size (DANNEMANN et al. 1991) or a high number of transfers to the other therapy group (KATLAMA ET AL. 1994) might have influenced these results. In fact, in the European multicenter study (KATLAMA ET AL. 1994), even if the number of patients who were transferred to the other regimen was similiar in the two treatment groups (37 in the pyrimethamine-clindamycin group and 45 in the pyrimethamine-sulfadiazine group), reasons for the change of therapy were clearly different. A change of therapy due to absence of clinical response at day 7 was more frequent in patients assigned to pyrimethamine-clindamycin than in those assigned to pyrimethamine-sulfadiazine ($p < 0.0001$). Based on this fact, the combination therapy with pyrimethamine-sulfadiazine has been defined as the treatment of choice for TE in HIV-infected patients. On the other hand, in patients randomized to this last regimen, the incidence of adverse reactions requiring treatment discontinuation was significantly higher than that observed for pyrimethamine-clindamycin ($p = 0.0001$) (DEWIT ET AL. 1994). Overall occurrence of adverse reactions was similiar in the two treatment groups. Cutaneous rash was frequently observed in patients treated with pyrimethamine combined with either sulfadiazine (39%–60%) or clindamycin (29%–43%). A sulfadiazine dose-desensitization protocol has been proposed, allowing 62% of patients with a known intolerance to sulfa compounds to continue first-line therapy (TENANT-FLOWERS et al. 1991). Hematologic toxicity, although frequently observed in both treatment groups, was rarely the reason for discontinuation. Administration of folinic acid at 10–20 mg/daily is useful in order to prevent myelotoxicity due to pyrimethamine. Gastrointestinal disturbances like diarrhea, sometimes due to *Clostridium difficile* colitis, occurred more commonly in patients treated with pyrimethamine-clindamycin. Furthermore, crystal-induced nephrotoxicity represents a well-recognized adverse reaction during sulfadiazine administration. Probability of survival was comparable in the two treatment groups, even though a trend towards an increased survival time in patients randomized to pyrimethamine plus sulfadiazine therapy was observed in one of the studies (DANNEMANN et al. 1991).

When therapy is initiated, it should be continued for 6 weeks followed by lifelong suppressive maintenance treatment. Anticonvulsants and a short course of corticosteroids should be added to the therapeutic regimen if seizures have occurred and CT/MRI show mass effects of FBL causing cerebral edema. Further drugs scheduled for the treatment of acute TE are under investigation at the moment. Atovaquone (3 g/daily) has been employed as salvage therapy in prospective noncomparative studies; a clinical improvement or stability was documented in 54%–88% after 6 weeks of treatment (KOVACS 1992; TORRES et al. 1993). However, an extremely high reccurrence rate of TE in patients treated with atovaquone as induction therapy has been shown (KATLAMA ET AL. 1994).

3.2 Maintenance Therapy

If treatment after primary therapy is discontinued, relapse of TE occurs in 53%–100% of patients (HAVERKOS 1987; COHN et al. 1989; PEDROL et al. 1990; DE GANS et al. 1992; PORTER and SANDE 1992). The time until TE recurs ranges between a few days and several months (COHN et al. 1989; PORTER and SANDE 1992). Therefore, in order to prevent relapse of TE, it is necessary that patients receive lifelong suppressive treatment with *Toxoplasma*-specific drugs. Recurrence of TE has been reported in up to 20%–30% of patients, who receive maintenance therapy and is often related to low compliance. Combination treatment with pyrimethamine and sulfadiazine showed a lower incidence rate of relapses compared with pyrimethamine combined with clindamycin (PEDROL et al. 1990; LEPORT et al. 1991; PORTER and SANDE 1992). A large, randomized, multicenter clinical trial showed a relapse rate of 7% in patients treated with pyrimethamine (25 mg/daily) plus sulfadiazine (2 g/daily) and of 28% in those receiving pyrimethamine combined with clindamycin (1.2 g/daily) (p=0.0007) (KATLAMA ET AL. 1994). Consequently, the most commonly used regimens are 25–50 mg/daily of pyrimethamine combined with 2 g/daily of sulfadiazine or 1.2–2.4 g/daily of clindamycin in patients intolerant to sulfa compounds. Daily administrations are recommended, since a twice-weekly schedule of pyrimeth-amine-sulfadiazine showed a significantly higher rate of TE recrudescence (POD-ZAMCZER et al. 1995). As in acute therapy, cutaneous rash represents the most common drug-limiting adverse reaction in patients treated with sulfadiazine, whereas diarrhea is frequently observed in those receiving clindamycin (KATLAMA ET AL. 1994). Administration of folinic acid is essential in patients who are treated with pyrimethamine in order to prevent hematologic toxicity (DE GANS et al. 1992). Although a decreased risk of *P. carinii* pneumonia was observed in patients on suppressive therapy consisting of pyrimethamine and sulfadiazine (HEALD et al. 1991; PODZAMCZER et al. 1995), inhalation of pentamidine, which does not have a systemic toxicity, should also be administered.

Both pyrimethamine alone and pyrimethamine-sulfadoxine showed some protective effect on TE relapse. However, breakthrough episodes have been observed, especially when pyrimethamine was administered at a low dose (PEDROL et al. 1990; DE GANS et al. 1992; RUF et al. 1993). In a small, open, uncontrolled study, a relapse rate of 25% was observed in patients who received atovaquone for maintenance therapy. This rate increased to 50% in those patients, who also had received atovaquone for initial acute therapy. Atovaquone, possibly combined with pyrimethamine, appears to be an alternative maintenance therapy in patients with severe intolerance to first-line treatments only (KATLAMA ET AL. 1994; MOUTHON et al. 1994). The long-term efficacies of combination therapies which use pyrimethamine with macrolides, such as azithromycin or clarithromycin, or with dapsone are currently under study.

3.3 Primary Prophylaxis

Cotrimoxazole has been shown to be effective in preventing TE in HIV-infected patients in retrospective as well as in prospective evaluations. In a retrospective study, none of the 60 patients taking cotrimoxazole as prophylaxis for *P. carinii* pneumonia developed toxoplasmosis when compared with 13% of those who received aerosolized pentamidine (CARR et al. 1992). Furthermore, in a trial that investigated secondary prophylaxis of *P. carinii* pneumonia (HARDY et al. 1992), nine out of the ten patients who developed TE were receiving aerosolized pentamidine and only one was taking cotrimoxazole at the time of diagnosis. Prospective observational studies found that the risk of developing TE ranged from 0.16 to 0.22 in patients who were treated with cotrimoxazole (JACOBSON et al. 1994; OKSENHENDLER et al. 1994). Dapsone, when combined with pyrimethamine, also showed a highly protective effect on the occurrence of TE. In a large randomized trial comparing dapsone-pyrimethamine (50 mg daily and 50 mg weekly, respectively) with aerosolized pentamidine for the combined primary prophylaxis of *P. carinii* pneumonia and TE, the adjusted relative risk of toxoplasmosis in the pentamidine group was 2.3–5.4 times higher than that observed in the dapsone-pyrimethamine group (GIRARD et al. 1993). In another trial, the adjusted relative risk of developing TE was 0.14 for patients who received dapsone-pyrimethamine once weekly (200 mg and 75 mg, respectively) (OPRAVIL et al. 1995).

We conducted a randomized, open-labeled trial to investigate the efficacy of dapsone-pyrimethamine at a low dosage of 100 mg weekly and 25 mg biweekly, cotrimoxazole (1 double-strength tablet every other day), and monthly aerosolized pentamidine as primary prophylactic regimens for *P. carinii* pneumonia and TE in HIV-infected patients with a CD4 cell count <200µl (ANTINORI et al. 1995). Compared with the pentamidine group, the incidence of TE was significantly lower in patients who were assigned to cotrimoxazole or dapsone-pyrimethamine, even though in this last treatment group a significantly higher risk of *P. carinii* pneumonia and mortality was observed than in patients receiving cotrimoxazole (ANTINORI et al. 1992). Almost equal effectiveness was confirmed in another trial for dapsone-pyrimethamine (100 mg and 50 mg twice weekly, respectively) and cotrimoxazole (1 double-strength tablet twice daily or three times weekly, respectively) (PODZAMCZER et al. 1995).

Pyrimethamine (25 mg three times weekly) as a single agent has been shown to be ineffective in the prevention of TE (MORLAT et al. 1993; JACOBSON et al. 1994). In one study, a significantly higher death rate was observed among patients who received pyrimethamine (JACOBSON et al. 1994). Since folinic acid was not simultaneously given in this trial, pyrimethamine-induced hematologic toxicity appears as a plausible explanation for the negative effect on survival. Several other agents, such as pyrimethamine plus sulfadiazine, atovaquone, azithromycin, and clarithromycin, which all have an activity against *T. gondii*, should be further investigated in order to evaluate their ability to protect against TE in humans with AIDS.

Based on these results and on the superior efficacy of cotrimoxazole in prevention of *P. carinii* pneumonia, the Centers for Disease Control (CENTERS FOR DISEASE CONTROL 1995) presently advises this drug as the first choice for the combined prophylaxis for *P. carinii* pneumonia and TE in HIV-positive individuals, who are characterized by positive *Toxoplasma* serology and CD4 cell counts <100µl. A regimen including dapsone-pyrimethamine should be administered in patients, who are intolerant to cotrimoxazole. Seronegative patients should be advised not to eat raw or undercooked meat, to wash fruits and vegetable well, and to wash their hands after touching meat, cats, or soil. In order to assess the future risk of TE and to facilitate diagnostic approach to patients with FBL, it appears useful to screen all newly observed HIV-infected persons for *T. gondii* antibodies. Seronegative subjects should be retested for *T. gondii* antibodies when their CD4 cell count falls to below 100/µl and afterwards every 3 months in order to detect seroconverted patients.

Acknowledgement. This work has been supported by the Istituto Superiore di Sanità, VIII Progetto AIDS 1995, grant n. 9305–30.

References

Ammassari A, Scoppettuolo G, Murri R, et al (1995) Temporal trends of toxoplasmic encephalitis (TE) and primary central nervous system lymphoma (PCNSL) in a prospective study on brain focal lesions. Fifth European Conference on Clinical Aspects and Treatment of HIV Infection, 26–29 September 1995, Copenhagen, Denmark, abstract 442

Anson JA, Glick RP, Reyes MR (1992) Diagnostic accuracy of AIDS-related CNS lesions. Surg Neurol 37:432–440

Antinori A, Murri R, Tamburrini E, De Luca A, Ortona L (1992) Failure of low-dose dapsone-pyrimethamine in primary prophylaxis of Pneumocystis carinii pneumonia. Lancet 340:788

Antinori A, Ammassari A, Murri R, Tumbarello M, Ortona L, Scerrati M, Roselli R, Cefaro GA (1993) Primary central nervous system lymphoma and brain biopsy in AIDS. Lancet 341:1411–1412

Antinori A, Murri R, Ammassari A, et al (1995) Aerosolized pentamidine, cotrimoxazole, and dapsone-pyrimethamine for primary prophylaxis of P. carinii pneumonia and toxoplasmic encephalitis. AIDS 12:1343–1350

Aurelius E, Johansson B, Skoldenberg B, Staland A, Forsgren M (1990) Rapid diagnosis of herpes simplex encephalitis by nested polymerase chain reaction of cerebrospinal fluid. Lancet 337:189–192

Bacellar H, Munoz A, Miller EN, Cohen BA, Besley D, Selnes OA, Becker JT, McArthur JC (1994) Temporal trends in the incidence of HIV-1-related neurologic diseases: multicenter AIDS cohort study, 1985–1992. Neurology 44:1892–1900

Baumgartner JE, Rachlin JR, Beckstead JH, Meeker TC, Levy RM, Wara WM, Rosenblum MC (1990) Primary central nervous system lymphomas: natural history and response to radiation therapy in 55 patients with acquired immunodeficiency syndrome. J Neurosurg 73:206–211

Bretagne S, Costa JM, Fleury-Feith J, et al (1995) Quantitative competitive PCR with bronchoalveolar lavage fluid for diagnosis of toxoplasmosis in AIDS patients. J Clin Microbiol 33:1662–1664

Burg JL, Grover CM, Pouletty P, Boothroyd JC (1989) Direct and sensitive detection of a pathogenic protozoan, Toxoplasma gondii, by polymerase chain reaction. J Clin Microbiol 27:1787–1792

Carr A, Tindall B, Penny R, Cooper DA (1992) Trimethoprim sulfamethoxazole appears more effective than aerosolized pentamidine as secondary prophylaxis against Pneumocystis carinii pneumonia in patients with AIDS. AIDS 6:165–171

Centers for Disease Control (1993) 1993 revised classification system for HIV infection and expanded surveillance case definition for AIDS among adolescents and adults. MMWR 41(RR17):1–19

Centers for Disease Control (1995) USPHS/IDSA guidelines for the prevention of opportunistic infections in persons infected with human immunodeficiency virus: a summary. MMWR 44:1–34

Chappell ET, Guthrie BL, Orenstein J (1992) The role of stereotactic biopsy in the management of HIV-related focal brain lesions. Neurosurgery 30:825–829

Cimino C, Lipton R, Williams A, Feraru E, Harris C, Hirschfeld A (1991) The evaluation of patients with human immunodeficiency virus-related disorders and brain mass lesions. Arch Intern Med 151:1381–1384

Cingolani A, De Luca A, Larocca LM et al (1995) EBV-DNA detection in cerebrospinal fluid for the in vivo diagnosis of AIDS-associated primary CNS lymphoma. Fifth European Conference on Clinical Aspects and Treatment of HIV Infection, 26–29 September 1995, Copenhagen, Denmark, abstract 339

Cinque P, Brytting M, Vago L, Castagna A, Parravicini C, Zanchetta N, d'Arminio-Monforte A, Wahren B, Lazzarin A, Linde A (1993) Epstein-Barr virus DNA in cerebrospinal fluid from patients with AIDS-related primary lymphoma of the central nervous system. Lancet 342:398–402

Cirillo S, Rosenblum ML (1991) Use of CT and MR imaging to distinguish intracranial lesions and to define the need for biopsy in AIDS patients. J Neurosurg 73:720–724

Cohn JA, McKeeking A, Cohen W, Jacobs J, Holzman RS (1989) Evaluation of the policy of empiric treatment of suspected Toxoplasma encephalitis in patients with the acquired immunodeficiency syndrome. Am J Med 86:521–527

Dannemann BR, Israelski DM, Leoung GS, McGraw T, Mills J, Remington JS (1991) Toxoplasma serology, parasitemia and antigenemia in patients at risk for toxoplasmic encephalitis. AIDS 5:1363–1365

de Gans J, Portegies P, Reiss P, Troost D, van Gool T, Lange JM (1992) Pyrimethamine alone as maintenance therapy for central nervous system toxoplasmosis in 38 patients with AIDS. J Acquir Immune Defic Syndr 5:137–142

De Luca A, Antinori A, Cingolani A, et al (1995) Evaluation of cerebrospinal fluid EBV-DNA and IL-10 as markers for in vivo diagnosis of AIDS-related primary central nervous system lymphoma. Br J Haematol 90:844–849

DeWit S, Katlama C, Clumeck N and the ENTA toxoplasmosis study group (1994) Tolerance of pyrimethamine-clindamycin and pyrimethamine-sulfadiazine during acute therapy of toxoplasmic encephalitis in AIDS patients in a randomized European multicentre study (ENTA 004). Fourth European Conference on Clinical Aspects and Treatment of HIV Infection, 16–18 March 1994, Milan, Italy, abstract P192

Dupon M, Cazenave J, Pellegrin JL Ragnaud JM, Cheyron A, Fischer I, Leng B, Lacut JY (1995) Detection of Toxoplasma gondii by PCR and tissue culture in cerebrospinal fluid and blood of human immunodeficiency virus-seropositive patients. J Clin Microbiol 33:2421–2426

Eggers C, Gross U, Klinker H, Schalke B, Stellbrink HJ, Kunze K (1995) Limited value of cerebrospinal fluid for direct detection of Toxoplasma gondii in toxoplasmic encephalitis associated with AIDS. J Neurol 242:644–649

Forsyth PA, Yahalom J, DeAngelis LM (1994) Combined-modality therapy in the treatment of primary central nervous system lymphoma in AIDS. Neurology 44:1473–1479

Girard PM, Landman R, Gaudebout C, Olivares R, Saimont AG, Jelazko P, Gaudebout C, Certain A, Boue F, Bouvet E, Lecompte T, Couland JP (1993) Dapsone-pyrimethamine compared with aerosolized pentamidine as primary prophylaxis against Pneumocystis carinii pneumonia and toxoplasmosis in HIV infection. N Engl J Med 328:1514–1520

Grant IH, Gold JWM, Rosenblum M, Niedzwiecki A, Armstrong D (1990) Toxoplasma gondii serology in HIV-infected patients: the development of cerebral nervous system toxoplasmosis in AIDS. AIDS 4:519–521

Grover CM, Thulliez P, Remington S, Boothroyd JC (1990) Rapid prenatal diagnosis of congenital toxoplasmosis from amniotic fluid by polymerase chain reaction. J Clin Microbiol 28:2297–2301

Hardy WD, Feinberg J, Finkelstein DM, Power ME, He W, Kaczka C, Frame PT, Holmes M, Waskin H, Fass RJ, et al (1992) A controlled trial of trimethoprim-sulfamethoxazole or aerosolized pentamidine for secondary prophylaxis of Pneumocystis carinii pneumonia in patients with the acquired immunodeficiency syndrome. N Engl J Med 327:1842–1848

Haverkos HW (1987) Assessment of therapy for toxoplasmic encephalitis. Am J Med 82:907–914

Heald A, Flepp M, Chave J-P, Malinverni R, Ruttimann S, Gabriel V, Renold C, Sugar A, Hirschel B (1991) Treatment of cerebral toxoplasmosis protects against Pneumocystis carinii pneumonia in patients with AIDS. Ann Intern Med 115:760–763

Iacoangeli M, Roselli R, Antinori A, Ammassari A, Murri R, Pompucci A, Scerrati M (1994) Experience with brain biopsy in acquired immune deficiency syndrome-related focal lesions of the central nervous system. Br J Surg 81:1508–1511

Israelski DM, Chmiel JS, Poggensee L, Phair JP, Remington JS (1993) Prevalence of Toxoplasma infection in a cohort of homosexual men at risk of AIDS and toxoplasmic encephalitis. J Acquir Immune Defic Syndr 6:414–418

Jacobson MA, Besch CL, Child C, Hafner R, Matts JP, Muth K, Wentworth DN, Neaton JD, Abrams D, Rimland D (1994) Primary prophylaxis with pyrimethamine for toxoplasmic encephalitis in patients with advanced human immunodeficiency virus disease: results of a randomized trial. J Infect Dis 169:384–394

Katlama C, De Wit S, Guichard A, Van Pottelsberghe C, Van Glabeke M, Clumeck N, and the ENTA Toxoplamosis Study Group (1994) Efficacy of pyrimethamine-clindamycin (P-C) for the long term suppressive therapy of toxoplasmosis encephalitis (TE) in AIDS patients (ENTA 04 study). Fourth European Conference on Clinical Aspects and Treatment of HIV Infection, 16–18 March 1994, Milan, Italy, abstract 043

Kovacs JA (1992) Efficacy of atovaquone in treatment of toxoplasmosis in patients with AIDS. Lancet 340:637–638

Lebech M, Lebech AM, Nelsing S, Vuust C, Mathiesen L, Petersen E (1992) Detection of Toxoplasma gondii-DNA by polymerase chain reaction in cerebrospinal fluid from AIDS patients with cerebral toxoplasmosis. J Infect Dis 165:982–983

Leport C, Raffi F, Matheron S, Katlama C, Regnier B, Saimot AG, Marche C, Vedrenne C, Vilde JL (1988) Treatment of central nervous system toxoplasmosis with pyrimethamine/sulfadiazine combination in 35 patients with the acquired immunodeficiency syndrome. Am J Med 84:94–100

Leport C, Tournerie C, Raguin G, Fernandez-Martin J, Nijongabo T, Vilde JL (1991) Long-term follow-up of patients with AIDS on maintenance therapy for toxoplasmosis. Eur J Clin Microbiol Infect Dis 10:191–193

Levy RM, Russell E, Yungbluth M, Hidvegi DF, Brody BA, Dal-Canto MC (1992) The efficacy of image-guided stereotactic brain biopsy in neurologically symptomatic acquired immunodeficiency syndrome patients. Neurosurgery 30:186–190

Luft BJ, Remington JS (1988) Toxoplasmic encephalitis. J Infect Dis 157:1–6

Luft BJ, Hafner R, Korzun AH, Leport C, Antoniskis D, Bosler EM, Bourland DD, Uttamchandani R, Fuhrer J, Jacobson J, et al (1993) Toxoplasmic encephalitis in patients with the acquired immunodeficiency syndrome. N Engl J Med 329:995–1000

Morgello S, Petito CK, Mouradian JA (1990) Central nervous system lymphoma in the acquired immunodeficiency syndrome. Clin Neuropathol 9:205–215

Morlat P, Chene G, Leport C, Rousseau F, Luft B, Aubertin J, Hafner R, Salamon R, Vilde JL (1993) Prevention primaire de la toxoplasmose cerebrale chez le sujet infecté par le VIH:resultats d'un essai randomisé en double insu, pyrimethamine versus placebo. Rev Med Interne 14:1002

Mouthon B, Katlama C, Caumes E, Bricaire F, Gentilini M (1994) Atovaquone (ATQ) as long-term suppressive therapy in toxoplasmosis. Fourth European Conference on Clinical aspects and Treatment of HIV Infection, 16–18 March 1994, Milan, Italy, abstract 043

Navia BA, Petito CK, Gold JWM, Cho ES, Jordan BD, Price RW (1986) Cerebral toxoplasmosis complicating the acquired immune deficiency syndrome:clinical and neuropathological findings in 27 patients. Ann Neurol 19:224–238

Novati R, Castagna A, Morsica G, Vago L, Tambussi G, Ghezzi S, Gervasoni C, Bisson C, d'Arminio-Monforte A, Lazzarin A (1994) Polymerase chain reaction for Toxoplasma gondii DNA in cerebrospinal fluid of AIDS patients with focal brain lesions. AIDS 8:1691–1694

Oksenhendler E, Charreau I, Tournerie C, Azihary M, Carbon C, Aboulker JP (1994) Toxoplasma gondii infection in advanced HIV infection. AIDS 8:483–487

Opravil M, Heald A, Lazzarin A, Heald A, Ruttimann S, Iten A, Furrer H, Oertle D, Praz G, Vuitton DA et al (1995) Dapsone-pyrimethamine versus aerosolized pentamidine for combined prophylaxis of PCP and toxoplasmic encephalitis. Clin Infect Dis 20:531–541

Pedrol E, Gonzalez-Clemente JM, Gatell JM, Mallolas J, Miro JM, Grans F, Alvarez R, Mercorder JM, Berenguer J, Jimenez de Anta MT et al (1990) Central nervous system toxoplasmosis in AIDS patients: efficacy of an intermittent maintenance therapy. AIDS 4:511–517

Pell MF, Thomas DGT, Whittle IR (1991) Stereotactic biopsy of cerebral lesions in patients with AIDS. Br J Neurosurg 5:585–589

Podzamczer D, Miro JM, Bolao F, et al (1995) Twice-weekly maintenance therapy with sulfadiazine-pyrimethamine to prevent recurrent toxoplasmic encephalitis in patients with AIDS. Ann Intern Med 123:175–180

Porter SB, Sande MA (1992) Toxoplasmosis of the central nervous system in the acquired immunodeficiency syndrome. N Engl J Med 327:1643–1648

Ruf B, Schürmann D, Bergmann F, Schüler-Maué W, Grünewald T, Gottschalk HJ, Witt H, Pohle HD (1993) Efficacy of pyrimethamine/sulfadoxine in the prevention of toxoplasmic encephalitis relapses and Pneumocystis carinii pneumonia in HIV-infected patients. Eur J Clin Microbiol Infect Dis 12:325–329

Schoondermark-van de Ven E, Galama J, Kraaijeveld C, van Druten J, Meuwissen J, Melchers W (1993) Value of polymerase chain reaction for the detection of Toxoplasma gondii in cerebrospinal fluid from patients with AIDS. Clin Infect Dis 16:661–666

Steinmetz H, Arendt G, Hefter H, Neuen-Jacob E, Dorries K, Aulich A, Kahn T (1995) Focal brain lesions in patients with AIDS:aetiologies and corresponding radiological patterns in a prospective study. J Neurol 242:69–74

Tenant-Flowers M, Boyle MJ, Carey D, Marriott DJ, Harkness JL, Penny R, Cooper DA (1991) Sulfadiazine desensitization in patients with AIDS and cerebral toxoplasmosis. 5:311–315

Torres R, Weinberg W, Stansell J, et al (1993) Multicenter clinical trial of atovaquone (ATQ) for salvage treatment and suppression of toxoplasmic encephalitis (TE). Ninth International Conference on AIDS, 6–11 June 1993, Berlin, Germany, abstract PO-B10-1453

Wallace MR, Rossetti RJ, Olson PE (1993) Cats and toxoplasmosis risk in HIV-infected adults. JAMA 269:76–77

Weiss LM, Udem SA, Salgo M, Tanowitz HB, Wittner M (1991) Sensitive and specific detection of Toxoplasma DNA in an experimental murine model: use of T.gondii-specific cDNA and the polymerase chain reaction. J Infect Dis 193:180–186

Zangerle R, Allerberger F, Pohl P, Fritsch P, Dierich MP (1991) High risk of developing toxoplasmic encephalitis in AIDS patients seropositive to Toxoplasma gondii. Med Microbiol Immunol 180:59–66

Zimmer C, Marzheuser S, Patt S (1992) Stereotactic brain biopsy in AIDS. J Neurol 239:394–400

Co-infection of *Toxoplasma gondii* with Other Pathogens: Pathogenicity and Chemotherapy in Animal Models

C. Lacroix[1], M. Brun-Pascaud[1], C. Maslo[2], F. Chau[1], S. Romand[2], and F. Derouin[2]

1 Introduction

Toxoplasma gondii is a protozoan parasite that is normally controlled by the host immune system and results in an asymptomatic chronic infection maintained by dormant cysts. Actually, animals and humans are simultaneously infected with *T. gondii* and by various micro-organisms during their life, and raise a specific protective immune response against each pathogen. The advent of immunosuppression related to concurrent viral infection such as human immunodeficiency virus (HIV) in humans or to the use of intensive immunosuppressive therapies is responsible for a profound alteration of protective immunity, and this often results in the recurrence of latent infections. Cerebral toxoplasmosis is one of the most common infections occurring in AIDS patients, especially in countries where *T. gondii* is highly endemic. However, other opportunistic infections due to protozoa, bacteria, and fungi are also commonly observed in these patients. Thus, co-infection with various pathogens is usual in patients with acquired immunodefiency syndrome (AIDS), and several lines of evidence suggest that interactions between the immune responses generated by each pathogen, due to cross-regulation of Th1- and Th2-activated CD4$^+$ T lymphocytes, macrophages, neutrophils and natural killer (NK) cells can result in an aggravation of one or more infections (MOSIER 1994).

[1]INSERM Unité 13, Hôpital Bichat-Claude Bernard, 170 bd Ney, 75018 Paris, France
[2]Laboratoire de Parasitologie-Mycologie, Faculté de Médecine, 15 rue de l'Ecole de Médecine, 75006 Paris, France

Furthermore, several of the pathogens infecting AIDS patients are intracellular (including HIV, *T. gondii, Mycobacterium avium-intracellulare*, MAI, and cyto-megalovirus, CMV), giving rise to the hypothesis that possible interactions between these pathogens within the cell may occur and result in alterations of cell functions.

Experimental studies in vitro or in vivo remain indispensable tools for analyzing such interactions, which may affect not only the replication of each pathogen but also the immune response of the host. In addition, a better understanding of the consequences of co-infections in immunocompromised hosts also represents a base for pharmacological studies, since new approaches for therapy and prophylaxis with broad spectrum activities are critically needed.

This review will focus on animal models of concurrent infections with *T. gondii* and several immunosuppressive viruses: murine cytomegalovirus (MCMV), feline immunodeficiency virus (FIV) and LP-BM5 murine leukemia virus (LP-BM5), and two of the most frequent opportunistic agents in AIDS patients: *Pneumocystis carinii* and MAI complex.

2 Co-infections with *Toxoplasma gondii* and Immunosuppressive Viruses

In vitro experiments with *T. gondii* and viruses show that co-infection may lead to modifications of *T. gondii* intracellular replication. Qualitative changes have been reported with CMV (GELDERMAN et al. 1968), and significantly enhanced replication of *T. gondii* is seen in HIV-infected cells (IKUTA et al. 1989; BIGGS et al. 1995). Although WELKER and collaborators (1993) did not find any interaction between *T. gondii* and HIV in concurrently infected monocytoid cells, BALA and coworkers (1994) showed that a soluble antigen of *T. gondii* can enhance HIV replication through induction of cytokine production. However, HIV-1 infection of human macrophages does not render macrophages unresponsive to cytokine activation for microbicidal activity against *T. gondii* (REED et al. 1992). Despite the limitations and drawbacks of in vitro studies, these results clearly indicate that significant alterations of cell functions occur in the dually infected cells, with possible consequences in vivo.

Thus, animal models of co-infection with *T. gondii* and an immunosuppressive virus have been developed to understand the pathogenesis of toxoplasmosis in HIV-infected hosts and provide a model of *Toxoplasma* reactivation (Tables 1 and 2). The aim of these models would also be to study new therapies with either antiparasitic, antiviral or immunomodulating drugs.

Since reactivation of *T. gondii* infection often occurs concurrently with active CMV disease in immunocompromised patients, and since CMV infection is known to increase immunosuppression, POMEROY et al. (1989) developed a

Table 1. Animal models of concurrent *T. gondii* and virus infections

Model	Strain of *T. gondii*	Viral strain	Symptoms and pathology
Cytomegalovirus (MCMV) POMEROY et al. 1992			
Mice	C56 6×10^4 tachyzoites ip	Smith strain 5×10^4 PFU ip	Pulmonary reactivation. Influx of CD8$^+$ lymphocytes with suppressor function
Feline immunodeficiency virus (FIV) LAPPIN et al. 1992, 1993			
Cats Germ free	ME 49 10^3 cysts po	Petaluma 0.5 ml blood iv	Absence of clinical reactivation. Depressed lymphocyte response to antigen
LIN et al. 1992			
Kittens Germ free	T264 10^3 cysts po	Augusta 2 ml plasma iv	Increase of parasitic burdens (brain and lymph nodes) during primary infection
DAVIDSON et al. 1993			
Cats Germ free	ME 49 10^4 tachyzoites intra-carotid	FIV-NCSU1 6×10^5 TCID$_{50}$ iv	Disseminated toxoplasmosis (primary infection). Defective lymphocyte response to antigen

PFU, plaque-forming units; TCID$_{50}$, 50% tissue culture infecting dose; po, peroral; ip, intraperitoneal; iv, intravenous

mouse model to study the effect of acute MCMV disease on latent *T. gondii* infection (Table 1). Mice were chronically infected with *T. gondii*, then challenged by MCMV. Toxoplasmosis reactivated and manifested as pneumonia. It is important to note that the severity of pneumonia was dependent on the infecting dose of MCMV. Although the mechanism of reactivation is still unknown, this model allowed a better understanding of the pathogenesis of *Toxoplasma* pneumonia. During reactivation, no change in macrophage functions has been noted, and the study of lung lymphocyte phenotypes has shown that the initial fall in the number of lung CD4$^+$ cells observed in MCMV infection may have induced the *T. gondii* reactivation. Furthermore, pneumonia appears to be related to a massive influx of CD8$^+$ T lymphocytes into the lungs (POMEROY et al. 1992). Interestingly, mice with latent toxoplasmosis were able to survive doses of MCMV that were lethal to control mice. The second major application of this model was in the study of the effect of antiviral agents on the occurrence of the pneumonia due to reactivation of toxoplasmosis, and in demonstrating the preventive efficacy of ganciclovir (BANISTER and POMEROY 1993). This drug could attenuate, but not eliminate, MCMV-induced reactivation of *T. gondii* pneumonia, but only when administered before or early after the viral challenge. Finally, this model of reactivation induced by the immunosuppressive virus MCMV revealed the major role of pulmonary infection in *Toxoplasma* pathogenesis, and demonstrated that the two pathogens interact with each other, since antiviral therapy can influence the intensity of *Toxoplasma* reactivation.

HIV infection is one of the major causes of *T. gondii* reactivation, thus animal models for human AIDS have been proposed to study the interaction of the two pathogens. Macaques infected with simian immunodeficiency virus (SIV) would represent the best animal model to investigate AIDS pathogenesis, but experimental co-infection with *T. gondii* has not been completed. Some rare cases of spontaneous reactivation of toxoplasmosis have been reported in SIV-infected macaques (LOWENSTINE et al. 1992), due probably to the low prevalence of *T. gondii* (SASSEVILLE et al. 1995), and no conclusion can be drawn from these individual reports. Belonging to the same family as HIV and SIV (lentiviridae), the FIV has been used to develop a model of AIDS in cats. Since cats are definitive hosts of *T. gondii,* such a model could be especially appropriate to investigate the effects of co-infection with *T. gondii*. The first retrospective epidemiological study on cats allowed the influence of FIV on *T. gondii* infection to be assessed (O'NEIL et al. 1991). Among cats with a positive serology for FIV, 57.1% had serological evidence of co-infection with *T. gondii*, which is not significantly different from the overall seroprevalence of *T. gondii* in cats. The most common clinical manifestation of toxoplasmosis in co-infected animals was ocular disease (77.7%), similar to that reported in FIV-naive cats, and no cat had central nervous system disease. However, while initial signs of disease were responsive to anti-*Toxoplasma* drugs, recurrences of symptoms were common. Finally, this report shows that under natural conditions of infection, it remains uncertain whether immunosuppression resulting from FIV infection predisposes cats to clinical toxoplasmosis. There are a few experimental studies of co-infection with FIV and *T. gondii* in cats which attempt to obtain a parasitic reactivation, or to study the effect of a previous FIV infection on the resistance to a primary infection with *T. gondii* (Table 1). In one study, LAPPIN et al. (1992) used cats which had been infected perorally with *T. gondii* cysts and had developed an asymptomatic infection. These animals were challenged 6 months later with FIV; after viral infection, the follow-up consisted in clinical examination, count of oocyst excretion and *T. gondii* serology. This co-infection did not result in any significant change, except for an increase in anti-*T. gondii* specific immunoglobulin M (IgM), which could be suggestive of an active toxoplasmosis, and a decrease of antigen-induced lymphocyte transformation. Two other studies examined the impact of a prior FIV-induced immunosuppression on the course of a primary infection with *T. gondii* (LIN et al. 1992; DAVIDSON et al. 1993), and concluded that FIV infection could favor the establishment of toxoplasmosis. In one study, FIV-infected cats were challenged perorally with cysts of *T. gondii* and co-infection resulted in a mild increase in parasite burdens in the brain and mesenteric lymph nodes of co-infected cats. In the second study, cats were challenged with *T. gondii* by intra-carotid injection of tachyzoites. *Toxoplasma* infection in the FIV-infected cats caused an acute systemic disease in all animals, with 75% mortality, even in those receiving a specific antitoxoplasmic therapy administered from the onset of symptoms. The predominant lesions responsible for death were interstitial pneumonitis, hepatic necrosis, and meningoencephalitis. All animals developed ocular lesions. Co-infection with *T. gondii* had apparently no effect on FIV-induced alterations of the macrophage activity, and the lymphocyte response to *T. gondii* antigen was not affected by FIV infection (LIN and BOWMAN 1992a). On the other hand, LIN and BOWMAN (1992b) noted that through an increased production of tumor necrosis factor (TNF), infection with *T. gondii* could increase immune dysfunctions caused by FIV in co-infected cats and thus favors a more rapid progression of the viral disease. However, this observation needs to be confirmed by quantification of the viral burden and assessment of immune functions during co-infection. Finally, under the experimental conditions that were used, these models of co-infection with *T. gondii* and FIV were not relevant to the study of toxoplasmic reactivation, confirming the observations that were made under natural conditions of co-infection. However, these models revealed significant interactions between the two pathogens, and the possible enhancing role of *T. gondii* on the course of the viral disease.

The murine model of AIDS (MAIDS) has been extensively used to study the pathogenesis of virus-induced immunosuppression. This model is based on infection of mice with the LP-BM5 MuLv mixture resulting, within 6–20 weeks, in a progressive immunodepression and an increased susceptibility to infection. Although this model is open to criticism, the induced syndrome is considered similar to that observed in AIDS (MORSE et al. 1992). This model has thus been used to examine dual infections of LP-BM5 with the most frequent AIDS-associated pathogens, i.e., *Cryptosporidium* (DARBAN et al. 1991), *Mycobacterium avium* (ORME et al. 1992), *Candida albicans* (COLE et al. 1992), *Plasmodium berghei* (ECKWALANGA et al. 1994), *Leishmania major* and *L. amazonensis* (BARRAL-NETTO et al. 1995; DOHERTY et al. 1995), *Trypanosoma cruzi* (SILVA et al. 1993), and *T. gondii* (GAZZINELLI et al. 1992; WATANABE et al. 1993; LACROIX et al. 1994).

The results obtained in three studies of co-infection with *T. gondii* markedly varied according to the experimental conditions, the strain and inoculum size of *T. gondii* and LP-BM5 (Table 2). Although WATANABE et al. (1993) could induce toxoplasmic reactivation which manifested as encephalitis in 100% of co-infected mice, GAZZINELLI et al. (1992) found reactivation of the parasitic infection, but only 30%–40% of mortality was seen. In this second study, in spite of an impaired production of interferon gamma (IFN-γ) in co-infected animals, additional administration of antibodies against IFN-γ or CD8+ T cells was necessary to induce a reactivation in all animals. In the same way, LACROIX et al. (1994) observed only a weak increase in pulmonary parasitic loads in co-infected mice, but no difference in mortality compared with mice solely infected with *T. gondii* or LP-BM5. In the two latter studies, the susceptibility of LP-BM5-infected mice to *T. gondii* infection was also examined. In both studies, mice infected with the viral complex were more susceptible to *T. gondii*, clinical manifestations were a disseminated and fatal toxoplasmosis (GAZZINELLI et al.

Table 2. T. gondii and LP-BM5 infection in mice (murine AIDS)

Mice	Strain of T. gondii	Viral dose	Symptoms and pathology
GAZZINELLI et al. 1992 C57Bl/6	C56 10^5 tachyzoites ME49 20 cysts ip	$10^{3.2}$ PFU ip	Inconstant cerebral reactivation (30% of cases, 5–16 weeks post infection with LP-BM5. Decrease of IFN-γ production. Increased mortality if injection of anti-CD4$^+$ or anti-IFN-γ mAb
WATANABE et al. 1993 C57Bl/6	Fukaya 10 cysts ip	3.8×10^4 PFU ip	Cerebral reactivation (100%) 9–14 weeks post LP-BM5 infection. Focal lesions. Depressed lymphocyte response to antigen
LACROIX et al. 1994 C57Bl/6	C 10 cysts po	4×10^3 PFU iv (retro-orbital)	Pulmonary reactivation during acute infection. No reactivation of chronic infection

mAb, monoclonal antibody; for other abbreviations see Table 1.

1992) or pulmonary toxoplasmosis assessed by a re-increase of pulmonary parasitic burdens after an initial clearance of the parasites (LACROIX et al. 1994). This enhanced susceptibility was associated with decreased IFN-γ production and a failure to activate CD4$^+$ and CD8$^+$ T cells (GAZZINELLI et al. 1992). These studies also showed that T. gondii infection could influence the natural course of MAIDS (GAZZINELLI et al. 1992; LACROIX et al. 1994). Indeed, while LP-BM5 infection induces progressive splenomegaly and lymphadenopathy, the weight of spleens and lymph nodes of co-infected mice was significantly lower than that of mice infected with the virus alone. These changes were associated with a reduced expression of the murine retrovirus (GAZZINELLI et al. 1992). Thus, T. gondii infection could interact with progression of the viral disease in dually infected mice. Mechanisms of interaction remain hypothetical, but possibly involve cross-regulation between the Th1 cytokines, that are largely predominant during T. gondii infection and the Th2 cytokines that are mainly involved in the progression of the viral disease.

Although these models of co-infection with LP-BM5 and T. gondii have proved to be of great interest in studying the pathogenesis of T. gondii reactivation and the in vivo interaction of parasitic infection on the course of a viral disease, no pharmacological applications have been developed to examine the preventive efficacy of antiviral or antiparasitic drugs on the occurrence of reactivated toxoplasmosis. This is due in part to the complexity of these models which are dependant on multiple factors and thus may lack reproducibility for large-scale studies of antimicrobial agents.

3 Co-infection with *Toxoplasma gondii* and Other Opportunistic Agents

Among the opportunistic infections that predominantly occur in patients with AIDS, pneumocystosis and disseminated MAI infection are the most frequent, and are almost invariably encountered at a late stage of the viral disease. As with toxoplasmosis, these infections are contained by immunity and remain latent for several years, with subsequent clinical manifestation as virus-induced immusuppression progresses. Since co-infection with these three pathogens is a common feature in AIDS patients, factors responsible for their reactivation or dissemination need to be studied concurrently in models of double or triple infection. For obvious technical reasons, such in vitro or in vivo models are difficult to establish and have mainly been developed for pharmacological studies, in an attempt to identify drugs or drug combinations which would be effective for treatment and prophylaxis of at least two opportunistic infections. Two animal models of co-infection, one with *P. carinii* and one with MAI have been developed in our laboratory and are currently being used for this purpose.

3.1 Co-infection with *Toxoplasma gondii* and *Pneumocystis carinii*

Because *P. carinii* cannot be maintained and grown in vitro, experimental studies of co-infection with this pathogen have only been performed in vivo. A model of concurrent infection has been developed in immunocompromised rats, associating a *P. carinii* pneumonia and an acute toxoplasmosis (BRUN-PASCAUD et al. 1994). For many years, rats have been extensively used for experimental pneumocystosis, but their use for studying toxoplasmosis has been limited because of their "natural" resistance to the *T. gondii* strains that are usually highly virulent in mice. The administration of corticosteroids during 5–7 weeks can induce a reactivation of the naturally acquired *P. carinii* infection, causing a progressive pneumonia, and also makes the rats susceptible to *T. gondii* infection. Thus, challenge with tachyzoites of the virulent RH strain of *T. gondii* can result in an acute toxoplasmosis which develops concurrently with *P. carinii* pneumonia. Without treatment or prophylaxis, rats die within 5 days of challenge with *T. gondii* and present typical *P. carinii* interstitial pneumonia associated with disseminated toxoplasmosis, involving spleen, lungs, liver, and brain. This model, in which a partially resistant host becomes susceptible to infection because of immunosuppression is closely comparable to the situation observed in humans. This model was found relevant for clinical treatment since it offers the unique opportunity to assess the effects of antimicrobial agents against two pathogens infecting an immunocompromised host. It has been used to confirm the efficacy of the combinations trimethoprim plus sulfamethoxazole and pyrimethamine plus dapsone against *P. carinii* and *T. gondii* and to show that other drugs like atovaquone and roxithromycin administered alone offer only a partial protection against *P. carinii* and *T. gondii,* respectively (BRUN-PASCAUD et al. 1994).

3.2 Co-infection with *Toxoplasma gondii* and *Mycobacterium avium-intracellulare* Complex

Both *T. gondii* and MAI are obligate intracellular pathogens which may interact within the same cell. BLACK et al. (1990) observed that in macrophages co-infected with *T. gondii* and MAI in vitro, the inhibition of intracellular growth of MAI that is usually observed within TNF-treated macrophages was abrogated by the presence of the parasite despite successful phagolysosomal fusion. This suggests that at a cellular level, co-infection could result in an exacerbation of one infection by the other. Such interaction has not been observed in vivo in a mouse model of concurrent chronic *Toxoplasma* and MAI infection (MASLO et al. 1995). On the contrary, co-infected mice showed less clinical symptoms than those infected by each agent alone, despite the fact that the kinetics of parasite burdens in brain and lungs and bacterial loads in lungs and spleen showed no difference between singly or dually infected mice (C. MASLO, unpublished). This was especially noted within 2–4 weeks post infection, a period at which *T. gondii* is predominantly found in the lungs, suggesting a possible interaction between the two pathogens on local pulmonary immune or inflammatory responses. Characterization of cytokines and cells involved in this response are in progress. The murine model of chronic co-infection with *T. gondii* and MAI has also been used for pharmacological studies, mainly to identify drugs that could be used for primary prophylaxis of both infections. The efficacy of rifabutin (which is effective against MAI and *T. gondii*) alone or in combination with atovaquone (which is effective against *T. gondii* cysts and tachyzoites) has been examined. These studies showed a synergistic effect of both drugs on *T. gondii* brain cysts, with a greater reduction of parasitic burdens in concurrently infected mice, compared to mice infected with *T. gondii* alone. Furthermore, a reduction of pulmonary MAI loads was observed in mice infected with MAI alone, and this reduction was more pronounced and prolonged in co-infected mice. These results imply that interactions between the two infections could markedly interfere with the mode of action and efficacy of drugs. These models of concurrent infection with *T. gondii* and other opportunistic agents are of real interest in the development of new strategies for combined prophylaxis or therapy. Our future objective will be to extend these models to a triple infection, by associating a MAI infection to the model of co-infection with toxoplasmosis and pneumocystosis.

4 Conclusions

The pathogenesis of reactivation of latent *T. gondii* infection in hosts with viral-induced immunosuppression remains poorly understood. Animal models of concurrent infection with *T. gondii* and a virus appear difficult to standardize because of the number of factors involved in the model, and are anyway distant from the human disease. Nevertheless, they are of major interest for studying the interactions between *T. gondii* and viral infections, and have highlighted the important regulatory role of the lungs in acute toxoplasmosis. In addition, these studies have revealed that *T. gondii* may influence the progression of viral disease, either in exacerbating the course of FIV infection or in reducing MAIDS progression. Additional studies are needed to assess the role of cytokines and different cell types of the immune system, as well as to quantify viral loads in infected animals. In the model of concurrent *T. gondii* and MAI infection, such interactions have also been seen to have significant effects on the efficacy of drugs. Taken together, these observations show that the pathogenesis and therapy of toxoplasmosis cannot be considered as an individual problem and should be examined within the context of multiple infections and combined therapies, a situation occurring in most patients and especially during AIDS. The models of co-infection which have been developed represent a first approach for such studies. Despite their drawbacks and limitations, these models offer a unique opportunity to analyze the individual response to each pathogen and to identify new combined therapies against toxoplasmosis and other common opportunistic infections.

Acknowledgements. This work was supported by a grant from the Agence National de la Recherche sur le Sida (ANRS) and from the Fondation pour la Recherche Médicale (FRM).

References

Bala S, Englund G, Kovacs J, Wahl L, Martin M, Sher A, Gazzinelli RT (1994) Toxoplasma gondii soluble products induce cytokine secretion by macrophages and potentiate in vitro replication of a monotropic strain of HIV. J Euk Microbiol 41:7S

Banister S, Pomeroy C (1993) Effect of Ganciclovir on murine cytomegalovirus-induced reactivation of toxoplasma pneumonia. J Lab Clin Med 122:576–580

Barral-Netto M, Santana da Silva J, Barral A, Reed S (1995) Up-regulation of T helper 2 and down-regulation of T helper 1 cytokines during murine retrovirus-induced immunodeficiency syndrome enhances susceptibility of a resistant mouse strain to Leishmania amazonensis. Am J Pathol 146:635–642

Biggs BA, Hewish M, Kent S, Hayes K, Crowe SM (1995) HIV-1 infection of human macrophages impairs phagocytosis and killing of Toxoplasma gondii. J Immunol 154:6132–6139

Black CM, Bermudez LE, Young LS, Remington JS (1990) Co-infection of macrophages modulates interferon γ and tumor necrosis factor-induced activation against intracellular pathogens. J Exp Med 172:977–980

Brun-Pascaud M, Chau F, Simonpoli AM, Girard PM, Derouin F, Pocidalo JJ (1994) Experimental evaluation of combined prophylaxis against murine pneumocystosis and toxoplasmosis. J Infect Dis 170:653–658

Cole GT, Saha K, Seshan KR, Lynn KT, Franco M, Wong PKY (1992) Retrovirus-induced immuno-deficiency in mice exacerbates gastrointestinal candidiasis. Infect Immun 60:4168–4178

Darban H, Enriquez J, Sterling CR, Lopez MC, Chen G, Abbaszadegan M, Watson R (1991) Crypto-sporidiosis facilitated by murine retroviral infection with LP-BM5. J Infect Dis 164:741–745

Davidson MG, Rottman JB, English RV, Lappin MR, Tompkins MB (1993) Feline immunodeficiency virus predisposes cats to acute generalized toxoplasmosis. Am J Pathol 143:1486–1497

Doherty TM, Morse HC, Coffman RL (1995) Modulation of specific T cell responses by concurrent infection with Leishmania major and LP-BM5 murine leukemia viruses. Int Immunol 7:131–138

Eckwalanga M, Marussig M, Dias Tavares M, Bouanga JC, Hulier E, Pavlovitch JH, Minoprio P, Portnoi D, Renia L, Mazier D (1994) Murine AIDS protects mice against experimental cerebral malaria: down-regulation by interleukin 10 of a T-helper type 1 CD4$^+$ cell-mediated pathology. Proc Natl Acad Sci 91:8097–8101

Gazzinelli RT, Hartley JW, Fredrickson TN, Chattopadhyay SK, Sher A, Morse HC (1992) Opportunistic infections and retrovirus-induced immunodeficiency: studies of acute and chronic infections with Toxoplasma gondii in mice infected with LP-BM5 murine leukemia viruses. Infect Immun 60:4394–4401

Gelderman AH, Grimley PM, Lunde MN, Rabson AS (1968) Toxoplasma gondii and cytomegalovirus: mixed infection by a parasite and a virus. Science 160:1130–1133

Ikuta K, Omata Y, Ramos MI, Nakabayashi T, Kato S (1989) Amplified replication of Toxoplasma gondii parasites in a human T-cell line persistently infected with HIV-1. AIDS 3:669–675

Lacroix C, Levacher-Clergeot M, Chau F, Sumuyen MH, Sinet M, Pocidalo JJ, Derouin F (1994) Interactions between murine AIDS (MAIDS) and toxoplasmosis in co-infected mice. Clin Exp Immunol 98:190–195

Lappin MR, Gasper PW, Rose BJ, Powell CC (1992) Effect of primary phase feline immunodeficiency virus infection on cats with chronic toxoplasmosis. Vet Immunol Immunopathol 35:121–131

Lappin MR, Marks A, Greene CE, Rose BJ, Gasper PW, Powell CC, Reif JS (1993) Effect of feline immunodeficiency virus infection on Toxoplasma gondii-specific humoral and cell-mediated im-mune responses of cats with serologic evidence of toxoplasmosis. J Vet Intern Med 7:95–100

Lin DS, Bowman DD (1992a) Macrophage functions in cats experimentally infected with feline immunodeficiency virus and Toxoplasma gondii. Vet Immunol Immunopathol 33:69–78

Lin DS, Bowman DD (1992b) Toxoplasma gondii: an AIDS enhancing cofactor. Med Hypotheses 39:140–142

Lin DS, Bowman DD, Jacobson RH (1992) Immunological changes in cats with concurrent Toxo-plasma gondii and feline immunodeficiency virus infections. J Clin Microbiol 30:17–24

Lowenstine LJ, Lerche NW, Yee JL, Uyeda A, Jennings MG, Munn RJ, McClure HM, Anderson DC, Fultz PN, Gardner MB (1992) Evidence for a lentiviral etiology in an epizootic of immune deficiency and lymphoma in stump-tailed macaques (Macaca arctoides). J Med Primatol 21:1–14

Maslo C, Perrone C, Della Bruna C, Pocidalo JJ, Derouin F (1995) Synergistic activity of rifabutin and atovaquone against chronic Toxoplasma gondii and Mycobacterium avium-intracellulare infection. ICAAC, San Francisco, Sept.17–20, Abstr. B63

Morse HC, Chattopadhyay SK, Makino M, Fredrickson TN, Hügin AW, Hartley JW (1992) Retro-virus-induced immunodeficiency in the mouse:MAIDS as a model of AIDS. AIDS 6:607–621

Mosier DE (1994) Consequences of secondary or co-infections for immunity. Curr Opin Immunol 6:539–544

O'Neil SA, Lappin MR, Reif JS, Marks A, Greene CE (1991) Clinical and epidemiological aspects of feline immunodeficiency virus and Toxoplasma gondii coinfections in cats. Am Anim Hosp Assoc 27:211–220

Orme IM, Furney SK, Roberts AD (1992) Dissemination of enteric Mycobacterium avium infections in mice rendered immunodeficient by thymectomy and CD4 depletion or by prior infection with murine AIDS retroviruses. Infect Immun 60:4747–4753

Pomeroy C, Kline S, Jordan MC, Filice GA (1989) Reactivation of Toxoplasma gondii by cytomega-lovirus disease in mice: antimicrobial activities of macrophages. J Infect Dis 160:305–311

Pomeroy C, Filice GA, Hitt JA, Jordan MC (1992) Cytomegalovirus-induced reactivation of Toxo-plasma gondii pneumonia in mice: lung phenotypes and suppressor functions. J Infect Dis 166:677–681

Reed SG, da Silva JS, Ho JL, Koehler J, Russo DM, Pihl DL, Coombs RW (1992) Cytokine activation of human macrophages infected with HIV-1 to inhibit intracellular protozoa. J Acquir Immune Defic Syndr 5:666–675

Sasseville VG, Pauley DR, MacKey JJ, Simon MA (1995) Concurrent central nervous system tox-oplasmosis and simian immunodeficiency virus-induced AIDS encephalomyelitis in a barbary macaque (Macaca sylvana). Vet Pathol 32:81–83

Silva JS, Barral-Netto M, Reed S (1993) Aggravation of both Trypanosoma cruzi and murine leukemia virus by concomitant infections. Am J Trop Med Hyg 49:589–597

Watanabe H, Suzuki Y, Makino M, Fujiwara M (1993) Toxoplasma gondii: Induction of toxoplasmic encephalitis in mice with chronic infection by inoculation of a murine leukemia virus inducing immunodeficiency. Exp Parasitol 76:39–45

Welker Y, Molina JM, Poirot C, Ferchal F, Decazes JM, Lagrange P, Derouin F (1993) Interaction between human immunodeficiency virus and Toxoplasma gondii replication in dually infected monocytoid cells. Infect Immun 61:1596–1598

Influence of Antimicrobial Agents on Replication and Stage Conversion of *Toxoplasma gondii*

U. Gross and F. Pohl

1 Introduction

As yet, only a few drugs are available that are frequently used for the treatment of human toxoplasmosis. Pyrimethamine and sulfadiazine are regularly used in combination and are the first choice for most clinical conditions. However, due to severe side effects that are observed especially with sulfadiazine, other drugs, such as clindamycin, have successfully been used as alternative treatment. Since pyrimethamine might be teratogenic when used in the first weeks of pregnancy, spiramycin has been introduced as first-choice treatment during the first 16–20 weeks of pregnancy. In addition to these four well-established drugs, others such as tetracycline or cotrimoxazole have been studied as therapeutics against toxoplasmosis. Although their effectiveness has been demonstrated to some extent, they have not been accepted as general treatment (Joss 1992). None of these drugs has been shown to completely eliminate the parasite in the human host. Since persistent infection with cysts is the source of reactivated toxoplasmosis, new drugs have been developed that

Institute of Hygiene and Microbiology, University of Würzburg, Josef-Schneider-Str. 2, 97080 Würzburg, Germany

might even be more effective against the cyst stage. The hydroxynaphtho-quinone atovaquone and macrolides such as azithromycin, clarithromycin, and roxithromycin are thought to inhibit the tachyzoite and the cyst stage as well, as has been shown in vitro or/and in vivo (ARAUJO et al. 1991,1992a,1992b; BLAIS et al. 1993a, HUSKINSON-MARK et al. 1991).

2 Mode of Action of Chemotherapeutics Against *Toxoplasma gondii*

2.1 Pyrimethamine

Pyrimethamine inhibits the enzyme dihydrofolate reductase, which is an im-portant key enzyme in the folic acid metabolism, and acts by converting di-hydrofolate to tetrahydrofolate. This reaction occurs in both mammalian and protozoan cells. Since there is little homology between this enzyme in both cell types, dihydrofolate reductase of the parasite possesses a significant higher affinity for pyrimethamine than the enzyme of the host cells, resulting in a toxicity for the parasite that is more than 1000 times greater than for the human cell. The combination of pyrimethamine with sulfadiazine, which acts synergistically, is currently the most effective therapy against the tachyzoite stage of *T. gondii* (JOSS 1992).

2.2 Sulfadiazine

Sulfadiazine, which is a chemically synthesized analog of para-aminobenzoic acid, competitively inhibits de novo synthesis of folic acid. Like pyrimethamine, sulfadiazine inhibits nucleic acid synthesis in *T. gondii*. Whereas the effect of pyrimethamine is thought to be parasiticidal, sulfadiazine is thought to inhibit replication rather than kill the parasite (DEROUIN and CHASTANG 1989). It has been shown that sulfadiazine induces more adverse side effects than any other drug routinely used for toxoplasmosis, and hypersensitivity results most often from long-term treatment (JOSS 1992).

2.3 Clindamycin

Long-term exposure to clindamycin seems to be important for the effectiveness of this drug against *T. gondii* (DEROUIN et al. 1988). This lincosamide inhibits protein synthesis of the parasite by binding to ribosomes, but has no effect on RNA synthesis of free tachyzoites (BLAIS et al. 1993b). In addition to reducing the level of replication of *T. gondii*, clindamycin impairs the ability of tachyzoites to infect host cells (BLAIS et al. 1993b). However, it is not understood why it presents selective toxicity to parasitic cells. It is possible that clindamycin acts on the multimembranous organelle that has recently been discovered in *T. gondii*. This organelle, which is located near the Golgi apparatus and the nucleus, harbors the 35-kb extrachromosomal element containing prokaryotic-type ribosomal genes. These genes might be the target for clindamycin and would thus explain the selectiveness of this drug (BECKERS et al. 1995). Since clindamycin is highly lipid-soluble and readily penetrates the eye, it is the drug of choice for treatment of ocular toxoplasmosis (TABBARA and O'CONNOR 1980). In combination with pyrimethamine, clindamycin has recently received more attention with treatment of cerebral toxoplasmosis in immunosuppressed patients, such as those suffering from the acquired immunodeficiency syndrome (AIDS) (HAVERKOS 1987).

2.4 Spiramycin

Spiramycin, a macrolide antibiotic, also acts by inhibiting protein synthesis of *T. gondii*. Although its antiparasitic activity seems to be rather low, the in vivo effect of spiramycin is explained by its exceptional persistence in tissues, especially in the placenta (SCHOONDERMARK-VAN DE VEN et al. 1994). Therefore, spiramycin is primarily used in the treatment of women in their first months of pregnancy. It is thought that spiramycin prevents spread of infection to the fetus by inhibiting the proliferation of parasites released from cysts which reside in the placenta (SCHOONDERMARK-VAN DE VEN et al. 1994).

2.5 Hydroxynaphthoquinones

Hydroxynaphthoquinones, such as atovaquone (566C80), target the parasite's mitochondrial bc_1 complex, in which electrons are transferred from ubiquinone to cytochrome c (HUDSON 1993). Investigations using *T. gondii* mutants resistant to atovaquone suggested that de novo pyrimidine synthesis is not the major biochemical target of this drug (PFEFFERKORN et al. 1993). Atovaquone seems to have a parasiticidal effect not only on tachyzoites, but also on bradyzoites. However, the drug is not able to totally eliminate the parasite from its infected host (FERGUSON et al. 1994)

2.6 New Macrolides

The new macrolides azithromycin, clarithromycin and roxithromycin are currently being evaluated for their effect on *T. gondii* (CHANG and PECHÈRE 1988). Azithromycin and clarithromycin seem to be parasitostatic (DEROUIN and CHASTANG

1990). Azithromycin, which concentrates inside infected host cells, specifically inhibits protein synthesis of *T. gondii*, without affecting RNA synthesis (BLAIS et al. 1993a).

3 Evaluation of the Effect of Drugs in Animal Experiments

In a murine model where mice were previously infected intraperitoneally with 10^4 tachyzoites of the RH strain, a dose which leads to early lung involvement and subsequent death in untreated control animals, sulfadiazine was shown to be the most effective therapy. All sulfadiazine-treated mice survived the period of treatment, and parasitic load significantly decreased after treatment. Pyrimethamine alone was less effective, protecting only 36.4% of mice. When sulfadiazine was combined with pyrimethamine, 100% of mice survived and parasites were undetectable during and after treatment. However, when the start of treatment was delayed 4 days after infection, mice were protected, with a complete clearance of parasites from blood and organs during treatment, but relapses were observed after cessation of therapy (PIKETTY et al. 1990). The efficacy of the combination of pyrimethamin and sulfadiazine was also confirmed for treatment of congenital toxoplasmosis in a rhesus monkey model. Following intravenous infection at day 90 of pregnancy (resembling the second trimester of organogenetic development in humans) with 5×10^6 RH tachyzoites, the monkeys with infected fetuses were treated throughout pregnancy with pyrimethamin, sulfadiazine, and folinic acid. No parasites could be detected in the newborns whose mothers were treated, whereas 75% of newborns whose mothers were untreated were *T. gondii* positive. The drugs, which were well-tolerated, crossed the placenta very efficiently (SCHOONDER-MARK-VAN DE VEN et al. 1995). In a similar study, spiramycin was evaluated with congenital toxoplasmosis in rhesus monkeys, also resulting in the absence of parasites in neonatal organs. Spiramycin accumulated mainly in maternal tissue, and in contrast to pyrimethamine, it could not be detected in the brains of fetuses (SCHOONDERMARK-VAN DE VEN et al. 1994).

Clindamycin was not protective in a murine model, where mice were previously infected intraperitoneally with 10^4 tachyzoites of the RH strain (PIKETTY et al. 1990).

The hydroxynaphthoquinone atovaquone (566C80) has been shown to significantly reduce the number of brain tissue cysts in mice that were intraperitoneally infected with 10–20 cysts of *T. gondii* strain ME49. Since a marked increase in the number of lysed or degenerate bradyzoites was observed upon treatment, it was postulated that atovaquone is more directed against the metabolically active immature bradyzoites than the mature cyst-stage parasites. In addition, the inflammatory response was significantly reduced in atovaquone-treated animals compared with untreated control animals (ARAUJO et al. 1992a; FERGUSON et al. 1994). When atovaquone was combined either with pyrimethamine, sulfadiazine, or clarithromycin, the survival rate of mice that had previously been infected intraperitoneally with 10^4 RH tachyzoites could be improved, whereas parasite load was not affected in comparison to mice treated with any of the agents alone (ROMAND et al. 1993).

Macrolide antibiotics affect acute toxoplasmosis differently in the murine animal model. Azithromycin protected 100% of mice which were intraperitoneally infected with 2.5×10^3 RH strain tachyzoites, compared to 70% protection with roxithromycin and 30% protection with clarithromycin. All untreated, infected mice and all mice that were treated with spiramycin died (ARAUJO et al. 1991, 1992b). No cysts could be detected in the brains of azithromycin-treated mice, indicating that this drug might be a promising alternative to agents currently in use. The effectiveness of clarithromycin can be improved by combining it with either pyrimethamine, sulfadiazine, or minocycline (ARAUJO et al. 1992b; DEROUIN et al. 1992). Like atovaquone, clarithromycin alone or in combination with minocycline reduced the total number of cysts within the brains of infected animals. However, a wide variation in the effectiveness against different *T. gondii* strains with either clarithromycin or azithromycin alone was observed (ARAUJO et al. 1991, 1992b). Therefore, *Toxoplasma* strain-dependent efficacy of antibiotics should always be considered when chemotherapeutics are under evaluation.

Doxycycline has been shown to protect 100% of mice from death. These animals had been previously infected intraperitoneally with 10^5 tachyzoites of the RH strain. However, although the mice were clinically asymptomatic, autopsies performed at termination of the experiments revealed the presence of the parasites in the brains of all animals, indicating that none of the animals were cured from *T. gondii*. When pyrimethamine was combined with doxycycline, a cure rate of 60% was achieved (CHANG et al. 1990).

4 Methods for Analyzing the Effect of Drugs on Parasite Replication in Cell Cultures

As with animal models, a method of standardization for analyzing in vitro drug activity with *T. gondii* does not exist. Although most studies used the same *T. gondii* strain, different types of host cells were used, as well as different methods for determining the rate of replication and different incubation periods. In a study performed by MACK and McLEOD (1984), murine peritoneal macrophages infected with RH strain parasites were incubated for 20 h in the presence of several drugs. The replication rate was determined by measuring uptake of [^{3}H]uracil. This incorporation is highly specific for *T. gondii* parasites because the eukaryotic host cells lack the required salvage enzyme, uracil phosphoribosyltransferase (PFEFFERKORN 1978). In this study, sulfadoxine (100 µg/ml) alone had no significant effect on parasite replication, whereas sulfadiazine (100 µg/ml) significantly inhibited parasite growth. The most effective monotherapy was with pyrimethamine (0.2 µg/ml). This effect could be enhanced by combining pyrimethamine (0.1 µg/ml) with sulfadiazine (100 µg/ml). Clindamycin had no effect on replication of either intracellular or extracellular parasites under these experimental conditions (MACK and McLEOD 1984). Using a similar experimental design, different macrolides and *T. gondii* RH tachyzoites added to monolayers of bovine turbinate cells were incubated for 20 h, and replication was determined by [^{3}H]uracil uptake (CHAMBERLAND et al. 1991). Whereas roxithromycin had no detectable activity and spiramycin showed only limited activity, with 20.16 µg/ml as the 50% inhibitory concentration (IC$_{50}$), erythromycin and azithromycin demonstrated better anti-*Toxoplasma* activity, with IC$_{50}$ of 14.38 and 8.61 µg/ml, respectively. Azithromycin was the only macrolide demonstrating inhibitory activity on the replication of intracellular tachyzoites, an effect that could be enhanced when combined with pyrimethamine (CANTIN and CHAMBERLAND 1993).

Recently performed studies indicate that onset of inhibition of replication is delayed when clindamycin and macrolide antibiotics, such as azithromycin or spiramycin, are used. It has been shown that clindamycin has little effect on parasite multiplication until many cell divisions have occurred (PFEFFERKORN and BOROTZ 1994). This might explain why clindamycin failed to demonstrate any effect in the short-term study performed by MACK and McLEOD (1984). When RH strain-infected human fibroblasts were incubated with clindamycin for at least 3 days, parasite growth could be reduced by 50% at 1 ng/ml clindamycin. At 6 ng/ml, this drug is even parasiticidal (PFEFFERKORN et al. 1992). A very recently performed study utilizing RH strain-infected human foreskin fibroblasts, demonstrated that clindamycin-treated parasites establish intracellular vacuoles in which they divide every 7 h, and the replication rate is not inhibited while they are in the first-cycle parasitophorous vacuole. When their clonal progeny ultimately rupture the host cell after 48–56 h, drug-treated parasites establish parasitophorous vacuoles inside new host cells as efficiently as untreated controls. However, replication in this second-cycle vacuole is significantly inhibited, and seems to be be dependent solely on the concentration and duration of drug treatment in the first (previous) vacuole. Efficient concentrations of clindamycin cause the parasites to emerge from the second vacuole, eventually leading to parasite death. Those treated less extensively, however, may infect a third cell where they might revert to normal replication rates (FICHERA et al. 1995). This peculiar kinetics of drug action differs from other drugs such as pyrimethamine, which inhibits parasite replication within the first parasitophorous vacuole.

An enzyme-linked immunosorbent assay (ELISA), utilizing a monoclonal antibody directed against the tachyzoite surface antigen SAG1, was used as another approach for determing replication rates of *T. gondii* (ROMAND et al. 1993). In this assay, MRC5 fibroblasts were infected with tachyzoites of the

RH strain for 4 h, followed by incubation with antibiotics for 72 h. The hydroxynaphthoquinone atovaquone was shown to be effective, with an IC_{50} of 23 ng/ml. Atovaquone has only limited absorption rates when administered perorally, demonstrated by the fact that this hydroxynaphthoquinone has remarkable in vitro activity at a very low concentration compared to necessity of high-dose treatment in vivo. The combination of atovaquone with either sulfadiazine, clarithromycin, or minocycline had no synergistic effect. Surprisingly, a significant antagonistic effect was observed with the combination of atovaquone and pyrimethamine (ROMAND et al. 1993). Since immunosuppressed patients might experience reactivation of latent *T. gondii* cysts, the effect of atovaquone against this parasite stage has also been evaluated. ME49 strain cysts isolated from the brains of infected mice were treated in vitro with 100 µg atovaquone per ml for 72 h, resulting in loss of viability of the cysts, which was determined by staining with either acridine orange and ethidium bromide or by survival of mice inoculated with treated cysts (ARAUJO et al. 1992a). In a similar study, in which cysts were incubated for 4–8 days in the presence of antibiotics, atovaquone and azithromycin separately, both at a concentration of 100 µg/ml, proved to be active against the cyst stage, whereas the same concentration of sulfadiazine had only limited effect. In contrast, clindamycin at 100 µg/ml and pyrimethamine at 5 µg/ml showed no activity against the cyst stage in vitro (HUSKINSON-MARK et al. 1991).

Other drugs that are currently not in routine therapeutic use against human toxoplasmosis were also evaluated in vitro. Doxycycline, but not tetracycline, has been shown to be active, with an IC_{50} of 6.4 µg/ml, when used to treat RH strain-infected monolayers of murine peritoneal macrophages (CHANG et al. 1990). At least 0.5 µg/ml of the antimalarial compound quinghaosu (artemisinin) was determined, by [^{3}H]uracil uptake, to inhibit parasite replication in vitro, when enterozytes or murine peritoneal macrophages were infected with RH strain tachyzoites. No inhibitory effect on intracellular *T. gondii* parasites has been observed with primaquine, mefloquine, or quinine in this experimental design (HOLFELS et al. 1994). Pentamidine, another antimalarial drug, has also been shown to be active against RH tachyzoites in vitro at a concentration of 10 or 25 µg/ml. However, in this assay where Vero cells were the host cells, pyrimethamine was superior at a concentration of 10 µg/ml (LINDSAY et al. 1991).

These studies demonstrate that a broad range of antibiotics have anti-*Toxoplasma* activity in vitro, some of which are only rarely used or not routinely used as anti-*Toxoplasma* therapeutics, such as clindamycin, atovaquone, or artemisinin. However, it should be noted that the effect of all these antibiotics was primarily evaluated with RH strain tachyzoites of *T. gondii*. Since it is not clear whether strain-specific differences exist for the action of certain drugs, other strains of *T. gondii* should be further examined.

5 Induction of Bradyzoite-Specific Antigen Expression of *T. gondii* by Antibiotics

Nearly all studies that have analyzed the activity of antibiotics in vitro have been restricted to determing their effect on replication, rather than on stage conversion. However, since the dormant cyst stage is the cause of reactivation in immunocompromised patients, it is postulated that optimal drug therapy should not only inhibit replication, but should also prevent stage differentiation from tachyzoites to bradyzoites.

Stress factors such as alkaline cell culture medium conditions, heat shock, or nitric oxide (NO) have been shown to induce the expression of bradyzoite-specific antigens (BOHNE et al. 1994; GROSS et al. 1996; SOÊTE et al. 1994). However, inhibition of parasite replication does not always correlate with stage differentiation, as was demonstrated with tachyzoite-infected and interferon-gamma (IFN-γ)-activated human fibroblasts, in which parasite replication was inhibited, but no induction of bradyzoite-specific antigens occurred (GROSS and BOHNE 1994).

The importance of NO as a molecule which inhibits the mitochondrial respiratory chain of both the host cell and the parasite could be confirmed in human fibroblasts and an isogenic mutant host cell line lacking mitochondrial DNA. Treatment with either the NO donor sodium nitroprusside or with the respiratory chain inhibitor oligomycin resulted in a lower growth rate of *T. gondii* and induction of bradyzoite-specific antigen expression in both host cell types (BOHNE et al. 1994). Since physical or chemical stress seems to be equally effective in inducing stage differentiation, the effect of several antibiotics on induction of bradyzoite-specific antigens was analyzed in vitro (U. GROSS and F. POHL, unpublished). Monolayers of murine bone marrow-derived macrophages were infected with tachyzoites from strains RH, NTE, or C56. These strains represent each of the three groups of a *T. gondii* classification system that has been recently proposed by PARMLEY et al. (1994). Following incubation with several antibiotics at different concentrations for 96 h, the amount of parasitophorous vacuoles expressing bradyzoite-specific antigens was determined by their reactivity with a stage-specific monoclonal antibody in an immunofluorescence assay. Simultaneously, the replication rate of the parasites was semiquantitatively determined by counting the average number of parasites per parasitophorous vacuole. Pyrimethamine, sulfadiazine, and atovaquone each significantly reduced replication of *T. gondii*, whereas no effect was observed with spiramycin and clindamycin under these experimental conditions. By determining [^{3}H]uracil uptake, atovaquone was shown to be superior in inhibiting parasite replication. In addition, pyrimethamine, sulfadiazine, and atovaquone also significantly induced differentiation towards the bradyzoite stage (Fig. 1). Strain-specific differences, with respect to the efficacy of the antibiotics, were not observed in this study. The antibiotics that were analyzed in this study did not significantly reduce replication without inducing

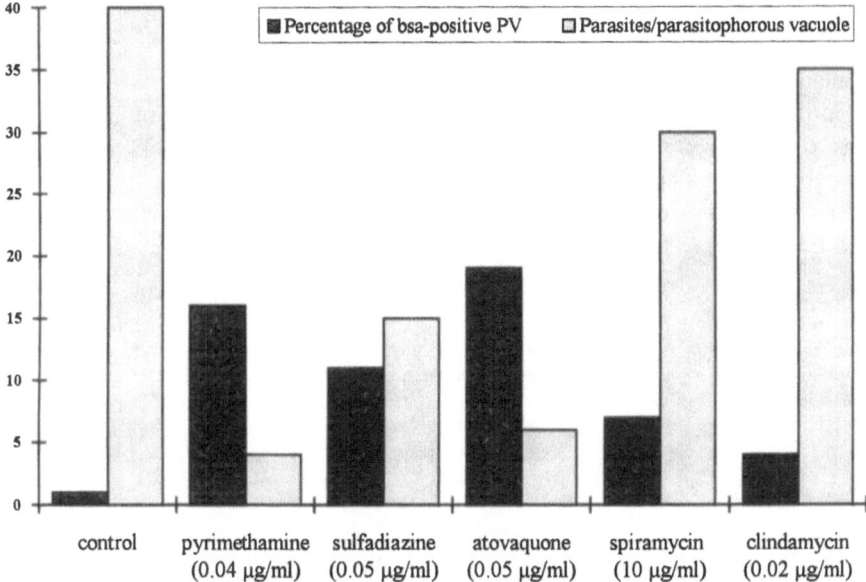

Fig. 1. Representative experiment demonstrating the in vitro activity of antibiotics on replication and tachyzoite-bradyzoite differentiation of *T. gondii*. Murine bone-marrow derived macrophages were infected with NTE-strain tachyzoites and incubated with antibiotics for 96 h. The ratio of parasitophorous vacuoles (*PV*) expressing bradyzoite-specific antigens (*bsa*) as an early indicator of tachyzoite-bradyzoite differentiation was analysed by using the bradyzoite-specific monoclonal antibody 4F8 (BOHNE et al. 1993) in an immunofluorescence assay. The parasite replication rate was determined by the absolute numbers of parasites per PV. Only the most efficient concentrations of antibiotics are compared. Numbers on y axis indicate percentage of bsa-positive PV and also absolute numbers of parasites per PV

stage differentiation. Therefore, under the conditions of this experimental design, no drug has been shown to fulfill the criteria of an optimal therapy to definitely eliminate the parasite.

6 Summary and Outlook

The combination of pyrimethamine and sulfadiazine, which is routinely used for therapy of human toxoplasmosis, has been shown to be most efficient in reducing tachyzoite replication in *T. gondii* infected animal models and cell culture. It has been demonstrated in vivo and in vitro that the recently developed antibiotics azithromycin and atovaquone are also active against the cyst stage of this protozoan parasite. Although significantly inhibiting tachyzoite replication, pyrimethamine, sulfadiazine, and atovaquone each induce the expression of bradyzoite-specific antigens and thereby probably initiate the persistence of *T. gondii*. Since optimal *Toxoplasma*-specific treatment should inhibit both replication and tachyzoite–bradyzoite stage conversion, new developments of therapeutics are necessary. One future approach could be the use of antisense oligonucleotides that specifically inhibit translation of essential and parasite-specific genes (SARTORIUS and FRANKLIN 1991), for example, genes encoding rhoptry antigens or ribosomal genes of the 35-kb extrachromosomal element (BECKERS et al. 1995). Alternatively, modulation of cytokine production which selectively inhibits parasite replication and tachyzoite–bradyzoite differentiation, could be another promising therapeutical approach killing the parasite within the host.

Acknowledgements. We would like to thank Christiana Cooper for critical reading of the manuscript. Studies performed by the authors have been supported by the Bundesministerium für Bildung und Forschung (01 KI 9454), the Deutsche Forschungsgemeinschaft (Gr 906/5–2) and the European Union (BMH1-CT92-1535).

References

Araujo FG, Shepard RM, Remington JS (1991) In vivo activity of the macrolide antibiotics azithromycin, roxithromycin and spiramycin against Toxoplasma gondii. Eur J Clin Microbiol Infect Dis 10:519–524

Araujo, FG, Huskinson-Mark J, Gutteridge WE, Remington JS (1992a) In vitro and in vivo activities of the hydroxynaphoquinone 566C80 against the cyst form of Toxoplasma gondii. Antimicrob Agents Chemother 36:326–330

Araujo FG, Prokocimer P, Lin T, Remington JS (1992b) Activity of clarithromycin alone or in combination with other drugs for treatment of murine toxoplasmosis. Antimicrob Agents Chemother 36:2454–2457

Beckers CJ, Roos DS, Donald RG, Luft BJ, Schwab JC, Cao Y, Joiner KA (1995) Inhibition of cytoplasmic and organellar protein synthesis in Toxoplasma gondii. Implications for the target of macrolide antibiotics. J Clin Invest 95:367–376

Blais J, Garneau V, Chamberland S (1993a) Inhibition of Toxoplasma gondii protein synthesis by azithromycin. Antimicrob Agents Chemother 37:1701–1703

Blais J, Tardif C, Chamberland S (1993b) Effect of clindamycin on intracellular replication, protein synthesis, and infectivity of Toxoplasma gondii. Antimicrob Agents Chemother 37:2571–2577

Bohne W, Heesemann J, Gross U (1993) Induction of bradyzoite-specific Toxoplasma gondii antigens in gamma interferon-treated mouse macrophages. Infect Immun 61:1141–1145

Bohne W, Heesemann J, Gross U (1994) Reduced replication of Toxoplasma gondii is necessary for induction of bradyzoite-specific antigens:a possible role for nitric oxide in triggering stage conversion. Infect Immun 62:1761–1767

Cantin L, Chamberland S (1993) In vitro evaluation of the activities of azithromycin alone and combined with pyrimethamine against Toxoplasma gondii. Antimicrob Agents Chemother 37:1993–1996

Chamberland S, Kirst HA, Current WL (1991) Comparative activity of macrolides against Toxoplasma gondii demonstrating utility of an in vitro microassay. Antimicrob Agents Chemother 35:903–909

Chang HR, Pechère JC (1988) In vitro effects of four macrolides (roxithromycin, spiramycin, azithromycin (CP-62, 993), and (A-56268) on Toxoplasma gondii. Antimicrob Agents Chemother 32:524–529

Chang HR, Comte R, Pechère JC (1990) In vitro and in vivo effects of doxycycline on Toxoplasma gondii. Antimicrob Agents Chemother 34:775–780

Derouin F, Chastang C (1989) In vitro effects of folate inhibitors on Toxoplasma gondii. Antimicrob Agents Chemother 33:1753–1759

Derouin F, Chastang C (1990) Activity in vitro against Toxoplasma gondii of azithromycin and clarithromycin alone and with pyrimethamine. J Antimicrob Chemother 25:708–711

Derouin F, Nalpas J, Chastang C (1988) Mesure in vitro de l'effet inhibiteur de macrolides, lincosamides et synergestines sur la croissance de Toxoplasma gondii. Path Biol 36:1204–1210.

Derouin F, Caroff B, Chau F, Prokocimer P, Pocidalo JJ (1992) Synergistic activity of clarithromycin and minocycline in an animal model of acute experimental toxoplasmosis. Antimicrob Agents Chemother 36:2852–2855

Ferguson DJP, Huskinson-Mark J, Araujo FG, Remington JS (1994) An ultrastructural study of the effect of treatment with atovaquone in brains of mice chronically infected with the ME49 strain of Toxoplasma gondii. Int J Exp Path 75:111–116

Fichera ME, Bhopale MK, Roos DS (1995) In vitro assays elucidate peculiar kinetics of clindamycin action against Toxoplasma gondii. Antimicrob Agents Chemother 39:1530–1537

Gross U, Bohne W (1994) Toxoplasma gondii: strain- and host cell dependent induction of stage conversion. J Euk Microbiol 41:10S–11S

Gross U, Bohne W, Soête M, Dubremetz JF (1996) Developmental differentiation between tachy-zoites and bradyzoites of Toxoplasma gondii. Parasitol Today 12:30–33

Haverkos HW (1987) Assessment of therapy for toxoplasma encephalitis. Am J Med 82:907–914

Holfels E, McAuley J, Mack D, Milhous WK, McLeod R (1994) In vitro effects of artemisinin ether, cycloguanil hydrochloride (alone and in combination with sulfadiazine), quinine sulfate, meflo-quine, primaquine phosphate, trifluoperazine hydrochloride, and verapamil on Toxoplasma gondii. Antimicrob Agents Chemother 38:1392–1396

Hudson AT (1993) Atovaquone – a novel broad-spectrum anti-infective drug. Parasitol Today 9:66–69

Huskinson-Mark J, Araujo FG, Remington JS (1991) Evaluation of the effect of drugs on the cyst form of Toxoplasma gondii. J Infect Dis 164:170–177

Joss AWL (1992) Treatment. In :Ho-Yen DO, Joss AWL (eds) Human toxoplasmosis. Oxford University Press, Oxford, pp 119–143

Lindsay DS, Blagburn BL, Hall JE, Tidwell RR (1991) Activity of pentamidine and pentamidine analogs against Toxoplasma gondii in cell cultures. Antimicrob Agents Chemother 35:1914–1916

Mack DG, McLeod R (1984) New micromethod to study the effect of antimicrobial agents on Toxoplasma gondii: comparison of sulfadoxine and sulfadiazine individually and in combination with pyrimethamine and study of clindamycin, metronidazole, and cyclosporin A. Antimicrob Agents Chemother 26:26–30

Parmley SF, Gross U, Sucharczuk A, Windeck T, Sgarlato GD, Remington JS (1994) Two alleles of the gene encoding surface antigen P22 in 25 strains of Toxoplasma gondii. J Parasitol 80:293–301

Pfefferkorn ER (1978) Toxoplasma gondii: the enzymatic defect of a mutant resistant to 5-fluo-rodeoxyuridine. Exp Parasitol 44:26–35

Pfefferkorn ER, Borotz SE (1994) Comparison of mutants of Toxoplasma gondii selected for re-sistance to azithromycin, spiramycin, or clindamycin. Antimicrob Agents Chemother 38:31–37

Pfefferkorn ER, Nothnagel RF, Borotz SE (1992) Parasiticidal effect of clindamycin on Toxoplasma gondii grown in cultured cells and selection of a drug-resistant mutant. Antimicrob Agents Chemother 36:1091–1096

Pfefferkorn ER, Borotz SE, Nothnagel RF (1993) Mutants of Toxoplasma gondii resistant to atova-quone (566C80) or decoquinate. J Parasitol 79:559–564

Piketty C, Derouin F, Rouveix B, Pocidalo JJ (1990) In vivo assessment of antimicrobial agents against Toxoplasma gondii by quantification of parasites in the blood, lungs, and brain of infected mice. Antimicrob Agents Chemother 34:1467–1472

Romand S, Pudney M, Derouin F. (1993) In vitro and in vivo activities of the hydroxynaphthoquinone atovaquone alone or combined with pyrimethamine, sulfadiazine, clarithromycin, or minocycline against Toxoplasma gondii. Antimicrob Agents Chemother 37:2371–2378

Sartorius C, Franklin RM (1991) The use of antisense oligonucleotides as chemotherapeutic agents for parasites. Parasitol Today 7:90–94

Schoondermark-van de Ven E, Melchers W, Camps W, Eskes T, Meuwissen J, Galama J (1994) Effectiveness of spiramycin for treatment of congenital Toxoplasma gondii infection in rhesus monkeys. Antimicrob Agents Chemother 38:1930–1936

Schoondermark-van de Ven E, Galama J, Vree T, Camps W, Baars I, Eskes T, Meuwissen J, Melchers W (1995) Study of treatment of congenital Toxoplasma gondii infection in rhesus monkeys with pyrimethamine and sulfadiazine. Antimicrob Agents Chemother 39:137–144

Soête M, Camus D, Dubremetz JF (1994) Experimental induction of bradyzoite-specific antigen expression and cyst formation by the RH strain of Toxoplasma gondii in vitro. Exp Parasitol 78:361–370

Tabbara KF, O'Connor GR (1980) Treatment of ocular toxoplasmosis with clindamycin and sulfa-diazine. Ophthalmology 87:129–134

Molecular Genetic Tools for the Identification and Analysis of Drug Targets in *Toxoplasma gondii*

D.S. Roos

1 Introduction

Chemotherapy for acute toxoplasmosis has classically relied on inhibitors of the folate metabolic pathway (BROOKS et al. 1987), typically a synergistic combination of sulfonamides (inhibitors of dihydropteroate synthase, which produces folic acid) and inhibitors of dihydrofolate reductase (e.g., pyrimethamine). Unfortunately, the combination of pyrimethamine and sulfonamides has been less successful against toxoplasmic encephalitis associated with acquired immunodeficiency syndrome (AIDS) (LUFT and REMINGTON 1992). Chronic treatment must be maintained to guard against the re-emergence of parasites from latent tissue cysts, as the bradyzoite forms are insensitive to most metabolic inhibitors. Long-term sulfonamide administration often produces a severe hypersensitivity response, however, and pyrimethamine alone is usually insufficient to prevent relapse; prolonged pyrimethamine therapy may also result in bone marrow depression (HAVERKOS 1987; LEPORT et al. 1988; TENANT-FLOWERS et al. 1991). Moreover, reliance on chronic treatment raises the fear that drug-

Department of Biology, University of Pennsylvania, 415 S. University Avenue, Philadelphia PA 19104–6018, USA

resistant parasites may emerge. These concerns have lent renewed impetus to the development of improved treatment protocols, the identification of novel parasiticidal agents, and studies on the metabolism of the parasite, with an eye toward more effective drug therapy (LAUGHON et al. 1991).

Several alternatives to antifolate therapy have been reported in recent years, but no generally applicable treatment for AIDS-associated toxoplasmosis has yet been developed (ARAUJO and REMINGTON 1992). In conjunction with pyrimethamine, antibiotics of the macrolide/lincosamide/azalide class (such as clindamycin and azithromycin) have been used successfully for initial and maintenance therapy of toxoplasmic encephalitis (KATLAMA 1991; DANNEMAN et al. 1992). Although these compounds are well-characterized as inhibitors of protein synthesis in prokaryotic systems (STEIGBIGEL 1990), they appear not to target the cytoplasmic ribosomes of *Toxoplasma* (BECKERS et al. 1995), and their mechanism of action against *Toxoplasma gondii* remains unclear (PFEFFERKORN and BOROTZ 1994; FICHERA et al. 1995). Other compounds have been identified which exhibit clear activity against *T. gondii* in vitro, including atovaquone (ARAUJO and REMINGTON 1992), cyclosporins (HIGH et al. 1994), and various suspected cytoskeletal inhibitors (STOKKERMANS et al. 1996), but the precise targets of these drugs are also unknown. A further source of potential anti-*Toxoplasma* agents are inhibitors known to be active against *Plasmodium* and other Apicomplexan parasites; agents developed for veterinary coccidiosis provide a particularly rich literature (RICKETTS and PFEFFERKORN 1993). Once again, however, development and optimization of these compounds for use in toxoplasmosis is compromised by the lack of known targets. This brief review explores the potential of recently developed molecular genetic technology for the elucidation of drug targets and resistance mechanisms in *T. gondii*.

2 Genetic Tools

Quite aside from its importance as a human pathogen, *T. gondii* has received considerable attention as a practical system for experimental manipulation (PFEFFERKORN 1990; ROOS et al. 1994; BOOTHROYD et al. 1995). The parasite is an obligate intracellular pathogen, but can be grown in vitro in a wide range of host cells and readily forms plaques, facilitating quantitation and clonal isolation. Its wide host range permits experiments conceptually similar to somatic cell genetic studies, taking advantage of mammalian cell mutants to ask what the parasite can do for itself and what it requires from its host (Pfefferkorn 1990). *T. gondii* parasites undergo a sexual cycle in cats, making classical genetic crosses feasible (PFEFFERKORN and PFEFFERKORN 1980) and permitting classical genetic mapping (SIBLEY et al. 1992).

Until recently, no molecular genetic system has been available for the investigation of *Toxoplasma* (or any other Apicomplexan parasite, for that matter), but the parasite's relatively undistinguished genetic structure – typical eukaryotic gene organization, codon usage, etc. (Roos 1993; ELLIS et al. 1993) – and small, haploid genome suggested that such an approach might be possible. Through the combined efforts of several laboratories, various transformation systems have now been developed (reviewed in Roos et al. 1994; BOOTHROYD et al. 1995). Explorations into the molecular genetics of *Toxoplasma* can now take advantage of the following strategies (among others):

- Transient expression of recombinant genes in parasite tachyzoites (with a choice of several promoters). Bradyzoite-specific expression may soon be possible as well (PARMLEY et al. 1994; BOHNE et al. 1995)
- Functional mapping of regulatory domains, using several reporter genes
- Stable transformation of parasites with recombinant material, using various selectable markers either linked to the selectable marker or co-transfected as distinct plasmids
- Overexpression of transgenes via amplification of co-integrated selectable markers
- Complementation cloning via introduction of DNA from mutant strains into wild-type parasites (or vice-versa)
- Insertional inactivation, using high-frequency transformation vectors which integrate into the genome by nonhomologous recombination
- 'Pseudodiploid' production (duplication of an endogenous locus in the haploid tachyzoite form), facilitating the study of essential genes as heterozygotes
- Genetic deletions (gene knockouts), by homologous recombination
- Allelic replacement, facilitating analysis of mutant alleles by removing the endogenous wild-type allele.

Of these many tools, complementation cloning and insertional mutagenesis are probably the most useful with respect to identification of drug-resistance genes. The former approach is more generally applicable, as it permits identification of the dominant mutations (positive selectable markers), which may be responsible for drug resistance. The latter approach permits identification of negative selectable markers (targets which can be selected against). Other tools on the above list provide assistance in validating the function of drug-resistance alleles once cloned, via gene knockouts and allelic replacements, transgenic expression of wild-type, mutant, and experimentally altered genes.

3 Identification of Drug Targets

In cases where a putative drug target has been identified and homologs are available from other organisms, it may be possible to identify the gene of interest using a variety of molecular genetic techniques (i.e., cross-hybridization with DNA or antibody probes, polymerase chain reaction (PCR) amplification based on primers designed from conserved sequences). Genes targeted by antifolate chemotherapy have been identified in this way (ROOS 1993; PASHLEY et al. 1995). Unfortunately, however, targets for many of the more recently developed drugs which show activity against *Toxoplasma* are unknown, as noted above (ARAUJO and REMINGTON 1992; RICKETTS and PFEFFERKORN 1993). Direct identification of drug targets typically involves careful biochemical and metabolic studies – often a laborious process, and one which genetic techniques cannot necessarily expedite (although molecular approaches can facilitate the validation of drug targets; see below). Molecular genetic techniques are well-suited to the identification of drug *resistance* mechanisms, however, as resistance mutations provide a powerful genetic selection. In some cases, mutations conferring resistance may arise within the gene encoding the drug target itself, but even where other resistance mechanisms are involved (e.g., transport mutations or detoxification mechanisms), identification of these genes is critical to the development of treatment strategies.

3.1 Complementation Cloning

Starting with a drug-resistant mutant (or any other identifiable phenotype) produced by chemical mutagenesis (PFEFFERKORN and PFEFFERKORN 1979) or other techniques, the high frequency of transformation reported for *Toxoplasma* (DONALD and ROOS 1993, 1994) should permit identification of the responsible gene by complementation. Electroporation of 10^7 extracellular *T. gondii* tachyzoites typically leaves about 10^6 viable parasites. If 5% of these transformants stably integrate transfected DNA (DONALD and ROOS 1994), this yields about 5×10^4 transgenic parasites. Assuming transfected DNA fragments of 40 kb in size (i.e., the size of cosmid inserts), the total amount of transgenic DNA in a single transfection would be 2×10^9 bp, or approximately 25 genome-equivalents. These calculations suggest that it should be possible to clone any dominant single-locus drug-resistance gene in a single electroporation.

To date, this approach has only been put to the test for mutations which confer resistance to pyrimethamine. In the case of known dihydrofolate reductase (DHFR) mutations, the transfer of drug-resistant alleles occurs at about five- to ten-fold lower frequency than predicted (i.e., 2–5 pyrimethamine-resistant parasites per transfection) – a frequency, which is still sufficient to clone this single copy gene by complementation (R.G. DONALD and D.S.ROOS, unpublished). Taking advantage of detailed maps available for the DHFR-TS locus, it has been possible to demonstrate that complementation cloning is independent of (i) the size of DNA fragments transfected (as long as the fragments are sufficiently large that they contain the entire resistance gene), (ii) the method used to generate DNA fragments (shearing versus partial or complete digestion with various different restriction enzymes), and (iii) the position of the resistance gene within the transfected fragment. Experience with transfection of cloned plasmid DNA suggests comparable frequencies of transformation using circular versus linear DNA, although this has not been confirmed under conditions suitable for complementation.

The generality of this approach remains untested, but two arguments suggest that experience with pyrimethamine-resistant DHFR genes will prove broadly applicable. First, mutant DHFR genes integrated at various different sites throughout the genome (by nonhomologous recombination) are as easily rescued as mutant genes at the endogenous DHFR locus. Second, preliminary results suggest that it is possible to rescue non-DHFR-based pyrimethamine-resistance genes (unpublished work in progress).

Having demonstrated the feasibility of transferring a drug-resistance (or other) gene into wild-type parasites by transfecting total DNA from a mutant strain, it is then necessary to identify the resistance locus. Several strategies have been used in other genetic systems. Perhaps the easiest approach is to co-transfect a plasmid containing a bacterial origin of replication and drug-resistance marker, and to hope that this plasmid will integrate at the same site as the transfected parasite DNA. Bacterial sequences then serve as a marker for plasmid rescue (see below), or for screening genomic libraries prepared from drug-resistant transgenic parasites. Recent experiments have shown that co-transfection is indeed feasible in *Toxoplasma* (BLACK et al. 1995).

Alternatively, a library can be prepared from drug-resistant transgenic parasites, introducing genomic sequences into a bacterial plasmid vector (which might also be engineered to include a parasite selectable marker). This method may prove more efficient in the long run, as it minimizes concerns about the linkage between bacterial sequences and the drug-resistance locus, and should facilitate plasmid rescue. Better still, a library could be prepared in a shuttle vector, which remains episomal in transfected parasites (PARENT et al. 1985). This method would make plasmid rescue a trivial matter, but awaits the development of stable episomal vectors for *Toxoplasma*. Ideally, both single-copy and high-copy episomal vectors could be used to identify both strong and weak resistance alleles, and to permit identification of high copy-number resistance genes or suppressor genes (COTRIM et al. 1994).

3.2 Insertional Mutagenesis

Using vectors which integrate throughout the parasite genome – apparently at random – by nonhomologous recombination (DONALD and ROOS 1993, 1995), it is also possible to produce insertional mutants in *Toxoplasma*. In contrast to transfection with total parasite DNA for complementation cloning (as outlined above), insertional mutagenesis employs a single cloned plasmid containing a fully functional pyrimethamine-resistant DHFR-TS allele (DONALD and ROOS 1993, 1995; DONALD et al. 1996). As for complementation strategies, however, the feasibility of insertional mutagenesis relies on high-frequency transformation of parasite tachyzoites (DONALD and ROOS 1993, 1994; BLACK et al. 1995).

Assuming that integration is indeed random throughout the parasite's 8×10^7 bp genome, the 5×10^4 transgenic parasites produced (see above) should represent an average of one insertion every 1.6 kb within the population of transfected parasites. In practice, targeting three independent loci has yielded frequencies of approximately one insertional mutant per transformation (DONALD and Roos 1995, and unpublished observations). It is not clear how closely this frequency corresponds to predictions, as gene target size is difficult to determine. The few insertion sites precisely identified to date all map to coding sequences, conserved splice junctions, or putative promoter domains, suggesting that insertion within introns or downstream of the coding sequence may not disrupt expression. Thus, while the uracil phosphoribosyl transferase (UPRT) locus (for example) spans about 6 kb, the effective target size for insertional mutagenesis is probably < 2 kb. Regardless, it is clear that the frequency and randomness of non-homologous recombination using insertional mutagenesis vectors is sufficiently close to predictions for practical application.

It is important to note that insertional mutagenesis is a rather blunt tool compared with chemical mutagenesis, as the technique is most likely to inactivate genes rather than to more subtly modify their function (in rare cases, however, insertion may alter gene/protein expression or stability). As a result, essential genes are not generally identifiable by insertional mutagenesis; the best targets for insertional mutagenesis are negative selectable markers, where the loss of genes produces a detectable phenotype. With respect to drug-resistance mechanisms, this means that insertional mutagenesis is only capable of identifying the enzymatic target for subversive or suicide substrates, and only in nonessential genes. While this bodes ill for the identification of drug targets themselves (for example, integration into the pyrimethamine target DHFR is expected to produce a dead cell, rather than a drug-resistant mutant), other resistance mechanisms are potentially accessible, such as transporters or detoxifying enzymes which interact with the drug in question (but whose action is not essential for the native substrate or other cellular functions). Despite its disadvantages, the ease of insertional mutagenesis (and identification of the tagged gene) makes this an appealing, if not guaranteed, approach.

Once a drug-resistant mutant has been identified by insertional mutagenesis and clonally rederived (by limiting dilution, plaque purification, or flow cytometry), identification of the tagged locus is greatly facilitated by the presence of bacterial sequences on the mutagenesis plasmid. Genomic DNA from the mutant, digested so as to produce a fragment containing both the bacterial vector and adjacent parasite gene, is recircularized by ligation at low concentration and rescued by transformation in *Escherichia coli* under ampicillin selection. Initial attempts at rescuing insertional mutants were complicated by disruption of essential bacterial sequences during the integration of circular plasmids, and by difficulties inherent in mapping and rescueing large DNA fragments including the entire transfected plasmid plus flanking DNA. These problems have now been solved by transfection with a linearized plasmid, producing transgene integrations of defined organization. This permits separation of the pyrimethamine-resistance marker from the bacterial sequences and rescue of the intact bacterial plasmid (DONALD and ROOS 1995; DONALD et al. 1996). Surprisingly little exonuclease activity has been observed when linearized plasmids are transfected into *T. gondii* parasites, and rearrangement of single plasmid insertions is often minimal (DONALD et al. 1996).

To minimize difficulties associated with the integration of multiple transgenes, it is important not to transfect with excess plasmid (optimum conditions have not yet been established, but transfecting >50 µg plasmid is clearly detrimental). A new generation of insertional mutagenesis vectors incorporating rare restriction sites has been designed to prevent rescue of the intact transfecting plasmid from tandem head-to-tail insertions (W.J. SULLIVAN and D.S. ROOS, unpublished). Perhaps the most important strategy for insertional mutagenesis is to generate multiple independent mutants which can be probed with rescued genomic fragments. Disruption of the same locus in independent mutants confirms the target prior to the more laborious work of cDNA and genomic cloning, targeted knockout, and retransformation (DONALD and ROOS 1995; DONALD et al. 1996). In our laboratory, approximately 50% of insertional tags have proved rescuable, although library construction provides an alternative approach for the identification of tagged loci (as discussed in Sect. 3.1, above).

4 Validation of Drug Targets

The ability to express recombinant material in transgenic parasites, coupled with techniques for generating targeted gene knockouts, provides the means to test putative drug-resistance genes identified by either complementation cloning or insertional mutagenesis. In the case of dominant mutations cloned by complementation, direct expression of the mutant gene is likely to confer drug resistance to mutant parasites. For further analysis, perfect allelic replacement of the wild-type locus with the mutant provides the ultimate proof; gene targeting can also be employed to determine whether the gene in question is essential. For negative selectable markers identified by insertional mutagenesis, this schedule is reversed: targeted gene knockouts provide validation of the candidate gene, and gene function can be assessed by the expression of wild-type or mutant transgenes in the knockout background.

4.1 Transgene Expression

Various transgenic expression systems are currently available for *Toxoplasma*, including vectors employing 5' and 3' flanking sequences derived from genes encoding the parasite's major surface antigen P30 (SAG1), ROP1 (a rhoptry antigen), β-tubulin, DHFR-TS, and hypoxanthine-guanine phosphoribosyl transferase (HXGPRT) (SOLDATI and BOOTHROYD 1993; DONALD and ROOS 1993, DONALD et al. 1996). Although limited deletion-mapping has been carried out for several genes (SOLDATI and BOOTHROYD 1995; A.D. SAGAR et al., unpublished), promoter sequences remain poorly defined in *Toxoplasma*. Sequences of the 3′ flanking site appear to be essential for certain promoters but not for others (A.L.C. MOULTON and D.S. ROOS, unpublished).

Transient transformation frequencies in *Toxoplasma* are sufficiently high that for many applications it is possible to assay function without isolating stable transformants. Expression of a chloramphenicol acetyltransferase (CAT) reporter gene under control of DHFR-TS 5′ and 3′ regulatory sequences is evident in at least 50% of transfected parasites, for example (and indirect evidence suggests that this may be a significant underestimate; unpublished observations). Populations of UPRT – or HXGPRT – knockout mutants transfected with transient expression vectors harboring wild-type UPRT or HXGPRT genes (respectively) express up to 50% of the enzyme activity observed in wild-type parasites (DONALD and ROOS 1995; DONALD et al. 1996). Stable parasite transformants expressing genes of interest can be produced by co-transfection with a variety of linked selectable markers (DONALD and ROOS 1993; KIM et al. 1993; SIBLEY et al. 1994). Alternatively, drug-resistance genes of interest can be used as selectable markers themselves.

4.2 Gene Knockouts and Allelic Replacements

In order to confirm that putative drug-resistance loci identified by insertional mutagenesis are indeed responsible for the observed phenotype, it is necessary to prepare a defined knockout mutant, deleting the genomic locus in question. Similarly, knockouts are also useful in characterizing targets identified by chemical mutagenesis techniques. Several strategies have been employed for generating gene knockouts in *T. gondii,* all of which rely on cloned fragments of the endogenous locus obtained from genomic libraries (KIM et al. 1993; DONALD and ROOS 1994, 1995). Although the mechanism by which long stretches of homologous DNA enhance the frequency of homologous recombination is not well understood (and frequencies appear to differ at different loci), experience with several genes suggests that homologous recombination occurs more frequently than nonhomologous recombination when ≥8 kb of genomic sequence is available, and less frequently when smaller fragments are used (DONALD and ROOS 1994, 1995, unpublished observations).

The most straightforward approach for gene targeting is a direct, single-step knockout, in which essential sequences are deleted from a cloned fragment of genomic DNA, which is then transfected into wild-type parasites and screened (or selected, when possible) to identify deletion mutants. Successful implementation of this approach relies on a high-efficiency screen/selection.

This is most easily accomplished for negative selectable markers, which provide direct selection for loss of the endogenous locus (DONALD and ROOS 1995). For loci whose loss cannot be directly selected, a marker may be introduced into the cloned genomic DNA derived from the locus in question, preferably in such a way as to delete essential sequences from the target locus (thereby preventing any possibility of reversion to wild-type). In this case, positive selection for the introduced marker identifies transgenic parasites, which are further screened to distinguish recombinants (which have altered the target locus) from nonhomologous recombinants (in which the targeted locus remains intact). This approach has been employed successfully at some loci (KIM et al. 1993; W. BOHNE, unpublished; S. TOMAVO, personal communication), but has been difficult at others, for reasons which remain unclear.

One strategy to minimize nonhomologous recombination is to introduce a negative selectable marker adjacent to the cloned genomic fragment, in addition to the positive selectable marker at an internal site (MORTENSEN 1993). In this case, simultaneous positive/negative selection might be expected to eliminate transgenics which integrate the entire construct at a single site (by either homologous or nonhomologous recombination), leaving only recombinants which integrated by double cross-over (removing the flanking negative selectable marker). Experience has shown that double cross-overs occur predominantly by homologous recombination. The identification of negative selectable markers by insertional mutagenesis (DONALD and ROOS 1995) makes this strategy (and others described below) feasible.

Perhaps the most elegant approach for generating targetted gene knockouts is a sequential positive/negative selection procedure (MORTENSEN 1993), in which positive and negative selectable markers are *both* introduced adjacent to (not within) the cloned (and suitably mutated) locus. This construct is transfected as a circular plasmid, and positive selection applied to yield a single-site homologous recombinant. In the resulting 'pseudodiploid', wild-type and mutant alleles flank the selectable marker and other vector sequences (DONALD and ROOS 1994). In the second step, parasites are removed from positive selection, permitting recombination between the duplicated loci, an event which appears to occur at a frequency of about 2×10^{-6} per cell generation (R.G. DONALD and D.S. ROOS, unpublished). These recombinants are then isolated under negative selection, and screened to distinguish those which have recombined so as to delete the mutant locus (yielding a wild-type revertant) from those which deleted the wild-type gene to leave a perfect allelic replacement.

This 'hit-and-run' approach, although somewhat time-consuming, offers several distinct advantages over other gene knockout strategies. First, because gene replacement occurs by two sequential single cross-overs instead of one double cross-over (a very rare event), it is more likely to be successful. Second, because selectable marker(s) are situated outside of the targeted gene itself, experiments are not limited to gene knockouts – – a variety of more subtle point mutations may be introduced as allelic replacements. Third, this strategy provides a means of distinguishing essential genes from those which cannot be deleted for purely technical reasons: if the 'hit-and-run' mutagenesis procedure yields only wild-type revertants instead of the expected 1:1 ratio of wild-type:mutant, this provides positive evidence that the locus in question is essential. Finally, because the introduced plasmid sequences and selectable markers are removed in the final product, the same selectable markers can be re-used for targeting a second gene (or for other applications in the same parasite). Moreover, the elimination of vector sequences is helpful for regulatory reasons (e.g., production of vaccine strains).

5 Parasite Reagents and Safety Considerations

The *Toxoplasma* field has proved remarkably congenial, and reagents are widely shared between laboratories. In most cases, parasites, libraries, and vectors cited in this review may be obtained directly from the investigators responsible for their development. To facilitate distribution, our laboratory has recently made a variety of reagents available through the AIDS Research and Reference Reagent program (operated by Ogden BioServices Corporation for the US National Institutes of Health, 685 Lofstrand Land, Rockville, MD 20850, USA, phone: +1–301–340–0245, Fax: +1–301–340–9245, Internet: obcaids@ix.netcom.com). Useful reagents that are currently available to all investigators free of charge (except for shipping costs) from this program include the following:

5.1 Libraries

- RH strain cDNA library in λZAPII (Stratagene). 0.4–2.2 kb inserts, 2×10^6 independent recombinants. Catalog number 1896.
- RH strain genomic library in λDASHII (Stratagene). Partial *Mbo* I digest, average insert size approximately 15 kb, 6×10^5 independent recombinants. Cat. no. 2862.
- RH strain genomic library in SuperCos (Stratagene). Partial *Mbo* I digest, average insert size approximately 40 kb, 3×10^4 independent recombinants. Cat. no. 2864.
- P(LK) strain genomic library in λDASHII. Partial *Sau* 3A digest, average insert size approximately 15 kb, 6×10^5 independent recombinants. Cat. no. 2863.

5.2 Plasmids

- pminCAT/HXGPRT+. Transformation vector suitable for transient or stable expression of chloramphenicol acetyltransferase (CAT) in transfected *T. gondii* parasites, under control of 5′ and 3′ flanking sequences derived from the parasite's DHFR-TS gene. The CAT reporter may be replaced with other genes of interest. Stable transgenic parasites may be selected in HXGPRT-deficient mutants (see below) using mycophenolic acid. Cat. no. 2850.

- pminiHXGPRT. Contains *T. gondii* HXGPRT gene under control of DHFR-TS 5' and 3' flanking sequences. Functions as either positive or negative selectable marker (using 6-thioxanthine or mycophenolic acid, respectively) in suitable host strains. Cat. no. 2855.
- pRHΔHXGPRT. HXGPRT targeting plasmid, suitable for generating HXGPRT-deficient mutants in any *T. gondii* host strain of interest. Cat. no. 2856.
- pDHTR-TSc3/M3. Low level pyrimethamine-resistance vector, suitable for amplification of linked genes under pyrimethamine selection (but see note on safety, below). Cat. no. 2853.
- pDHTR-TSc3/M2M3. High level pyrimethamine-resistance vector expressing mutant DHFR-TS enzyme; suitable for insertional mutagenesis or as a selectable marker for co-transformation (but see note on safety, below). Cat. no. 2854.

5.3 Parasite Strains

- RH(EP). Wild-type host strain RH (highly pathogenic in mice). Cat. no. 2859.
- RH(EP)ΔHXGPRT. HXGPRT knockout mutant of RH strain (above). Suitable for use with HXGPRT-containing vectors. Cat. no. 2857.
- P(LK). Wild-type host strain P, (clonal isolate of strain ME49; produces brain cysts in mice). Cat. no. 2858.
- P(LK)HXGPRT–. HXGPRT-deficient mutant of P strain (above). Suitable for use with HXGPRT-containing vectors. Cat. no. 2860.

5.4 Safety Issues

It should be noted that reagents containing mutant DHFR genes render transfected *T. gondii* parasites completely resistant to pyrimethamine therapy. Pyrimethamine-resistant parasites should therefore be used with extreme caution, and only when absolutely necessary (such as for insertional mutagenesis, or studies on pyrimethamine-resistance). Genes initially identified using pyrimethamine-resistance markers should be deleted (or otherwise altered) using alternative markers prior to distribution. Any possibility of infection with pyrimethamine-resistant *T. gondii* should be treated immediately, using an alternative therapy (such as clindamycin and sulfadiazine). In both in vitro assays and in vivo studies in mice, pyrimethamine-resistant parasites are hypersensitive to sulfonamides, and remain sensitive to clindamycin, azithromycin, atovaquone, or arprinocid (B.J. Luft and D.S. Roos, unpublished).

Acknowledgements. I would like to acknowledge members of my research laboratory and the international community of *T. gondii* researchers for critical discussions. Work discussed in this review was supported by grants from the National Institutes of Health. D.S.R. is a Burroughs Wellcome New Investigator in Molecular Parasitology, and a Presidential Young Investigator of the National Science Foundation (with support from Merck Research Laboratories and the MacArthur Foundation).

References

Araujo FG, Remington JS (1992) Recent advances in the search for new drugs for treatment of toxoplasmosis. Int J Antimicrob Agents 1:153–164

Beckers CJ, Roos DS, Donald RG, Luft BJ, Schwab JC, Cao Y, Joiner KA (1995) Inhibition of cytoplasmic and organellar protein synthesis in Toxoplasma gondii. Implications for the target of macrolide antibiotics. J Clin Invest 95:367–376

Black M, Seeber F, Soldati D, Kim K, Boothroyd JC (1995) Restriction enzyme-mediated integration elevates transformation frequency and enables cotransfection of Toxoplasma gondii. Mol Biochem Parasitol 74:55–63

Bohne W, Gross U, Ferguson DJP, Heesemann J (1995) Cloning and characterization of a brady-zoite-specifically expressed gene (HSP30/BAG1) of Toxoplasma gondii, related to genes encoding small heat-shock proteins of plants. Mol Microbiol 16:1221–1230

Boothroyd JC, Black M, Kim K, Pfefferkorn ER, Seeber F, Sibley LD, Soldati D (1995). Forward and reverse genetics in the study of the obligate, intracellular parasite Toxoplasma gondii. In: Adolph K (ed) Methods in Molecular Genetics. Academic, New York, pp 3–29

Brooks RG, Remington JS, Luft BJ (1987) Drugs used in the treatment of toxoplasmosis. Antimicrob Agents Annu 2:297–306

Cotrim PC, Garrity LK, Beverly SM (1994) Isolation of drug resistance genes in Leishmania major by transfection using a shuttle cosmid vector. Molecular Parasitology Meeting, 18–22 September 1994, Woods Hole, MA, abstract 125C

Danneman B, McCutchan JA, Israelski D, Antoniskis D, and the California Collaborative Treatment Group (1992) Treatment of toxoplasmic encephalitis in patients with AIDS. A randomized trial comparing pyrimethamine plus clindamycin to pyrimethamine plus clindamycin. Ann Int Med 116:33–43

Donald RGK, Roos DS (1993) Stable molecular transformation of Toxoplasma gondii: a selectable dihydrofolate reductase-thymidylate synthase marker based on drug-resistance mutations in malaria. Proc Natl Acad Sci USA 90:11703–11707

Donald RGK, Carter D, Ullman B, Roos DS (1996) Insertional tagging, cloning and expression of the Toxoplasma gondii hypoxanthine -xanthine -guanine phosphoribosyltransferase gene: use as a selectable marker for stable transfection. J Biol Chem (in press)

Donald RGK, Roos DS (1994) Homologous recombination and gene replacement at the DHFR-TS locus in Toxoplasma gondii. Mol Biochem Parasitol 63:243–253

Donald RGK, Roos DS (1995) Insertional mutagenesis in a protozoan parasite: Direct cloning of the uracil phosphoribosyl transferase gene from Toxoplasma gondii. Proc Natl Acad Sci USA 92:5749–5753

Ellis J, Griffin H, Morrison D, Johnson AM (1993) Analysis of dinucleotide frequency and codon usage in the phylum Apicomplexa. Gene 126:163–170

Fichera ME, Bhopale MK, Roos DS (1995) In vitro assays elucidate peculiar kinetics of clindamycin action against Toxoplasma gondii. Antimicrob Agents Chemother 39:1530–1537

Haverkos HW (1987) Assessment of therapy for toxoplasma encephalitis. Am J Med 82:907–914

High KP, Joiner KA, Handschumacher RE (1994) Isolation, cDNA sequences, and biochemical characterization of the major cyclosporin-binding proteins of Toxoplasma gondii. J Biol Chem 269:9105–9112

Katlama C (1991) Evaluation of the efficacy and safety of clindamycin and pyrimethamine for induction and maintenance therapy of toxoplasmic encephalitis in AIDS. Eur J Clin Microbiol Infect Dis 10:189–191

Kim K, Soldati D, Boothroyd JC (1993) Gene replacement in Toxoplasma gondii with chloramphenicol acetyltransferase as selectable marker. Science 262:911–914

Laughon BE, Allaudeen HS, Becker JM, Current WL, Feinberg J, Frenkel JK, Hafner R, Hughes WT, Laughlin CA, Meyers JD, Schrager LK, Young LS (1991) Summary of the workshop on future directions in discovery and development of therapeutic agents for opportunistic infections associated with AIDS. J Infect Dis 164:244–251

Leport C, Raffi F, Katlama C, Regnier B, Saimot AG, Marche C, Vedrenne C, Vilde JL (1988) Treatment of central nervous system toxoplasmosis with pyrimethamine/sulfonamide combination in 35 patients with the acquired immunodeficiency syndrome. Am J Med 84:94–100

Luft BJ, Remington JS (1992) Toxoplasmic encephalitis in AIDS. Clin Infect Dis 15:211–222

Mortensen R (1993) Overview of gene-targeting by homologous recombination. In: Ausubel FM (ed) Current protocols in molecular biology. Wiley,Cambridge, Chap. 9.15.1–9.15.6

Parent SA, Fenimore CM, Bostian KA (1985) Vector systems for the expression, analysis and cloning of DNA sequences in S. cerevisiae. Yeast 1:83–138

Parmley SF, Yang S, Harth G, Sibley LD, Sucharczuk A, Remington JS (1994) Molecular characterization of a 65-kilodalton Toxoplasma gondii antigen expressed abundantly in the matrix of tissue cysts. Mol Biochem Parasitol 66:283–296

Pashley TV, Delves CJ, Hyde JE, Sims PFG (1995) Molecular cloning and sequence analysis of the Toxoplasma gondii dihydropteroate synthase gene. Molecular Parasitology Meeting, 17–21 September 1995, Woods Hole, MA, abstract 442

Pfefferkorn ER (1990) Cell biology of Toxoplasma gondii. In: Wyler DJ (ed) Modern parasite biology. Freeman, New York, pp 26–50

Pfefferkorn ER, Borotz SE (1994) Comparison of mutants of Toxoplasma gondii selected for resistance to azithromycin, spiramycin, or clindamycin. Antimicrob Agents Chemother 38:31–37

Pfefferkorn ER, Pfefferkorn LC (1979) Quantitative studies on the mutagenesis of Toxoplasma gondii. J Parasitol 65:364–370

Pfefferkorn ER, Pfefferkorn LC (1980) Toxoplasma gondii: Genetic recombination between drug-resistant mutants. Exp Parasitol 50:305–316

Ricketts AP, Pfefferkorn ER (1993) Toxoplasma gondii: Susceptibility and development of resistance to anticoccidial drugs in vitro. Antimicrob Agents Chemother 37:2358–2363

Roos DS (1993) Primary structure of the fused dihydrofolate reductase/thymidylate synthase gene of Toxoplasma gondii. J Biol Chem 268:6269–6280

Roos DS, Donald RGK, Morrissette NS, Moulton ALC (1994) Molecular tools for genetic dissection of the protozoan parasite Toxoplasma gondii. Methods Cell Biol 45:27–63

Sibley LD, LeBlanc AJ, Pfefferkorn ER, Boothroyd JC (1992) Generation of a restriction fragment length polymorphism linkage map for Toxoplasma gondii. Genetics 132:1003–1015

Sibley LD, Messina M, Niesman IR (1994) Stable DNA transformation in the obligate intracellular parasite Toxoplasma gondii by complementation of tryptophan auxotrophy. Proc Natl Acad Sci USA 91:5508–5512

Soldati D, Boothroyd JC (1993) Transient transfection and expression in the obligate intracellular parasite Toxoplasma gondii. Science 260:349–352

Soldati D, Boothroyd JC (1995) A selector of transcription initiation in the protozoan parasite Toxoplasma gondii. Mol Cell Biol 15:87–93

Steigbigel NH (1990) Erythromycin, lincomycin and clindamycin. In: Mandel GL, Douglas RG Jr, Bennet JE (eds) Principles and practice of infectious diseases. Churchill Livingstone, New York

Stokkermans TJW, Schwartzman JD, Keenan K, Morisette NS, Tilney LG, Roos DS. Inhibition of Toxoplasma gondii replication by dinitroaniline herbicides. Exp Parasitol (in press)

Tenant-Flowers M, Boyle MJ, Carey D, Marriott DJ, Harkness JL, Penny R, Cooper DA (1991) Sulfadiazine desensitization in patients with AIDS and cerebral toxoplasmosis. AIDS 5:311–315

Studies of the Effect of Various Treatments on the Viability of *Toxoplasma gondii* Tissue Cysts and Oocysts

V. Kuticic[1] and T. Wikerhauser[2]

1 Introduction

In the prevention of *Toxoplasma gondii* infection in man and animals, the inactivation of the parasite's cysts in animal tissues and its oocysts in environment is essential. The objective of our studies was to extend our knowledge of the effects of various physical and chemical treatments on the viability of two stages of *T. gondii*. In several separate studies, we examined the effects of freezing, heating, and irradiation on the viability of tissue cysts, as well as the effects of freezing, heating, drying, four disinfectants [Aldesol, tincture of Hibisept, Izosan-G (all manufactured by PLIVA, Zagreb, Croatia) and Virkon-S (KRKA, Novo Mesto, Slovenia)], two alcohols, and 10% formalin on the viability of oocysts. In all our experiments, the viability of the treated and control tissue cysts and oocysts was assessed by bioassays on cats and/or mice. The cats were inoculated following the principles described by DUBEY and STREITEL (1976) and the mice were inoculated according to a protocol by DUBEY at al. (1984).

---

[1]Department of Parasitology and Parasitic Diseases, Faculty of Veterinary Medicine, University of Zagreb, 10000 Zagreb, P.O.Box 190, Croatia
[2]Croatian Academy of Sciences and Arts, Zrinski Trg 11, 10000 Zagreb, Croatia

2 Tissue Cysts

For the production of tissue cysts, specific pathogen free (SPF) mice were subcutaneously infected with approximately 100 sporulated oocysts per mouse and young pigs were inoculated orally with 1000, 5000 or 10 000 sporulated oocysts per pig. Mice were killed 4–6 weeks after infection, and their brains were examined microscopically for the presence of tissue cysts. Pigs were killed 6–9 weeks after oral infection and the presence of cysts in their tissues was demonstrated by bioassays in subsequent trials.

2.1 Freezing and Heating

To study the effects of various treatments on the viability of tissue cysts, minced and pooled tissues of infected pigs (skeletal muscles, heart, and brain), and pooled brains of infected mice were used. Porcine samples were divided into 50 g samples. Infected murine brains were added to the 50 g porcine samples to ensure sufficient numbers of tissue cysts in each sample. The samples were placed in plastic jars and frozen at –7°C and –12°C. Control samples were kept at 4°C. Four days later, the frozen samples were thawed and, as with the controls, homogenized by means of HCl-pepsin (Dubey et al. 1984), and subcutaneously injected into mice. Each sample was divided among six mice. Six weeks later, the mice were killed, and their brains examined for the presence of *T. gondii* tissue cysts. The results indicated that the parasites did not survive a 4-day freezing at –7°C or –12°C (Kuticic 1992). These results are in accordance with earlier reports of Jacobs et al. (1960) and Dubey (1974), in which a relatively poor resistance of tissue cysts to freezing was demonstrated. However, Dubey and Frenkel (1973) observed that tissue cysts survived for 16 days at –16°C, which indicates the possibility of the existence of freeze-resistant strains of *T. gondii*.

In order to analyze the effect of heating on viability of *T. gondii* cysts, infected murine brains were pooled, diluted with saline and immersed in water at 50°C and 58°C for 15 and 30 min, respectively. The heated and control samples were bioassayed on mice, each sample being divided among six mice. According to examinations of the latter the parasites survived heating at 50°C for 30 min; however, there was no evidence of parasites when heated at 58°C (Kuticic and Wikerhauser 1994). In an earlier study by Dubey et al. (1970) tissue cysts were destroyed at 55°C in 30 min.

2.2 Irradiation

Tissue cysts were produced in murine brains and in edible porcine tissues by inoculation with oocysts isolated in Croatia, in America and in China. Cysts harbouring tissues were irradiated with X-rays or γ-rays at doses ranging from 0.3 to 1.0 kGy (30–100 krad). The source of irradiation was either Philips X-ray equipment (Philips 250/30, Philips, The Netherlands) or ^{60}Co (Institute Rudjer Boskovic, Zagreb, Croatia). The results were assessed by bioassays using cats and/or mice. Each irradiated and control sample was fed to one or two cats and/or injected in 6–12 mice. According to the results of bioassays, some slight differences in radiosensitivity of geographically different isolates were observed. Thus, a complete inactivation of the Croatian isolate was achieved only after irradiation with 0.7 kGy; however for the same effect on the American and Chinese isolates, 0.4 and 0.5 kGy, respectively, were sufficient. At sublethal doses, a sharp decrease of infectivity was observed (WIKERHAUSER et al. 1993). In an earlier study, DUBEY et al. (1986) reported that cysts in murine brains irradiated with 0.5 kGy (50 krad) were no longer infective to mice, and those in porcine tissues irradiated with 0.25 or 0.3 kGy (25 or 30 krad) were no longer infective to mice and cats.

3 Oocysts

Previously uninfected cats were fed porcine diaphragms and, when they started to shed oocysts, the parasites were collected from the feces by sedimentation and filtration, followed by multiple washings in water (DUBEY et al. 1970). After sporulation in 2% potassium bichromate, the oocysts were stored at 4°C until used.

3.1 Freezing and Heating

The oocysts, suspended in water, were maintained at –20°C. After weeks 2 and 3, their viability was bioassayed in mice, as was that of the unfrozen controls. According to the results, a period of 2 weeks was ineffective, whereas that of 3 weeks at –20°C resulted in complete inactivation of all oocysts (KUTICIC and WIKERHAUSER 1994). In an earlier study by FRENKEL and DUBEY (1973), the oocysts proved more resistant to low temperatures.

In order to analyze the effect of heating on viability, the oocysts, suspended in water, were maintained at 58°C for 15 min and for 30 min and then bioassayed in mice, as were the untreated controls. The results showed that 15 min at 58°C was sufficient to inactivate all oocysts (KUTICIC and WIKERHAUSER 1994). These results are in accordance with an earlier observation of DUBEY et al. (1970).

3.2 Drying

A few drops of a concentrated watery suspension of oocysts were poured onto strips of filter paper and left in uncovered Petri dishes at room temperature, in the range of 21–23°C, with the relative humidity in the range of 30%–68%. At weekly intervals, between weeks 3 and 7, the strips were immersed in saline to wash off the oocysts. The oocysts' viability, and that of the untreated controls, was bioassayed in mice. The results showed that at the end of the experiment (7 weeks), the viability of the dried oocysts remained comparable with that of the untreated controls (KUTICIC and WIKERHAUSER 1994). In an earlier study by FRENKEL and DUBEY (1972), dried oocysts had a shorter duration of viability.

3.3 Disinfectants

The effect of the following four disinfectants was studied: 33% watery solution of Aldesol (5 g benzalkonium chloride, 6 g glutaraldehyde, and 8 g glioxal in 100 g solution), tincture of Hibisept (0.5 g of chlorhexidine gluconate in 70% ethanol per 100 ml of tincture), 0.02% and 0.04% watery solutions of Izosan-G (sodium dichloroisocyanurate-dihydrate; $Cl_2Na/NCO_3.H_2O$), and 1% and 2% watery solutions of Virkon-S (potassium peroxisulfate 50%). The sporulated oocysts were kept at room temperature for 2 h in Aldesol and for 1, 2, 4 and 24 h in all other disinfectants. After four washings in sterile saline, the treated oocysts and the untreated controls were bioassayed in mice. After examination of the mouse tissue, cysts of *T. gondii* were demonstrated in all groups of mice, regardless of whether they had been injected with the treated or with the control oocysts (KUTICIC and WIKERHAUSER 1993a, 1994). Thus, the tested disinfectants failed to inactivate the oocysts. These results are in accordance with reports on the high resistance of oocysts against chemicals, such as 6% sodium hydroxide (DUBEY et al. 1970), 3% carbolic acid, 0.1% sublimate solution and 0.5% Hibitane solution in 70% ethanol (ITO et al. 1975).

3.4 Alcohols and Formalin

The effect of 95% and 75% ethanol, 100% methanol, and 10% formalin was studied. The sporulated oocysts were treated at room temperature for 1, 2, 4 and 24 h. After the treatment, the oocysts were repeatedly washed in sterile saline and subcutaneously injected into mice for bioassay. The results of the treated oocysts were comparable with those of the untreated controls (KUTICIC and WIKERHAUSER 1993b). Thus, the tested chemicals failed to inactivate all oocysts. Our results using formalin are in accordance with earlier reports by DUBEY et al. (1970), FRENKEL and DUBEY (1972), and ITO et al. (1975); however, our result did not concur with the results of ITO at al. (1975), who reported that 99% ethanol killed oocysts after 24 h and concentrated methanol killed oocysts after 12 h.

4 Conclusions

Complete inactivation of *T. gondii* tissue cysts was achieved by freezing at −7°C for 4 days, by heating at 58°C for 15 min or by irradiation with a dose of 0.7 kGy, whereas sporulated oocysts were completely inactivated by freezing at −20°C for 3 weeks or by heating at 58°C for 15 min. Drying of oocysts at room temperature for 7 weeks and separate treatments with disinfectants, alcohols, or formalin failed to kill all oocysts.

References

Dubey JP (1974) Effect of freezing on the infectivity of Toxoplasma cysts in cats. J Am Vet Ass 165:534–536
Dubey JP, Frenkel JK (1973) Experimental Toxoplasma infection in mice with strains producing cysts. J Parasitol 59:505–512
Dubey JP, Streitel RH (I976) Prevalence of Toxoplasma gondii infection in cattle slaughtered at an Ohio abattoir. J Am Vet Med Ass 169:1197–1199
Dubey JP, Miller NL, Frenkel JK (1970) Characterization of the new fecal form of Toxoplasma gondii. J Parasitol 56:447–456
Dubey JP, Murrell KD, Fayer R (1984) Persistance of encysted Toxoplasma gondii in tissues of pigs fed oocysts. Am J Vet Res 45:1941–1943
Dubey JP, Brake RJ, Murrell KD, Fayer R (1986) Effect of irradiation on the viability of Toxoplasma gondii in tissues of mice and pigs. Am J Vet Res 47:518-520
Frenkel JK, Dubey JP (1972) Toxoplasmosis and its prevention in cats and man. J Infect Dis 126:664–673
Frenkel JK, Dubey JP (1973) Effect of freezing on the viability of Toxoplasma gondii cysts and oocysts. J Parasitol 59:587–588
Ito SH, Tsunoda K, Shimada K, Taki T, Matsui T (1975) Disinfectant effects of several chemicals against Toxoplasma oocysts. Jpn J Vet Sci 37:229–234
Jacobs L, Remington JS, Melton ML (1960) The resistance of the encysted form of Toxoplasma gondii. J Parasitol 46:11-12
Kuticic V (1992) Otpornost tkivnih cista Toxoplasma gondii na smrzavanje. Vet Arh 62:213–216
Kuticic V , Wikerhauser T (1993a) Effects of three disinfectants on the viability of Toxoplasma gondii oocysts. Period Biol Zagreb 95:345–346
Kuticic V, Wikerhauser T (1993b) A study of the effects of ethanol, methanol and formalin on the infectivity of Toxoplasma gondii. Zb Vet Fac Univ Ljubljana 30:113–115
Kuticic V, Wikerhauser T (1994) Effects of some chemical and physical factors on the viability of Toxoplasma gondii. Vet Arh 64:89–93
Wikerhauser T, Kuticic V, Razem D, Orsanic L, Besvir J (1993) Irradiation to control infectivity of Toxoplasma gondii in murine brains and edible porcine tissues. In: Use of irradiation to control infectivity of food-borne parasites. Proceedings of the final research co-ordination meeting organized by the joint FAO/IAEA division of nuclear techniques in food and agriculture, Mexico City, Mexico 24–28 June 1991, pp 133–136. International Atomic Energy Agency, Vienna

Subject Index

Current Topics in Microbiology and Immunology

Volumes published since 1989 (and still available)

Vol. 199/I: **Doerfler, Walter; Böhm, Petra (Eds.):** The Molecular Repertoire of Adenoviruses I. 1995. 51 figs. XIII, 280 pp. ISBN 3-540-58828-0

Vol. 199/II: **Doerfler, Walter; Böhm, Petra (Eds.):** The Molecular Repertoire of Adenoviruses II. 1995. 36 figs. XIII, 278 pp. ISBN 3-540-58829-9

Vol. 199/III: **Doerfler, Walter; Böhm, Petra (Eds.):** The Molecular Repertoire of Adenoviruses III. 1995. 51 figs. XIII, 310 pp. ISBN 3-540-58987-2

Vol. 200: **Kroemer, Guido; Martinez-A., Carlos (Eds.):** Apoptosis in Immunology. 1995. 14 figs. XI, 242 pp. ISBN 3-540-58756-X

Vol. 201: **Kosco-Vilbois, Marie H. (Ed.):** An Antigen Depository of the Immune System: Follicular Dendritic Cells. 1995. 39 figs. IX, 209 pp. ISBN 3-540-59013-7

Vol. 202: **Oldstone, Michael B. A.; Vitković, Ljubiša (Eds.):** HIV and Dementia. 1995. 40 figs. XIII, 279 pp. ISBN 3-540-59117-6

Vol. 203: **Sarnow, Peter (Ed.):** Cap-Independent Translation. 1995. 31 figs. XI, 183 pp. ISBN 3-540-59121-4

Vol. 204: **Saedler, Heinz; Gierl, Alfons (Eds.):** Transposable Elements. 1995. 42 figs. IX, 234 pp. ISBN 3-540-59342-X

Vol. 205: **Littman, Dan R. (Ed.):** The CD4 Molecule. 1995. 29 figs. XIII, 182 pp. ISBN 3-540-59344-6

Vol. 206: **Chisari, Francis V.; Oldstone, Michael B. A. (Eds.):** Transgenic Models of Human Viral and Immunological Disease. 1995. 53 figs. XI, 345 pp. ISBN 3-540-59341-1

Vol. 207: **Prusiner, Stanley B. (Ed.):** Prions Prions Prions. 1995. 42 figs. VII, 163 pp. ISBN 3-540-59343-8

Vol. 208: **Farnham, Peggy J. (Ed.):** Transcriptional Control of Cell Growth. 1995. 17 figs. IX, 141 pp. ISBN 3-540-60113-9

Vol. 209: **Miller, Virginia L. (Ed.):** Bacterial Invasiveness. 1996. 16 figs. IX, 115 pp. ISBN 3-540-60065-5

Vol. 210: **Potter, Michael; Rose, Noel R. (Eds.):** Immunology of Silicones. 1996. 136 figs. XX, 430 pp. ISBN 3-540-60272-0

Vol. 211: **Wolff, Linda; Perkins, Archibald S. (Eds.):** Molecular Aspects of Myeloid Stem Cell Development. 1996. 98 figs. XIV, 298 pp. ISBN 3-540-60414-6

Vol. 212: **Vainio, Olli; Imhof, Beat A. (Eds.):** Immunology and Developmental Biology of the Chicken. 1996. 43 figs. IX, 281 pp. ISBN 3-540-60585-1

Vol. 213/I: **Günthert, Ursula; Birchmeier, Walter (Eds.):** Attempts to Understand Metastasis Formation I. 1996. 35 figs. XV, 293 pp. ISBN 3-540-60680-7

Vol. 213/II: **Günthert, Ursula; Birchmeier, Walter (Eds.):** Attempts to Understand Metastasis Formation II. 1996. 33 figs. XV, 288 pp. ISBN 3-540-60681-5

Vol. 213/III: **Günthert, Ursula; Schlag, Peter M.; Birchmeier, Walter (Eds.):** Attempts to Understand Metastasis Formation III. 1996. 14 figs. XV, 262 pp. ISBN 3-540-60682-3

Vol. 214: **Kräusslich, Hans-Georg (Ed.):** Morphogenesis and Maturation of Retroviruses. 1996. 34 figs. XI, 344 pp. ISBN 3-540-60928-8

Vol. 215: **Shinnick, Thomas M. (Ed.):** Tuberculosis. 1996. 44 figs. XII, 307 pp. ISBN 3-540-60985-7

Vol. 216: **Rietschel, Ernst Th.; Wagner, Hermann (Eds.):** Pathology of Septic Shock. 1996. 34 figs. X, 321 pp. ISBN 3-540-61026-X

Vol. 217: **Jessberger, Rolf; Lieber, Michael R. (Eds.):** Molecular Analysis of DNA Rearrangements in the Immune System. 1996. 43 figs. IX, 224 pp. ISBN 3-540-61037-5

Vol. 218: **Berns, Kenneth I.; Giraud, Catherine (Eds.):** Adeno-Associated Virus (AAV) Vectors in Gene Therapy. 1996. 38 figs. IX, 137 pp. ISBN 3-540-61076-6